# Switching Power Supply
# Design & Optimization

# Switching Power Supply Design & Optimization

**Sanjaya Maniktala**
*National Semiconductor Corp. USA*

Boston, Massachusetts    Burr Ridge, Illinois
Dubuque, Iowa    Madison, Wisconsin    New York, New York
San Francisco, California    St. Louis, Missouri

**Library of Congress Cataloging-in-Publication Data**

Maniktala, Sanjaya.
  Switching power supply design and optimization /
  Sanjaya Maniktala.
      p.    cm.
  ISBN 0-07-143483-6
  1. Switching power supplies. 2. Electronic apparatus and
  appliances—Power supply. 3. Electric current converters.
  I. Title.
  TK7868.P6M36      2004
  621.381'044–dc22

                                                    2004054664

    2 3 4 5 6 7 8 9   BKM BKM     0 9 8 7 6

ISBN 0-07-143483-6

*The sponsoring editor was Steve Chapman and the production
supervisor was Pamela Pelton. This book was set in New Century School-
book by International Typesetting and Composition. The art director for
the cover was Margaret Webster-Shapiro.*

# Contents

# Preface

Power Conversion is a fascinating area. Even after many years of intense involvement, not a day passes in which, at least in some small way, we learn a little more. This may come as a genuine technical nugget, a different perspective, or just an interesting elucidation. And it ultimately spans all corporate, geographic and even human boundaries. Though initially at least, it's engineers like us, everywhere, from Bombay to Finland, burning TO-220's with midnight oil, sometimes struggling simultaneously with an archetypal supervisor, who ultimately make this field evolve, gain meaning, and take form. This book is dedicated to that struggling persona: the committed power conversion engineer in each of us

This rather challenging engineering discipline and the resultant book have a rather personal meaning to me in many ways, since I have no formal education in electronics. I am therefore presenting information in much the same way as I taught myself over the years. Yes, certain things that may seem obvious to the trained engineer may still not be entirely so to me, and vice versa. So I do apologize in advance if I ever linger on the obvious or breeze past the complex. But the reader will notice that I have defined several new terms like the gap factor $z$ in magnetics, "LSD" cells in dc–dc converters and their configurations, $\alpha$ for helping in the discontinuous mode equations etc. My hope was that these terms really aid the explanations, and help demystify the more complex aspects of power converter design.

On the topic of demystification—as I was writing this book, I looked around in related literature, and was surprised to sense a "business strategy" being practiced by some companies. I had no other explanation considering that their applications information was so well written, but the accompanying schematics were always hopelessly confused. Maybe they just had really rotten luck for several years now in their CAD hires. Though I admit that I tended to believe that the general idea was to confuse a prospective customer, to the extent that he would think all the advantages would accrue only if he uses their part. I certainly once worked in a company where a similar attitude prevailed. So this observation is what made "demystification" my primary goal in this book. To that end I have invariably tried to reveal the underlying general principles, not part-specific (or seemingly part-specific) information. The discussion on dc–dc topologies and their configurations is an example of this approach.

Magnetics is the undoubtedly the holy grail of power conversion. To that end I have inserted four chapters covering magnetics. From basic concepts to the "mapping" of any off-the-shelf inductor to a specific application, to a detailed exploration of the proximity effect and also the fringing flux correction, I have tried to cover a lot of ground in the simplest possible language. Design charts abound, as in the rest of the book.

The chapter on thermal management presents a unifying study of the various forms any given equation of convection can be cast. The transformations may make the forms seem unrelated or unrecognizable, but they could very well represent the very same equation, or something very close to it. So the confusion created by so many apparently different equations abounding should be laid to rest. Besides natural convection, forced convection and corrections at high altitudes have also been discussed.

In the chapter "Things to Try" various ideas and practical circuits are presented which could stimulate further thought and lead to even better ideas. But it is up to the reader to check if there are any intellectual property, patent or licensing issues in force, before using them.

I have, on last count, worked in more than seven companies across four continents, building off-line power supplies for IT and telecom markets. More recently I migrated to the semiconductor side of power where I have dealt with switcher ICs, both for dc–dc and off-line applications. I think it has all probably added an unusual technical and business perspective. I would be tempted to claim that I have 'seen it all', but in power nothing can be captured in one lifetime. This is one area where something you learn in one sub-topic, can strike a match in another. How do you connect the dots within a hazy nebula? Every point is connected to every other. That is probably what makes Power such good fare for the ambitious and also scares away the faint-hearted.

But I did have more than my share of experiences. Different design approaches, perspectives, work cultures, attitudes, techniques, production technologies etc. I have seen companies with state-of-art production technologies. I have met brilliant colleagues from whom I learned. But I also saw companies in which an engineer was not allowed to make his own schematic or PCB layout because there were "assigned" specialists to do that. I have seen places where you could be fired if ever found looking into the pages of *Electronic Design* or *EDN* during "work hours." Then there were companies that would never let you even put your name on an Application Note you really toiled over. Sometimes they would even resort to declaring their in-house "visionary" as an abiding inventor on every patent application filed by the company, though we all knew he usually had nothing to do with it. Soon someone out there would be walking around with 35+ largely undeserved patents under the belt, something that even Edison would yearn for. And the talented but disheartened engineer would just have to try and try again.

The bottom line is that learning the art of power conversion, and then being recognized for it, are actually much harder than one can imagine. Not to mention the overwhelming sense of speed also required in the commercial arena. That is why I wanted to dedicate this book to the engineering community. I think I wrote it from their perspective because I am one of them and feel that I understand their aspirations and struggles. Hopefully they are the ones that will ultimately find this book useful and gain the most from it.

But I confess I certainly had a lot of fun writing this book (all in my spare time of course!). Hope you enjoy reading it too.

*Sanjaya Maniktala*

# Acknowledgments

Like almost everything else in this book, I couldn't help bulleting these out too:

- Dr G.T. Murthy, former vice-president in Bombay. For all the breaks he gave me, and his lasting inspiration. My one and only mentor.

- Herr Massat, former head of Siemens in Leipzig, Germany for all the faith they placed in me, and the awards of excellence and appreciation, monetary and otherwise.

- Dr. Karl Rinne, a brilliant German engineer and colleague, now a Professor in Ireland. Some of the work in this book is based on work we did together. Karl also showed me the power of Mathcad, for which I owe him.

- Mike Hongyang He, a former colleague of mine who until recently, worked as a Public Relations specialist at National Semiconductor. He helped my budding writing career and shared in the joy (and occasional bitterness too), like a true friend and professional.

- Steve Ohr, editor of Planet Analog, for his remarkable ability to spot talent and then to whole-heartedly encourage it. An endearing person who always makes it fun.

- My wonderful wife Disha and daughter Aartika for stoically managing without me, as I slaved late into the night for months while writing this book.

- Some others who gave support along the way: Ravi Sidhu (former journalist, The Tribune), Naveen Garewal (former journalist, Indian Express), and Dr. M.V. Pitke (former Director of C-DOT).

- Finally, at National Semiconductor I would like to thank Suneil Parulekar, Dennis Monticelli, Edward Lam, Paul Greenland, Stephen K. Lee, Fred Wise, Brian Ridgeway, Cole Reif and Ajithkumar Jain for their constant support.

# About the Author

**Sanjaya Maniktala** is a Principal Engineer with National Semiconductor. He holds two patents in power supply technology and has written numerous articles on power supply design, appearing in such magazines as *Power Electronics Technology, EDN, Electronic Engineering*, and *Planet Analog*. He lives in Fremont, California.

# Overview of Switching Power

## 1.1 Introduction

Switching power conversion spans several different areas. We know this clearly from the fact that there is almost nothing we can change at any given point in the power supply that will fail to affect some parameter or function at another point, either voluntarily or inadvertently. That is why this area is rightly considered by many as "difficult." The author remembers one of his experienced colleagues being reminded by a nervous manager as he handed him a power supply which a customer was having problems with—"fix it, but don't change anything." This underscores the dread that most companies have about the obvious interplay of several engineering disciplines they don't fully understand—and don't believe others do too well either. Their recurring nightmare seems to be about an overly creative engineer creating a web of changes—some intended and others hitherto unnoticed—until everyone from design to production gets involved, hopelessly entangled, while the customer gets increasingly impatient.

The key to understanding power is not to focus on one narrow aspect at a time but to be aware of the broader picture as we move ahead. In this book we are going to assume that the reader has some basic knowledge to start with—hopefully derived from the several excellent reference books available on the subject—and that his or her desire here is simply to acquire a deeper understanding. There is just too much to know, too much to cover in this exciting area of power electronics.

We will therefore also generally avoid exotic topologies in this book. In the commercial arena most people are involved in building only the well-known and basic workhorse topologies: the forward converter, the flyback, and their variants. There is a general perception that there is a large risk involved in putting out an untested technology into mass production. To most enthusiastic and creative engineers, this often sounds like their stuffy manager has caught the seasonal "Catch-22 flu." Though admittedly, this conservativism does have some anecdotal support to back it up. But we should not underestimate the skill and knowledge that goes into designing and producing a typical commercial power supply. Here we all are thoroughly tested in our ability to extract maximum performance at the lowest cost, without compromising reliability or safety. The author remembers the prolonged brainstorming in a company he worked for in Southeast Asia over the issue of whether to put in a 5-cent zener at the gate of the switching MOSFET (metal-oxide semiconductor field effect transistor). As the CEO (cum senior-most design engineer) put it: "I ship 50,000 power supplies a month. With that money I could hire another engineer here." Well, surprisingly the engineer finally *didn't* get hired, and it was certainly not because the proffered salary was too low. He had sadly just lost his position to a 500 mW zener diode. But more surprisingly perhaps, the zener *wasn't* placed to "protect" the MOSFET, but rather to satisfy safety norm UL1950. More on this intriguing discussion and explanation later in Chap.17!

Now, it would be nice if we could delve deep into the mathematics too. But not many power supply designers would swear that they feel very comfortable with Laplace transforms, Fourier transforms, or even Fourier series. Also, realizing that time is money in the commercial arena, we must accept that most designers are at any given moment just rushing to complete a prototype or debug a horrendous control or layout scheme. Therefore, although they may want to garner a better understanding that helps them design better, some would rather just have as many equations, formulas, and design graphs as possible, so that they can get it over with. With that perspective, the author has tried to do much of "the dirty work" himself, in the form of some detailed mathematical simulations that went into creating some of the curves presented herein.

The situation is much like the one concerning the several popular integrated switcher IC (integrated circuit) families available from most major semiconductor vendors, one of which the author had, until recently supported for several years. Some people instinctively presume that these switchers must be rather "simple." In fact they are, but only to their users. And that is mainly because we engineers constantly struggle behind the scenes to make them so. As they say: "never watch a sausage being made." Well, power certainly is one very large sausage. And you probably don't want to know, or need to know, what all may have gone into making this particular one.

## 1.2   The Voltseconds Law

Let's introduce the most basic concept in power conversion:

In normal *square-wave* (nonresonant) power conversion we always apply a certain *constant* voltage (denoted here by $V_{ON}$) during the switch on-time ($t_{ON}$). Then we will automatically get a constant voltage (of opposite sign, whose magnitude is denoted here as $V_{OFF}$) during the off-time ($t_{OFF}$). This leads to piecewise-linear current segments. So we can write (in terms of magnitudes)

$$V_{ON} = L\frac{\Delta I_{ON}}{t_{ON}}$$

$$V_{OFF} = L\frac{\Delta I_{OFF}}{t_{OFF}}$$

A "steady state" in power conversion can be defined as

$$\Delta I_{ON} = \Delta I_{OFF}$$

(again in terms of magnitudes). This equality in effect implies that the current at the end of a given switching cycle returns to the exact instantaneous value it had at the start of the same cycle, every cycle. Thus the entire current (and voltage) pattern becomes repetitive, and the operation is in that sense "steady."

This basically means that

$$\boxed{V_{ON} \times t_{ON} = V_{OFF} \times t_{OFF}}$$

The product of the applied voltage and the duration it is applied for is called *voltseconds*. The above equation therefore forms the *voltseconds law*.

Note that the off-time, $t_{OFF}$ above, is defined from the viewpoint of the inductor, not of the switch. It is equal to the time for which the current in the inductor continues to ramp down. This is not necessarily equal to the *entire available* off-time $T - t_{ON}$ [that is, $(1-D)/f$, where $T = 1/f$ is the time period of the switching cycle]. In *discontinuous conduction mode* (DCM) for example, the voltage reversal across the inductor lasts for a duration less than $T - t_{ON}$. During the remaining part of the cycle, the voltage across the inductor remains zero, and so does the current through it.

If for any reason the inductor current fails to return to zero every cycle, the voltage-reversal duration will increase, and ultimately span the entire available off-time, even adjusting the latter duration if necessary. As a result, either the output voltage will automatically get readjusted (open loop condition), or the feedback loop will not let that happen, and instead change the duty cycle to maintain the output voltage. Either way, the effort is to make the inductor current return to its original (now nonzero) starting value. And if this succeeds, the steady state condition we will now be operating in would be termed a *continuous conduction mode* (CCM).

It is easy to understand and confirm this voltseconds law for an inductor, but what about a multiwinding magnetic element like a transformer or coupled inductor? In fact, this law applies to *any winding* present on the magnetic structure. We can check this out—but while doing so, each winding should be considered *individually*, and without any regard as to whether any other winding is present or not, and even whether that is passing current or not. Therefore, we cannot for example combine the voltage applied during the on-time across one winding, with the voltage present across another winding during the off-time (when we write out our voltseconds equation).

But also note that this certainly doesn't imply that the multiple windings are independent in any way. In fact the currents flowing through each and every winding have all gone into establishing the very steady state (and duty cycle) that we are analyzing in the first place.

We must be clear that any topology that *exists* (discovered or yet to be), must tend to automatically move toward a steady state (or else how could it exist?). Regulation only maintains the output rail to a desired (and precise) value under various line and load variations. On its own, feedback (or in fact any signal-level control) cannot "enforce" a successful realization of any topology. Physics is not deceived by op-amps, multipliers, and the like. Eventually, our first responsibility is to transfer raw power from the input to the output, not to write a control theory manuscript. The voltseconds law being directly related to energy flow is therefore fundamental. Despite its seeming simplicity, it is the basic tool for confirming the viability or existence of any topology.

The input to output transfer function $V_O/V_{IN}$ for any topology thus follows simply from the voltseconds law. Alternatively, we can also express the duty cycle $t_{ON}/T$ (where $T = 1/f$) in terms of the input and output voltages (their magnitudes). See Table 1.1 for derivations for a buck, a boost and a buck-boost (all assumed to be in (CCM), and ignoring parasitic voltage drops).

## 1.3  Basic Waveform Analysis

We learn a lot from basic geometry as applied to converters. If we don't understand the waveforms in a switching converter well, we are not likely to understand anything more. In Fig. 1.1 we have tabulated the rms and average of several typical waveforms we may encounter.

**TABLE 1.1  Derivation of Input-Output Transfer Functions from Voltseconds Law.**

| | Buck | Boost | Buck-Boost |
|---|---|---|---|
| $V_{ON}$ | $V_{IN} - V_O$ | $V_{IN}$ | $V_{IN}$ |
| $V_{OFF}$ | $V_O$ | $V_O - V_{IN}$ | $V_O$ |
| $t_{ON}$ | $D/f$ | | |
| $t_{OFF}$ | $(1-D)/f$ | | |
| **Voltseconds** '$AB = CD$' | $(V_{IN} - V_O)D = V_O(1-D)$ | $V_{IN}D = (V_O - V_{IN})(1-D)$ | $V_{IN}D = V_O(1-D)$ |
| $\dfrac{A}{C} = \dfrac{D}{B}$ | $\dfrac{V_{IN} - V_O}{V_O} = \dfrac{1-D}{D}$ | $\dfrac{V_{IN}}{V_O - V_{IN}} = \dfrac{1-D}{D}$ | $\dfrac{V_{IN}}{V_O} = \dfrac{1-D}{D}$ |
| $\dfrac{A+C}{C} = \dfrac{D+B}{B}$ | $\dfrac{V_{IN} - V_O + V_O}{V_O} = \dfrac{1-D+D}{D}$ | $\dfrac{V_{IN} + V_O - V_{IN}}{V_O - V_{IN}} = \dfrac{1-D+D}{D}$ | $\dfrac{V_{IN} + V_O}{V_O} = \dfrac{1-D+D}{D}$ |
| **therefore** | $\dfrac{V_{IN}}{V_O} = \dfrac{1}{D}$ | $\dfrac{V_O}{V_O - V_{IN}} = \dfrac{1}{D}$ | $\dfrac{V_{IN} + V_O}{V_O} = \dfrac{1}{D}$ |
| **reciprocal** | $D = \dfrac{V_O}{V_{IN}}$ | $D = \dfrac{V_O - V_{IN}}{V_O}$ | $D = \dfrac{V_O}{V_{IN} + V_O}$ |

*(margin note: $D = $ DUTY cycle)*

But is there a more general rule for an arbitrary waveform? Yes, if it is "piecewise-linear" we do have such an equation. This is provided in Fig. 1.2. What we basically do here is as follows:

> We pick any portion of the waveform that is "repetitive." It does not matter whether we start at the moment the switch turns *on* or at some other point, provided at the end we return to exactly the same point we started with. That is by definition one complete cycle.

> We split the cycle into segments of *constant* slope; thus the ends of a segment are usually the breakpoints of slope. Though it is imperative that we ensure that no segment *encloses* any such breakpoint.

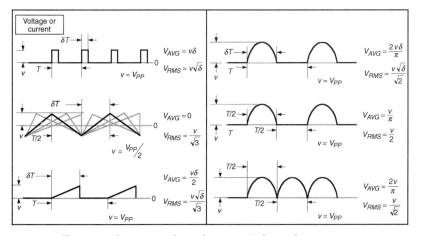

**Figure 1.1**  The rms and average values of some typical waveforms.

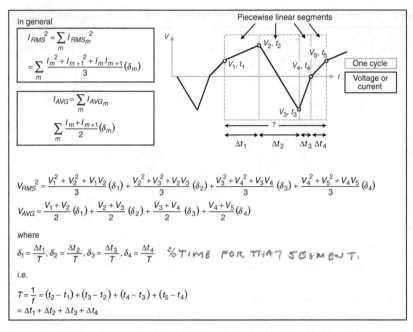

In general

$$I_{RMS}^2 = \sum_m I_{RMS_m}^2$$

$$= \sum_m \frac{I_m^2 + I_{m+1}^2 + I_m I_{m+1}}{3}(\delta_m)$$

$$I_{AVG} = \sum_m I_{AVG_m}$$

$$\sum_m \frac{I_m + I_{m+1}}{2}(\delta_m)$$

Piecewise linear segments

$V_2, t_2$

$V_5, t_5$

$V_1, t_1$    $V_4, t_4$

One cycle

Voltage or current

$V_3, t_3$

$T$

$\Delta t_1$    $\Delta t_2$    $\Delta t_3$  $\Delta t_4$

$$V_{RMS}^2 = \frac{V_1^2 + V_2^2 + V_1 V_2}{3}(\delta_1) + \frac{V_2^2 + V_3^2 + V_2 V_3}{3}(\delta_2) + \frac{V_3^2 + V_4^2 + V_3 V_4}{3}(\delta_3) + \frac{V_4^2 + V_5^2 + V_4 V_5}{3}(\delta_4)$$

$$V_{AVG} = \frac{V_1 + V_2}{2}(\delta_1) + \frac{V_2 + V_3}{2}(\delta_2) + \frac{V_3 + V_4}{2}(\delta_3) + \frac{V_4 + V_5}{2}(\delta_4)$$

where

$$\delta_1 = \frac{\Delta t_1}{T}, \delta_2 = \frac{\Delta t_2}{T}, \delta_3 = \frac{\Delta t_3}{T}, \delta_4 = \frac{\Delta t_4}{T} \quad \% TIME \ \ FOR \ THAT \ SEGMENT.$$

i.e.

$$T = \frac{1}{f} = (t_2 - t_1) + (t_3 - t_2) + (t_4 - t_3) + (t_5 - t_4)$$

$$= \Delta t_1 + \Delta t_2 + \Delta t_3 + \Delta t_4$$

**Figure 1.2**  RMS and average of an arbitrary piecewise linear waveform.

We calculate the average value of each segment (considered independently), and then sum over all segments to get the average for the entire waveform. Thus

$$I_{AVG} = I_{AVG\_1} + I_{AVG\_2} + I_{AVG\_3}\cdots$$

where $I_{AVG}$ of each segment is

$$I_{AVG\_n} = \frac{I_n + I_{n+1}}{2} \times \delta_n$$

Similarly, we calculate the rms values of each segment and then sum the *squares* of the rms of each segment to get the square of the rms for the entire waveform.
So,

$$I_{RMS}^2 = I_{RMS\_1}^2 + I_{RMS\_2}^2 + I_{RMS\_3}^2\cdots$$

where $I_{RMS}$ of each segment can be found from

$$I_{RMS\_n}^2 = \frac{I_n^2 + I_{n+1}^2 + I_n I_{n+1}}{3}\delta_n$$

Here $\delta$ is the *geometric duty cycle* of that segment, i.e., the ratio of its duration to the time period of the entire waveform.

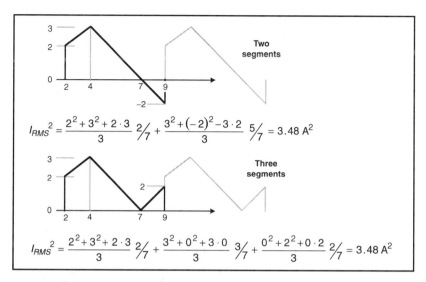

$$I_{RMS}^2 = \frac{2^2 + 3^2 + 2 \cdot 3}{3} \,{}^2\!\!\big/_{\!7} + \frac{3^2 + (-2)^2 - 3 \cdot 2}{3} \,{}^5\!\!\big/_{\!7} = 3.48 \text{ A}^2$$

$$I_{RMS}^2 = \frac{2^2 + 3^2 + 2 \cdot 3}{3} \,{}^2\!\!\big/_{\!7} + \frac{3^2 + 0^2 + 3 \cdot 0}{3} \,{}^3\!\!\big/_{\!7} + \frac{0^2 + 2^2 + 0 \cdot 2}{3} \,{}^2\!\!\big/_{\!7} = 3.48 \text{ A}^2$$

**Figure 1.3** Rectifying a waveform does not change its rms value.

The rms value of a waveform is not affected if the waveform goes below zero (ground). We could therefore take the entire part of the waveform that is below ground and fold it to be above ground. This is equivalent to rectifying it. See Fig. 1.3 for a worked example. Though the rms is unchanged in the process, the average value does change.

Though it may seem obvious, we should note that reflecting a waveform *horizontally* does not change either its rms or average value. So a switch waveform can change its familiar "up-ramp" shape and become a diode ("down-ramp") waveform, and the same basic equations for rms and average would still apply. However the duty cycle $\delta$ (of each segment in the equations above) is $D$ for the switch, but is $1 - D$ for the diode (assuming the converter is operated in CCM). We can do the same for, say, a transformer isolated flyback, but we will need to first reflect the currents to the same side of the transformer.

## 1.4 The *r* and *K* of the Current Waveform

In Fig. 1.4 we have shown a typical switch current waveform. We define one of the most fundamental design parameters of any switching power supply, the *current ripple ratio* r. We will show in Chap. 9, which details the inductor selection procedure for dc-dc converters, that setting *r* to about 0.4 represents the *most optimum value* in terms of the overall stresses in the power supply. By definition, *r* is related to the average of the inductor current (the geometric center, or the dc value) and to the

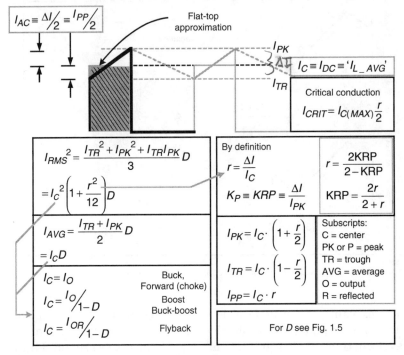

**Figure 1.4** A standard switch waveform and its components.

total swing in the current $\Delta I$ by

$$r = \frac{\Delta I}{I_C}$$

$\substack{I_C = I_{DC} \ \ MEASURED \\ = I_{Conduction}}$

*Selection of* r *should be the starting point of any converter design since once it is chosen, it impacts everything else.* For a given switching frequency, this is indirectly a statement of the value of inductance $L$. But since frequency can vary widely, it is not possible to give a general rule based on inductance. Note that for $r$ we have just provided such a rule. And this $r$-*rule* applies to any topology, any frequency, and virtually any input and output condition.

In the same figure we have used the subscript extension $R$ to mean "reflected." This bears relevance mainly to the flyback topology, and the reader should therefore read Chap. 6 for more details. Here it suffices to say that if voltage is being reflected from say the primary side to the secondary side, we need to *divide* it by the turns ratio $n = n_P/n_S$. For current, we *multiply* it by the same factor. And in going the other way, i.e., from secondary to primary, we just invert the respective factors.

Some engineers talk in terms of a $K$ factor as indicated in the figure. But they usually give it a subscript $P$ (for peak), since it is the ratio of the current swing to the *peak* value of the waveform, rather than to its center (as is $r$). But sometimes they also call it a *reflected peak* K and denote it by $KRP$. The relationship between $r$ and this $K$ factor (or whatever it is called) is provided in Fig. 1.4.

**Note:** Calling the $K$ factor $KRP$ or $K_{RP}$ ($R$ for reflected) is rather disingenuous because we must remember that if we look at the primary current waveform and calculate either its $r$ or its $K$ factor, we get the same result as if we had done the calculation based on the secondary-side current waveform. That follows from the scaling law of transformers. The peak, center, and $\Delta I$ all scale by the same factor from primary to secondary side and vice versa. So in fact both $r$ and $K$, which are ratios, are invariant across the sides of a transformer. There is therefore no reflected $K$ *factor* really, just $K$ or $K_P$.

## 1.5  Basic Design Procedure for Inductors

Without going into too much detail here, we now present a simplified but universal design starting point for all switching power converters. This is shown in Fig. 1.5. It starts by setting $r = 0.4$ at the appropriate

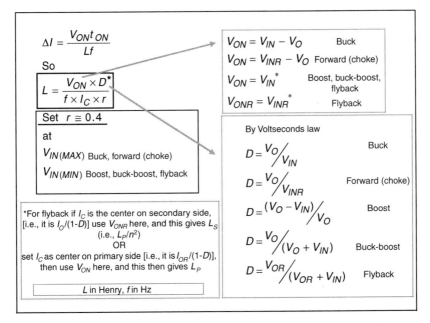

**Figure 1.5**  Basic inductor design procedure.

input voltage end. This applies to any frequency and for any topology. Then we can calculate the inductance as indicated. This is the reason why acquiring a mental picture based on $r$ is such a useful strategy. We couldn't have written any such general rule for $L$.

## 1.6   Calculating RMS Current for Capacitors

We will see a little later why it may be very important to know the input/output rms capacitor currents in a converter. In Fig. 1.6 we indicate the general procedure for calculating the rms current through a capacitor. We must remember that in a steady state the average current through a capacitor is zero. So the procedure is as follows:

> We take the original waveform (belonging to the switch, diode, or inductor current, whichever is associated with the capacitor). We calculate the average value and the rms value. These are called $I_{AVG\_old}$ and $I_{RMS\_old}$ respectively.

> Now to convert it into the associated capacitor current waveform, we need to reposition the horizontal axis (zero current level) at the calculated average current value. We can also look upon this as translating the original waveform vertically, so as to bring its (old) average level to line up with the horizontal axis. In doing so, we effectively force

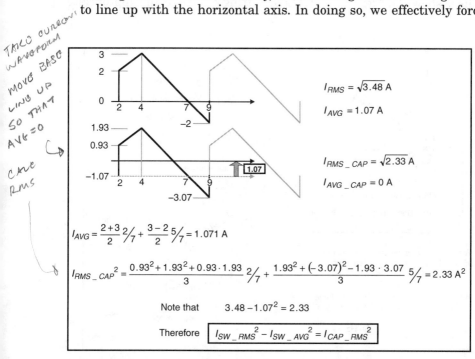

$$I_{AVG} = \frac{2+3}{2} \,{}^2\!/_7 + \frac{3-2}{2} \,{}^5\!/_7 = 1.071 \text{ A}$$

$$I_{RMS\_CAP}^2 = \frac{0.93^2 + 1.93^2 + 0.93 \cdot 1.93}{3} \,{}^2\!/_7 + \frac{1.93^2 + (-3.07)^2 - 1.93 \cdot 3.07}{3} \,{}^5\!/_7 = 2.33 \text{ A}^2$$

Note that    $3.48 - 1.07^2 = 2.33$

Therefore    $\boxed{I_{SW\_RMS}^2 - I_{SW\_AVG}^2 = I_{CAP\_RMS}^2}$

**Figure 1.6**  How to calculate the associated capacitor current of any waveform.

the average of the new (translated) waveform to be zero. For that is exactly what a capacitor will strive to do eventually.

The rms of the new (capacitor) current waveform can then be shown to be

$$I_{RMS\_new} = \sqrt{I_{RMS\_old}^2 - I_{AVG\_old}^2}$$

*USE THIS EQ'N.*

Note that we could have started off with a waveform with a different amount of dc offset, and we would still get the same capacitor rms current by the above procedure. Therefore we can also conclude that for any general waveform, the following quantity must always be preserved

$$\sqrt{I_{RMS}^2 - I_{DC}^2} = \text{constant}$$

This term is often called the *AC* rms of the waveform, $I_{AC\_RMS}$. This is considered to be an attribute of a waveform that does not change so long as we keep its basic shape unchanged, i.e., even if we move it "up or down" or "side to side."

The above procedure can be applied to derive the input capacitor rms current from the switch current waveform for the buck and for the buck-boost. For the boost topology, the input capacitor is associated with an inductor, so here we must first find the rms and average, *not of the switch current, but of the inductor current waveform*. Thereafter, the procedure is exactly the same as previously described. Also see Chap. 4 for a good exercise on this topic as applied to DCM.

As for the output capacitor, we have to use the diode current waveform as the "associated" waveform, but only for the boost and the buck-boost. For the buck, the output capacitor is connected to the inductor, and we therefore have to use the inductor current waveform for this purpose.

Finally, all the tools we need for calculating the capacitor currents in virtually every situation are available in Fig. 1.1 or Fig. 1.2. In Box 1.1 we have derived the rms of the input and output capacitor currents of a buck-boost as an example.

## 1.7  Topologies and Worst-Case Capacitor Currents

A fundamental difference between the topologies concerns the basic shape of the input and output waveforms. For a buck or forward converter for example, the output current into the capacitor is relatively smooth as it comes through an inductor. However, the input current is "chopped" (pulsating). For a boost the situation reverses and it is the output current that is pulsating. For a buck-boost or a flyback, both

| Input Capacitor:<br>From Figure 1.6 [using $I_C = I_{OR}/(1-D)$]<br><br>$I_{SW\_RMS} = \dfrac{I_{OR}}{1-D}\sqrt{\left(1+\dfrac{r^2}{12}\right)D}$<br><br>$I_{SW\_AVG} = \dfrac{I_{OR}}{1-D}D$<br><br>Using $\boxed{I_{CAP\_RMS} = \sqrt{I_{RMS}^2 - I_{AVG}^2}}$<br><br>$I_{CAP\_IN\_RMS}$<br>$= \dfrac{I_{OR}}{1-D}\sqrt{\left(1+\dfrac{r^2}{12}\right)D - D^2}$<br><br>$\boxed{I_{CAP\_IN\_RMS} = \dfrac{I_{OR}}{1-D}\sqrt{D\left[1-D+\dfrac{r^2}{12}\right]}}$ | Output Capacitor:<br>From Figure 1.6 [using $I_C = I_O/(1-D)$]<br><br>$I_{D\_RMS} = \dfrac{I_O}{1-D}\sqrt{\left(1+\dfrac{r^2}{12}\right)(1-D)}$<br><br>$I_{D\_AVG} = \dfrac{I_O}{1-D}(1-D) = I_O$<br><br>Using $\boxed{I_{CAP\_RMS} = \sqrt{I_{RMS}^2 - I_{AVG}^2}}$<br><br>$I_{CAP\_OUT\_RMS}$<br>$= \dfrac{I_O}{1-D}\sqrt{\left(1+\dfrac{r^2}{12}\right)(1-D)-(1-D)^2}$<br><br>$\boxed{I_{CAP\_OUT\_RMS} = I_O\sqrt{\dfrac{D+\dfrac{r^2}{12}}{1-D}}}$ |

**Box 1.1**  Flyback/buck-boost capacitor currents.

the input and output currents are pulsating. This contributes to the inability of the flyback to handle larger power (another restrictive factor is the leakage inductance). For the Cuk converter, which is essentially a composite of a boost stage input and a buck stage output, we get the best of two worlds in that sense. So both the input and output currents are smooth and this topology is therefore sometimes called an *ideal dc-dc converter*.

In Table 1.2, we have followed the general procedure for finding the capacitor currents of the three basic topologies.

We note that *for all topologies, a low* D *implies a high* $V_{IN}$ *and a high* D *implies a low* $V_{IN}$. We are of course only referring to the magnitudes of the voltages involved.

Thus for all the cases in the table where we have $1 - D$ in the denominator, we expect the corresponding terms to be maximum at low input voltages.

But for small $r$, the input capacitor of a buck has the following dependency on $D$

$$I_{RMS} \propto \sqrt{D(1-D)}$$

Since $D = 1 - D$ at $D = 0.5$, the worst-case input capacitor current for a buck occurs at about 50 percent duty cycle (input voltage roughly twice the output voltage). We have to be mindful of this aspect when we

**TABLE 1.2   RMS Capacitor Currents for the Three Main Topologies.**

| | Buck | Boost | Buck-Boost |
|---|---|---|---|
| RMS current in input cap (A) | $I_O \cdot \sqrt{D \cdot \left[1 - D + \dfrac{r^2}{12}\right]}$ | $\dfrac{I_O}{1-D} \cdot \dfrac{r}{\sqrt{12}}$ | $\dfrac{I_O}{1-D} \cdot \sqrt{D \cdot \left[1 - D + \dfrac{r^2}{12}\right]}$ |
| RMS current in output cap (A) | $I_O \cdot \dfrac{r}{\sqrt{12}}$ | $I_O \cdot \sqrt{\dfrac{D + \frac{r^2}{12}}{1-D}}$ | $I_O \cdot \sqrt{\dfrac{D + \frac{r^2}{12}}{1-D}}$ |
| $r$ | $\dfrac{V_O + V_D}{I_O \cdot L \cdot f} \cdot (1 - D) \cdot 10^6$ | $\dfrac{V_O - V_{SW} + V_D}{I_O \cdot L \cdot f} \cdot D \\ \cdot (1 - D)^2 \cdot 10^6$ | $\dfrac{V_O + V_D}{I_O \cdot L \cdot f} \cdot (1 - D)^2 \cdot 10^6$ |
| $V_{IN\_D=0.5}(V)$ | $(2 \cdot V_O) + V_{SW} + V_D$ | $\dfrac{1}{2} \cdot [V_O + V_{SW} + V_D]$ | $V_O + V_{SW} + V_D$ |
| | $\approx 2 \cdot V_O$ | $\approx V_O / 2$ | $\approx V_O$ |

design and test the input capacitor. Note that if we do not have $D = 0.5$ included within the input range of our application, we have to pick the closest input voltage end to this point.

**Example 1.1**   A buck with a 12 V output has an input of 15 V to 20 V. What input voltage represents the worst-case for the input capacitor current?

At $D = 0.5$ we require an input of 24 V for a 12 V output since $D = V_O/V_{IN}$. The point closest to 24 V in our input range is $V_{INMAX}$, that is, 20 V, and we must design and test the input capacitor at this voltage.

**Example 1.2**   A buck with a 12 V output has an input of 28 V to 50 V. What input voltage represents the worst-case for the input capacitor current?

At $D = 0.5$ we require an input of 24 V for a 12 V output since $D = V_O/V_{IN}$. The point closest to 24 V in our input range is $V_{INMIN}$, that is, 28 V, and we must design and test the input capacitor at this voltage.

## 1.8   Worst-Case Input Voltage for a Power Supply

In the section where we introduced $r$ we did not point out what happens when we have a wide input range. Do we set $r$ to the suggested 0.4 at $V_{INMIN}$ or at $V_{INMAX}$? For that we have to understand what constitutes the "worst-case" in a general sense for the entire power supply. A key to converter design is ensuring that the "magnetics" do not saturate. So that is the point where we should "ensure" our design. Clearly, at this point the inductor current is a maximum. However note from Fig. 1.4, that the center of the inductor current waveform (its average value) has the following form for the boost and the buck-boost (also flyback)

$$I_{L\_AVG} \equiv I_C \propto \frac{1}{1 - D}$$

For a buck

$$I_{L\_AVG} \equiv I_C = \text{constant}$$

For all topologies, a high $D$ corresponds to a low $V_{IN}$. So for the boost and the buck-boost, the 'worst-case' corresponds to high $D$, that is, low input. For the buck, the average inductor current is always equal to the load current $I_O$, so its worst-case cannot be determined by the average inductor current alone. We turn to its peak value and we see that it has a maximum at high input voltage. So for a buck (and for the forward converter choke) the design must start at high line. However, after we set $r$ at the appropriate input end, as we change the input (or $D$), the current ripple ratio changes. From Table 1.2 we see that

For a buck, having set $r = 0.4$ at $D_{MIN}$, as we increase $D$ (lower the input voltage), the current ripple ratio will *decrease* because it has the form $(1 - D)$

For the buck-boost, having set $r = 0.4$ at $D_{MAX}$, as we decrease $D$ (increase the input voltage), the current ripple ratio will *increase* as it has the form $(1 - D)^2$. This is attributable to the fact that $r = \Delta I / I_C$, and $I_C$ starts falling rapidly.

For the boost, having set $r = 0.4$ at $D_{MAX}$, as we decrease $D$ (increase the input voltage), the current ripple ratio will increase at first, but as $D$ becomes even smaller, it will eventually decrease. This is because it has the form $D(1 - D)^2$, and, in fact, this function has a maximum at $D = 0.33$.

Relating the variation of $r$ to the input capacitor current of a boost we see that

$$I_{RMS} \propto \frac{r}{1 - D} \propto \frac{D(1 - D)^2}{1 - D} = D(1 - D)$$

But we already know that the function $D(1 - D)$ has a maximum at $D = 0.5$. So we conclude that for a boost we must also design and test the input capacitor at $D = 0.5$ (or the closest point to it). The input voltage at which $D = 0.5$ is designated as $V_{IN\_D=0.5}$ in Table 1.2. So for the boost this corresponds to an input voltage equal to half the output voltage.

**Note:** For the sake of completeness we have included the forward drops across the switch and diode, $V_{SW}$ and $V_D$, respectively, in the table. But we can choose to ignore them for now.

## 1.9  Using Too High an Inductance (small *r*)

We know from practical experience that a very large inductor (small $r$) is not a cost-effective or optimum choice. But what are some other issues associated with it?

One major concern is the leading edge spike as shown in Fig. 1.7. This can cause jitter and, in severe cases, an inability to deliver full power. If we increase the inductance we could cause premature termination of the switching pulses in current mode controllers because in doing so we are inadvertently raising the pedestal on which the spike is riding. But since less than the required energy will be delivered for that (prematurely terminated) cycle, in the next cycle the converter will try to compensate with a larger duty cycle. In this process it gets some unexpected help because after the early termination of the previous pulse, the inductor current now has a longer time to slew down. Thus the pedestal on which the leading edge spike is riding comes down, probably enough to help it evade early pulse-limiting in the next cycle. Finally, what we may see on the oscilloscope are *alternate wide and narrow pulses*, which incidentally mimic what we get under "subharmonic instability" (see Chap. 14). On looking at the oscilloscope, we may be rather surprised by the pattern because we always thought that a high inductance should help us *avoid* subharmonic instability, but here it seems to be *aggravating* it. Such are the apparent mysteries (and perils) of practical power supply design! The leading edge spike also causes erratic responses from the current limit protection circuit for both current mode and voltage mode control. We obviously can't set an effective or trustworthy current limit based on a spike, especially since we will discover that this spike varies from unit to unit (as it is based on various

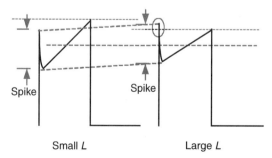

Small *L*             Large *L*

**Figure 1.7**  Large inductance increases jitter from leading edge spike.

uncontrolled and/or uncharacterized parasitics). We could of course set a large "blanking time" for current mode control, and/or we can add some delay to the current limit detection circuitry. But we also then run the danger of not being able to react fast enough to an actual abnormal load condition, especially if the inductor starts to saturate abruptly. Thus we may end up solving the problem by further oversizing our inductor in an effort to prevent saturation. But we see that this chain of design missteps was not really necessary, and probably came about only because we started off with an instinctive and possibly unjustified desire to somehow decrease $r$ beyond reasonable practical limits.

Though to be fair to the designer, he or she may have just been trying to reduce the output *voltage* ripple (which is clearly dependent on $r$). But in that case a simple LC postfilter should have been considered instead. In general, for any switching power stage it just doesn't pay to be "ripple-phobic," either from the point of view of current or voltage. For example, at the input of an off-line supply, if we increase the input capacitance too much in an effort to "smooth" the input dc rail, this could add tremendously to the cost of the EMI filter (see Chap. 5 for more details).

One advantage of a large inductance is that the converter will stay in continuous conduction mode down to extremely light loads. If we set a certain $r$ at maximum load condition and then start decreasing the load, *the converter will turn discontinuous (DCM) at $r/2$ times the maximum load*.

We also remember that an inductor takes time for current through it to change. So it cannot react fast enough to sudden changes in the load demand. The loop response may thus be poor with a large inductance.

An inductor working in steady state never parts with a certain minimum or residual energy that it stores every cycle (assuming it is in CCM). This energy level is $1/2 \times L \times I_{TR}^2$ ($I_{TR}$ is the trough of the current). But under a fault condition, even if the controller reacts by ceasing all switching action, the energy stored in the inductor will get dumped into the output capacitor. If the output capacitance is small, it can see a high spike in its voltage. And this spike can be even higher than the regulated level. So we have to be careful when using a large inductance combined with a low output capacitance.

Some think that by using a large inductance we are somehow reducing the heating in the FET. Actually that doesn't bear out either. To see this we take up the *flat-top approximation* next.

## 1.10  The Flat-Top Approximation

For a practicing engineer, there is always a constant struggle to work with less imposing equations and yet compromise little in terms of accuracy. One of these is the *flat-top approximation*. This approximation is in fact used extensively by engineers because it is easier to handle and also for its sheer availability. We will discuss various concerns and pitfalls when using this popular approximation, and also provide some correction curves to get more accurate results, despite the approximations. But we will also show why at some stage *we do need the more exact form of the equations*.

The flat-top approximation is equivalent not to actually using, but *assuming* an extremely large inductance. Let us see the possible issues with this.

If we need to set a current limit, we cannot afford to forget that *the peak value of the current is 50 × r percent higher than the flat-top assumption*. Therefore, if we set the current limit based on the flat-top approximation, we will likely have a power supply that can't deliver its expected power output.

By assuming that the current swing is zero, we are in effect assuming that the inductor has no ripple. So it has a pure dc current only. There is no current swing, and therefore we cannot estimate the core losses with this approximation either.

In peak current mode control we encounter subharmonic oscillations at about $D > 0.5$ unless we set "slope compensation." This compensating slope is typically set to a certain factor of the slope of the down-ramp of the inductor current. By assuming no down-ramp (zero slope) we are solving the subharmonic instability problem trivially, by not requiring any slope compensation at all. Clearly, the flat-top approximation is incompatible with the design of a practical slope compensation scheme.

The flat-top approximation does underestimate the switch rms current somewhat, but as we can see from the error curves in Fig. 1.8, the loss in accuracy is fairly insignificant under the normal range of selection of $r$. We must however keep in mind that *heating* is proportional to rms *squared* and so the error in that may be higher.

The rms of the capacitor current is also underestimated *and much more so*. In Fig. 1.8 we have plotted the error for a buck-boost/flyback. *We see that we get a large flat-top approximation error in the input capacitor current estimate for large* D *and a large error in the output capacitor for low* D.

As a shortcut in most cases we can suggest that the designer takes the flat-top equations, works with them, but then applies the corrections as indicated in Fig. 1.8.

Note:    In Fig. 1.8, the switch rms error curve applies to the boost and the buck-boost/flyback. The remaining curves are only for the buck-boost/flyback.

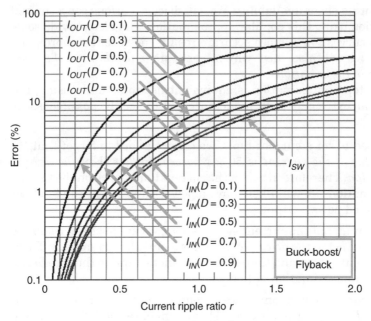

**Figure 1.8**  Error in rms estimate by using the flat-top approximation.

**Example 1.3**  By the flat-top approximation, the input capacitor rms current for a flyback at a given input voltage ($D = 0.5$) is estimated to be 1.5 A. What is a more exact value if the measured current ripple ratio is 1.2?

Eyeballing the curves in Fig. 1.8, we can see the error is 10 percent. The conversion for a known error of x percent is

$$I_{exact} = \frac{I_{flat-top}}{1 - \frac{x}{100}}$$

So

$$I_{IN} = \frac{1.5}{0.9} = 1.67 \text{ A}$$

So we have had to increase the flat-top rms estimate by 11 percent here. But in terms of the heating in the capacitor, which depends on rms$^2$, we need to increase our estimate by 24 percent. Note that stated in reverse, this means that we have underestimated the capacitor temperature by 19 percent by using the flat-top approximation. We can see that under the same conditions the error in the switch rms estimate is only *half* of this. So we have to be very cautious in applying the flat-top approximation to capacitor currents, though for the switch and diode waveforms it may work just fine.

One thing we have to be even more careful about when using the flat-top approximation are *cumulative errors*. For example, if we do a flat-top estimate of the switch current—starting from a measured (known) value of the input capacitor current (or vice versa)—the errors will build up. In that case the error curves presented in Fig. 1.8 do not apply.

As an exercise we will try to do just that and see where the problem lies. The input capacitor flat-top equation is

$$I_{IN} = \frac{I_{OR}}{1 - D} \bullet \sqrt{D \bullet (1 - D)}$$

Solving for the reflected output current ($I_{OR} = I_O \times n_S/n_P$) we get

$$I_{OR} = \frac{I_{IN} \bullet (1 - D)}{\sqrt{D \bullet (1 - D)}} = \frac{1.5 \bullet (1 - 0.5)}{\sqrt{0.5 \bullet (1 - 0.5)}} = 1.5 \text{ A}$$

The flat-top equation for the switch rms is then solved

$$I_{SW} = \frac{I_{OR}}{1 - D} \bullet \sqrt{D} = \frac{1.5}{1 - 0.5} \bullet \sqrt{0.5} = 2.12 \text{ A}$$

Now let us do an exact calculation to compare with. We have the input rms current

$$I_{IN} = \frac{I_{OR}}{1 - D} \bullet \sqrt{D \bullet \left[1 - D + \frac{r^2}{12}\right]}$$

Solving for the reflected output current we get

$$I_{OR} = \frac{(1 - D) \bullet I_{IN}}{\sqrt{D \bullet \left[1 - D + \frac{r^2}{12}\right]}} = \frac{(1 - 0) \bullet 1.5}{\sqrt{0.5 \bullet \left[1 - 0.5 + \frac{1.2^2}{12}\right]}} = 1.347 \text{ A}$$

So for the switch rms we get

$$I_{SW} = \frac{I_{OR}}{1 - D} \bullet \sqrt{D \bullet \left[1 + \frac{r^2}{12}\right]} = \frac{1.347}{1 - 0.5} \bullet \sqrt{0.5 \bullet \left[1 + \frac{1.2^2}{12}\right]}$$

$$I_{SW} = 2.016 \text{ A}$$

This time the flat-top equations being used *progressively*, actually *overestimated* the switch current. Clearly Fig. 1.8 cannot help. The reason is that we have $I_{OR}$ in both equations and we should have known this parameter directly rather than use one flat-top equation to provide an input to the other and so on. That way *the errors accumulate*. If we were dealing with a buck-boost, we would have had no problem since $I_{OR}$ is just $I_O$, the load current (and we better know that!). For a flyback we should have performed a sanity check using the turns ratio of the transformer since we know that $I_{OR} = I_O \times n_S/n_P$.

## 1.11  Tolerance of Set Output Voltage

All designers know the basics of how to set up a simple voltage divider. But there is much more to know here. Looking at Fig. 1.9 we see that assuming that the current through $R_2$ passes straight into $R_1$, we get from Ohm's law

$$\frac{V_o - V_{fb}}{V_{fb} - 0} = \frac{I \times R_2}{I \times R_1} = \frac{R_2}{R_1}$$

Solving, we get the standard equation

$$R_2 = R_1 \times \left( \frac{V_o}{V_{fb}} - 1 \right)$$

This basic equation seems to suggest that $R_2$ is always a certain fixed ratio to $R_1$, for a given output voltage. For example, if $V_{fb} = 2.42$ V we can set $R_2 = 806$ and $R_1 = 2.21$ kΩ for an output of 3.3 V. Now it may have been thought that using $R_2 = 8.06$ kΩ and $R_1 = 22.1$ kΩ would have been equally acceptable. But the simple divider equation is just that—an ideal equation that unrealistically assumes zero current into or out of the feedback pin. It can be easily shown that the effect of any minute current, say $I_{in}$, flowing into the feedback pin is to raise the output voltage slightly from the ideal. Now the current in the two resistors gets slightly unbalanced, so we have used $I_1$ and $I_2$ in the figure. This difference causes an output voltage error which needs to be understood and quantified. There is of course a basic tolerance band for the reference

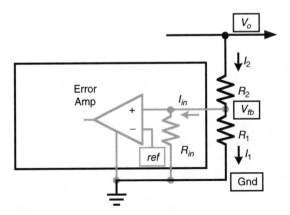

Figure 1.9  Simple voltage divider.

voltage too. There can also be a dc offset error between the two input pins of the error amplifier. So the feedback pin may settle down, under regulation, to a voltage slightly different from the reference applied on the other pin. Achieving overall output accuracy is not an easy or trivial task. But for the system-level engineer there is still a lot that can be done outside the IC. What we don't want to do is to compound the output error by not doing something to account for and control the effect of the current that flows into (sometimes out of) the feedback pin.

If the feedback pin "steals" $I_{in}$ away from $I$, we are left with $I_1 = I - I_{in}$ flowing through $R_1$. We assume that the resistors $R_2$ and $R_1$ have not changed, but $V_o$ has. Call it $V_{o'}$ now. So the equation becomes

$$\frac{V_{o'} - V_{fb}}{V_{fb} - 0} = \frac{I_2 \times R_2}{I_1 \times R_1}$$

Comparing it with the ideal equation and solving for the error we get

$$\Delta V = (V_o - V_{fb}) \times \frac{I_{in}}{I_1}$$

So for a given feedback voltage, *the clue to minimizing the error is to increase the current in the divider*, that is, $I_1$ and $I_2$. The results are plotted out in Fig. 1.10 for the most common case of $V_{fb} = 2.5$ V

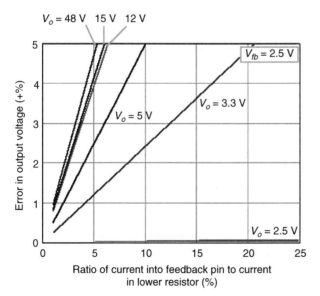

Figure 1.10 Minimizing the output error due to the feedback pin bias current.

(as when we use the popular TL431 voltage reference). Note that if the input bias current were fixed we could easily compensate it by tweaking the resistors of the divider. The reason we can't use this approach in production has to do with the *variability* of this bias current. It can vary anywhere between the "min" and "max" limits stated in the datasheet of the switcher IC. For the worst-case error estimate we must take the maximum value of the spread in the bias current, not its typical (or typ) value, nor its "min" value in the calculation above. Note also that for the *same bias current and same divider current we get higher output errors for high output voltages (absolute error and even expressed as a percentage)*. The small bias current clearly has a crowbar effect on the output voltage level.

**Example 1.4**   We have a 2.5 V reference and the lower resistor of the divider is set to 62.5 kΩ. We set up a divider with the ideal divider equation to get 12 V on the output. What voltage will we see instead, if the pin flowing into the feedback pin is 2 $\mu$A?

With 62.5 kΩ at the lower position, the divider current is

$$I_1 = \frac{2.5}{62.5k} = 0.04 \text{ mA}$$

So the ratio of the bias current to the divider current is

$$\frac{I_{fb}}{I_1} = \frac{2 \times 10^{-6}}{4 \times 10^{-5}} = 0.05$$

So

$$\Delta V = (12 - 2.5) \times 0.05 = 0.475 \text{ V}$$

The output voltage thus will be 12.475 V, an error of $+0.475/12 = +4$ percent. We could have seen this directly from Fig. 1.10. Under the same conditions, for a 3.3 V output we would have gotten about $+1.2$ percent error only.

Therefore for setting higher output voltages accurately, we must pass more current in the divider (decrease $R_1$ and $R_2$). This unfortunately works against minimizing the power dissipated in the divider (when we need it most). In particular the dissipation in the upper resistor $R_2$ will increase significantly and we should check it against the rating of the chosen component.

In geographical regions, where labor is cheap, it is a normal practice to leave large resistors in parallel to $R_2$ or $R_1$. These are calculated to be able to shift the output voltage by manually testing the output and then cutting the resistors according to the measured deviation $\Delta$ and a lookup table. We of course need to ensure that with no resistor cut, the power supply output will always be offset in one direction. For maximum flexibility and trimming range these resistors are mutually ordered as a *binary weighted* sequence $R$, $2R$, $4R$, and so forth. $R$ is chosen so that if we cut *all* the resistors, the initial worst-case deviation $\Delta$ will swing and become close to $-\Delta$. In theory, with $n$ trim resistors we can reduce the output spread by a factor of $2^n - 1$. In practice, even with 5 percent tolerance trim resistors (the standard series—we won't get exact binary values with these) we get closer to $1.9^n - 1$. For example, if we start with a 10 percent deviation, with two resistors we can reduce it to 3.8 percent (i.e., $\pm 1.9$ percent).

**Note:** Trimpots are sometimes used, effectively forming a small variable resistor in series with the larger lower fixed resistor of the divider. But generally speaking, trimpots are expensive and unreliable in the long run, so we should avoid them if possible.

### 1.12   Preferred Resistor values

In the past, circuit designers were able to precisely specify the value of the resistor required at a particular circuit location based on calculations. But this created difficulties for the resistor manufacturers since there was little or no standardization and hence all resistors were "custom made." Even today, in the author's experience, it is bewildering to discover that even some big American PC system manufacturers routinely ask for whatever theoretical value they calculate, and usually get it. The only explanation is that resistor vendors just didn't consider it prudent to risk losing a big account by "educating" the customer about "preferred values." But most of us are not so lucky. For example, especially in Europe, the E-series is fully known and widely used.

Manufacturing costs can go up for several reasons. For example, (1) if we use custom parts or nonpreferred values and (2) if we have a large number of possibly *unnecessary* part values. For example, if we look carefully at some power supply circuits we may find a 10 kΩ resistor at some point and for some odd reason (probably associated with blindly cutting and pasting a "nice" circuit block from a previous product) we may find 10.5 kΩ at another point. On closer analysis we will often see no reason why the 10.5 kΩ could not have been made into a 10 kΩ. Thus we forget that not only do we want to reduce the number of parts in our power supply, but we also want to reduce the number of types of different parts. This would serve us logistically, especially because each new part's tape and reel needs to be separately loaded onto a pick-and-place machine.

One of our design goals is *not* to cater to a circuit's sensitivity to tolerances but to diminish it. With some thought we could *make* the circuit accept a wider range of values in that location thus also increasing our

TABLE 1.3  Standard E-series Tolerances.

| Series | Tolerance (±%) | Values per decade |
|---|---|---|
| E3 (obs.) | 50 | 3 |
| E6 | 20 | 6 |
| E12 | 10 | 12 |
| E24 | 5, 2 | 24 |
| E48 | 2 | 48 |
| E96 | 1 | 96 |
| E192 | 0.5, 0.25, 0.1 | 192 |

manufacturing yields. We could also consider using a 10 kΩ in series with a 470 Ω if the latter value was being used elsewhere in the circuit. Not only would we reduce ordering costs in the process, but we would also get parts cheaper because we are using larger quantities of each. All told, we may even end up adding a couple of cents to our *bill of materials* (BOM), but it could save a lot of related costs and headaches in the long run. As even component manufacturer Vishay advises, "Use of standard values is encouraged because stocking programs are designed around them."

But what are *preferred values?* We all remember using values like 220 Ω, 470 Ω, and the like in the past. Today these are being replaced by 221 Ω, 464 Ω, and the like. There are reasons for this. Look at Table 1.3, which gives the usual tolerances of the standard E-series.

The preferred values (E-series) are a modern system for selecting nominal values within a given decade, *based on the accuracy with which they can be manufactured.* For example, when we had the capability to produce only 10 percent tolerance resistors, and we arbitrarily picked the first preferred value to be 100 Ω, then it made little sense to produce a 105 Ω resistor, since it falls within the 10 percent tolerance range of the 100 Ω resistor. The next *reasonable* value is 120 Ω. So on and so forth, and we will get the well-known E12 series. But as manufacturing capability improved, we progressed to 0.1 percent tolerances. So we have also added several new series. All the series start with the letter 'E' followed by a number denoting how many nominal values there are within a decade (a decade is, say, 10 to 100 Ω, or, say, 100 Ω to 1 kΩ).

The series from E3 through E24, however, comprises a separate and distinct set of values, and from E48 through E192 we have a different set. For example, our old 220 Ω, 470 Ω, and the like all belong to the E24 series. Though *metal oxide film* (MOF) resistors for current sense applications, or even *metal oxide* power resistors, or the various high wattage/high voltage types may still come as 0.33 Ω, 0.22 Ω, 22 Ω, it probably makes no commercial sense anymore to carry inventories in E24 signal-level resistors. So engineers who feel that they must place a 5 percent resistor somewhere because they think they are saving money or designing 'optimally' must be stopped. In the author's experience and opinion, many companies overlooked this for a few years and by then had already reached exploding inventories before they decided to act. But they couldn't push back the clock fully because several power supplies were already in the field with the older E24 values, and they had to continue to support them. Not only that, they couldn't "redesign" them with the 1 percent or 2 percent resistors because that could require requalification by the customer and possibly a fresh safety approval process too. We should note that any major changes after product release, particularly on the primary side of an off-line power supply, may require a fresh safety approval. In any case, safety agencies have always to be informed of subsequent changes.

Standard resistor values are provided in Table 1.4. Remember that E192 series are very expensive and rarely used in commercial power supplies. Also see that the E48 series is a subset of the given E96 series. All successive decade values can be generated by multiplying the previous decade by 10.

**Caution:** The designer should think twice before using any single resistance larger than about 0.5 MΩ in his or her power supply. In fact, some extremely quality-conscious power supply companies have internal rules prohibiting any value greater than 100 kΩ. Contamination on the printed circuit board (PCB) or moisture and humidity can cause a large change in the resistance. So we may need to use several 100 kΩ resistors in series rather than a single large-value resistor.

## 1.13   Optimum Divider Selection

While designing voltage dividers we must study the type of error amplifier it connects to. We can have a transconductance op-amp or a regular op-amp. These can be recognized as follows. The transconductance type will have no loop compensation components present between its output and its inverting pin. It will just have an RC type of network connected straight from the output (of the op-amp) to ground. But the regular/conventional op-amp will need direct feedback components between its output pint and its inverting pin. As far as the overall stability loop is concerned it has different rules for these two types of error amplifiers, and this will determine how we set up the divider.

TABLE 1.4  Standard Resistor Series.

| E6 | E12 | E24 | E48 | E96 | E192 | E6 | E12 | E24 | E48 | E96 | E192 | E6 | E12 | E24 | E48 | E96 | E192 |
|---|---|---|---|---|---|---|---|---|---|---|---|---|---|---|---|---|---|
| 100 | 100 | 100 | 100 | 100 | 100 | 220 | 220 | 220 | 215 | 215 | 215 | 470 | 470 | 470 | 464 | 464 | 464 |
|  |  |  |  |  | 101 |  |  |  |  |  | 218 |  |  |  |  |  | 470 |
|  |  |  |  | 102 | 102 |  |  |  |  | 221 | 221 |  |  |  |  | 475 | 475 |
|  |  |  |  |  | 104 |  |  |  |  |  | 223 |  |  |  |  |  | 481 |
|  |  |  | 105 | 105 | 105 |  |  |  | 226 | 226 | 226 |  |  |  | 487 | 487 | 487 |
|  |  |  |  |  | 106 |  |  |  |  |  | 229 |  |  |  |  |  | 493 |
|  |  |  |  | 107 | 107 |  |  |  |  | 232 | 232 |  |  |  |  | 499 | 499 |
|  |  |  |  |  | 109 |  |  |  |  |  | 234 |  |  |  |  |  | 505 |
|  |  | 110 | 110 | 110 | 110 |  |  | 240 | 237 | 237 | 237 |  |  | 510 | 511 | 511 | 511 |
|  |  |  |  |  | 111 |  |  |  |  |  | 240 |  |  |  |  |  | 517 |
|  |  |  |  | 113 | 113 |  |  |  |  | 243 | 243 |  |  |  |  | 523 | 523 |
|  |  |  |  |  | 114 |  |  |  |  |  | 246 |  |  |  |  |  | 530 |
|  |  |  | 115 | 115 | 115 |  |  |  | 249 | 249 | 249 |  |  |  | 536 | 536 | 536 |
|  |  |  |  |  | 117 |  |  |  |  |  | 252 |  |  |  |  |  | 542 |
|  |  |  |  | 118 | 118 |  |  |  |  | 255 | 255 |  |  |  |  | 549 | 549 |
|  |  |  |  |  | 120 |  |  |  |  |  | 258 |  |  |  |  |  | 556 |
|  | 120 | 120 | 121 | 121 | 121 |  | 270 | 270 | 261 | 261 | 261 |  | 560 | 560 | 562 | 562 | 562 |
|  |  |  |  |  | 123 |  |  |  |  |  | 264 |  |  |  |  |  | 569 |
|  |  |  |  | 124 | 124 |  |  |  |  | 267 | 267 |  |  |  |  | 576 | 576 |
|  |  |  |  |  | 126 |  |  |  |  |  | 271 |  |  |  |  |  | 583 |
|  |  |  | 127 | 127 | 127 |  |  |  | 274 | 274 | 274 |  |  |  | 590 | 590 | 590 |
|  |  |  |  |  | 129 |  |  |  |  |  | 277 |  |  |  |  |  | 597 |
|  |  |  |  | 130 | 130 |  |  |  |  | 280 | 280 |  |  |  |  | 604 | 604 |
|  |  |  |  |  | 132 |  |  |  |  |  | 284 |  |  |  |  |  | 612 |
|  |  | 130 | 133 | 133 | 133 |  |  | 300 | 287 | 287 | 287 |  |  | 620 | 619 | 619 | 619 |
|  |  |  |  |  | 135 |  |  |  |  |  | 291 |  |  |  |  |  | 626 |
|  |  |  |  | 137 | 137 |  |  |  |  | 294 | 294 |  |  |  |  | 634 | 634 |
|  |  |  |  |  | 138 |  |  |  |  |  | 298 |  |  |  |  |  | 642 |
|  |  |  | 140 | 140 | 140 |  |  |  | 301 | 301 | 301 |  |  |  | 649 | 649 | 649 |
|  |  |  |  |  | 142 |  |  |  |  |  | 305 |  |  |  |  |  | 657 |
|  |  |  |  | 143 | 143 |  |  |  |  | 309 | 309 |  |  |  |  | 665 | 665 |
|  |  |  |  |  | 145 |  |  |  |  |  | 312 |  |  |  |  |  | 673 |
| 150 | 150 | 150 | 147 | 147 | 147 | 330 | 330 | 330 | 316 | 316 | 316 | 680 | 680 | 680 | 681 | 681 | 681 |
|  |  |  |  |  | 149 |  |  |  |  |  | 320 |  |  |  |  |  | 690 |
|  |  |  |  | 150 | 150 |  |  |  |  | 324 | 324 |  |  |  |  | 698 | 698 |
|  |  |  |  |  | 152 |  |  |  |  |  | 328 |  |  |  |  |  | 706 |
|  |  |  | 154 | 154 | 154 |  |  |  | 332 | 332 | 332 |  |  |  | 715 | 715 | 715 |
|  |  |  |  |  | 156 |  |  |  |  |  | 336 |  |  |  |  |  | 723 |
|  |  |  |  | 158 | 158 |  |  |  |  | 340 | 340 |  |  |  |  | 732 | 732 |
|  |  |  |  |  | 160 |  |  |  |  |  | 344 |  |  |  |  |  | 741 |
|  |  | 160 | 162 | 162 | 162 |  |  | 360 | 348 | 348 | 348 |  |  | 750 | 750 | 750 | 750 |
|  |  |  |  |  | 164 |  |  |  |  |  | 352 |  |  |  |  |  | 759 |
|  |  |  |  | 165 | 165 |  |  |  |  | 357 | 357 |  |  |  |  | 768 | 768 |
|  |  |  |  |  | 167 |  |  |  |  |  | 361 |  |  |  |  |  | 777 |
|  |  |  | 169 | 169 | 169 |  |  |  | 365 | 365 | 365 |  |  |  | 787 | 787 | 787 |
|  |  |  |  |  | 172 |  |  |  |  |  | 370 |  |  |  |  |  | 796 |
|  |  |  |  | 174 | 174 |  |  |  |  | 374 | 374 |  |  |  |  | 806 | 806 |
|  |  |  |  |  | 176 |  |  |  |  |  | 379 |  |  |  |  |  | 816 |
|  | 180 | 180 | 178 | 178 | 178 |  | 390 | 390 | 383 | 383 | 383 |  | 820 | 820 | 825 | 825 | 825 |
|  |  |  |  |  | 180 |  |  |  |  |  | 388 |  |  |  |  |  | 835 |
|  |  |  |  | 182 | 182 |  |  |  |  | 392 | 392 |  |  |  |  | 845 | 845 |
|  |  |  |  |  | 184 |  |  |  |  |  | 397 |  |  |  |  |  | 856 |
|  |  |  | 187 | 187 | 187 |  |  |  | 402 | 402 | 402 |  |  |  | 866 | 866 | 866 |
|  |  |  |  |  | 189 |  |  |  |  |  | 407 |  |  |  |  |  | 876 |
|  |  |  |  | 191 | 191 |  |  |  |  | 412 | 412 |  |  |  |  | 887 | 887 |
|  |  |  |  |  | 193 |  |  |  |  |  | 417 |  |  |  |  |  | 898 |
|  |  | 200 | 196 | 196 | 196 |  |  | 430 | 422 | 422 | 422 |  |  | 910 | 909 | 909 | 909 |
|  |  |  |  |  | 198 |  |  |  |  |  | 427 |  |  |  |  |  | 920 |
|  |  |  |  | 200 | 200 |  |  |  |  | 432 | 432 |  |  |  |  | 931 | 931 |
|  |  |  |  |  | 203 |  |  |  |  |  | 437 |  |  |  |  |  | 942 |
|  |  |  | 205 | 205 | 205 |  |  |  | 442 | 442 | 442 |  |  |  | 953 | 953 | 953 |
|  |  |  |  |  | 208 |  |  |  |  |  | 448 |  |  |  |  |  | 965 |
|  |  |  |  | 210 | 210 |  |  |  |  | 453 | 453 |  |  |  |  | 976 | 976 |
|  |  |  |  |  | 213 |  |  |  |  |  | 459 |  |  |  |  |  | 988 |

*For the transconductance error amplifier.* It doesn't care about the values of the upper and lower resistors as long as their *ratio* is maintained. The divider is just a step-down gain block in its feedback control loop. Therefore we could use, say, 2.21 kΩ and 806 Ω for a 3.3 V output setting with a feedback voltage of 2.42 V, or we could also choose to use 22.1 kΩ and 8.06 kΩ instead. Here the only dominant concern may be concerning the bias current issue discussed earlier.

*For the regular op-amp.* The upper resistor (only) comes into play in the control loop. The lower resistor just acts as a dc biasing element and does not enter the stability calculations. So in this case we would rather first set the upper resistor (based on stability considerations), and then choose the lower resistor using the voltage divider equation. Note that in this case we may be able to do very little about the bias current variability issue, since we would usually want to give priority to the loop stability. Indirectly, this implies that better output accuracy is usually possible with transconductance type of error op-amps, unless of course a very tight control is maintained on the feedback pin bias current.

Hint: Feedback loop calculations are much easier with regular op-amps. So in this case, if we have a converter with a previously established and desirable loop response characteristic, we should keep the upper resistor (and all other associated compensation components) unchanged if we want to set it up for a different output voltage setting. In doing so, the Bode plot will not change, since it does not care about the value of the lower resistor.

If we have full flexibility in choosing the upper and lower resistors, we can ask the question: *What combination of standard* one *percent*

TABLE 1.5  Best Voltage Divider Combinations for Some Typical Cases.

| $R_2$ (upper) | 115 | 523 | 590 | 806 | 1.18E+03 |
|---|---|---|---|---|---|
| $R_1$ (lower) | 357 | 137 | 118 | 130 | 137 |
| Error (%) | 0.161 | 0.364 | 0 | 0 | 0.137 |
| Output rail (V) | 3.3 | 12 | 15 | 18 | 24 |
| Reference (V) | 2.5 | 2.5 | 2.5 | 2.5 | 2.5 |
| $R_2$ (upper) | 301 | 133 | 137 | 1.07E+03 | 1.02E+03 |
| $R_1$ (lower) | 1.37E+03 | 287 | 133 | 634 | 332 |
| Error (%) | 0.016 | 0 | 0.12 | 0.177 | 0.178 |
| Output rail (V) | 1.5 | 1.8 | 2.5 | 3.3 | 5 |
| Reference (V) | 1.23 | 1.23 | 1.23 | 1.23 | 1.23 |
| $R_2$ (upper) | 931 | 1.02E+04 | 1.50E+03 | 1.07E+04 | 5.23E+03 |
| $R_1$ (lower) | 107 | 909 | 110 | 576 | 137 |
| Error (%) | 0.569 | 0.213 | 0.015 | 0.328 | 0.385 |
| Output rail (V) | 12 | 15 | 18 | 24 | 48 |
| Reference (V) | 1.23 | 1.23 | 1.23 | 1.23 | 1.23 |

## E96 SERIES

ORIGIN $\equiv 1$

We have 96 values of resistor from 1 K$\Omega$ to 10 K$\Omega$ (in 1 decade) forming a geometric progression with the multiplying factor:

$10^{\frac{1}{96}} = 1.024$

Resistor values index:

$r := 1 .. 96$

Unrounded Values are:

$R96\_base_r := 100 \cdot \left(10^{\frac{1}{96}}\right)^{r-1}$

Rounding up to get the standard series:

$R96_r :=$ $\begin{vmatrix} f \leftarrow \text{floor}(R96\_base_r) \\ c \leftarrow \text{ceil}(R96\_base_r) \\ f \quad \text{if} \quad R96\_base_r - f < 0.5 \\ c \quad \text{otherwise} \end{vmatrix}$

4 decades of standard values are thus:

$R96\_extend := 10.R96$

$R96\_extend\_extend := 10^2.R96$

$R96\_extend\_extend\_extend := 10^3.R96$

We stack them into one table:

$R := \text{stack}[R96, (\text{stack}(R96\_extend, \text{stack}(R96\_extend\_extend,$
$R96\_extend\_extend\_extend)))]$

This is the E96 set of values which will be used for the best combination sweep below.

Q: What is the best upper and lower resistor combination for lowest error?

Set regulated voltage rail

$Vin := 3.3$

Set reference voltage

$Vfb := 2.42$

Create an index to generate all values from the same 'Solution' algorithm below: $p := 1 .. 3$

$\text{Solution}_p :=$ $\begin{vmatrix} Err \leftarrow 1 \\ \text{for} \quad a \in 1 .. 384 \\ \quad \text{for} \quad b \in 1 .. 384 \\ \qquad \begin{vmatrix} rl \leftarrow R_a \\ ru \leftarrow R_b \\ vf \leftarrow Vin \cdot \dfrac{rl}{rl + ru} \\ err \leftarrow \left| \dfrac{vf - vfb}{vfb} \right| \\ \text{continue} \quad \text{if} \quad err \geq Err \\ Err \leftarrow err \\ RL \leftarrow rl \\ RU \leftarrow ru \end{vmatrix} \\ RU \quad \text{if} \quad p = 1 \\ RL \quad \text{if} \quad p = 2 \\ (Err \cdot 100) \quad \text{if} \quad p = 3 \\ 0 \quad \text{otherwise} \end{vmatrix}$

Results are: $Rupper := \text{Solution}_1$  $Rlower := \text{Solution}_2$  $Error := \text{Solution}_3$
$Rupper = 115$  $Rlower = 316$  $Error = 0.021 \%$

**Box 1.2**  Mathcad file for best resistor combination.

*resistors will produce the lowest output error?* Here we are ignoring the feedback pin bias current issue for now and are only interested in achieving the least error using standard resistor values. In addition, we are hoping to avoid using parallel or series combinations to achieve the required accuracy. We want to use just two resistors in the divider.

We will need to scan every possible combination of standard resistors and minimize the error. In Box 1.2 we have a Mathcad file for the purpose. Note that in this file we have not explicitly considered the feedback pin bias current. The suggested resistor combinations are based on an ideal calculation, and can be further increased in decade steps (e.g., multiplying both resistors either by 10 or by 100). We can go as far as the bias current error analysis allows.

In Table 1.5 we have provided some typical divider solutions for quick reference. These are the best combinations using only two resistors in the divider, when using standard values.

NOTES:

- VOLT-SECOND   $V_{ON} \cdot I_{ON} = V_{OFF} \cdot I_{OFF}$

- DCM - DISCONTINUOUS CONDUCTION MODE - WHEN THE REVERSAL OF CURRENT IN THE INDUCTOR TAKES LESS THAN $I_{OFF}$. THE INDUCTOR STOPS CONDUCTING FOR A SPLIT PART OF THE CYCLE.

- THE GOAL OF CONTROL IS TO HAVE THE CURRENT IN THE INDUCTOR RETURN TO IT'S NO-ZERO STATE EVERY CYCLE. THE CONTROL CIRCUIT ADJUSTS THE DUTY CYCLE TO MAKE THIS HAPPEN; STEADY STATE.

CCM - STEADY STATE

r - RIPPLE CURRENT RATIO, RATIO OF CURRENT SWING TO CONT. CURRENT

$K_p$ - RATIO OF CURRENT SWING TO PEAK CURRENT

$$K = K_o = \frac{\Delta I}{I_p} \qquad r = \frac{\Delta I}{I_c}$$

# 2

# DC-DC Converters and Their Configurations

## 2.1  Introduction

One of the first things we must realize is that a switcher IC (integrated circuit) for a dc-dc converter application can actually be used across topologies quite freely. For example a buck IC can be used in a buck-boost application and vice-versa. This is not so puzzling once we understand that a switcher is just that—it basically switches a transistor between *on* and *off* states. How we use the transistor and configure our circuit to apply and switch the voltage across an associated inductor, and how we finally route the energy to the output is not necessarily preordained. Indeed, there are limitations on what all we can do, but that has more to do with how the transistor is actually driven, how the internal control of the IC is referenced, and the like. It is certainly important to understand the internal construction of the IC itself, but to unveil the other "hidden" applications, needs a better (and more abstract) understanding of switcher topologies themselves.

Related literature occasionally refers to some of these applications but in a rather scattered and unintelligible way. One major practical impediment to a clear understanding that we noticed was the way schematics were often being drawn. We saw schematics that required several right-angle turns and/or horizontal and/or vertical mirror reflections to make any sense of. To compound the confusion, there was rarely any attempt to even explicitly state the fact that the underlying topology itself had changed in the process. Therefore, in the accompanying text, it may just have been mentioned casually that the buck IC was now being used in an "inverting" configuration (or a "positive to negative" converter). Go figure! What is the maximum load current in this new configuration and what is the safe input operating voltage range? Why did these change? Is that the feedback path? And how does it really work now? Is the accuracy of the output regulation as good as it was for the original circuit? Questions like these just add to the general mystery surrounding switcher ICs. We will attempt to not only go past these impediments, but to understand the subtle chain of logic by which we can virtually predict, even before putting pen to paper, what all we can do with a certain switcher IC, other than its primary intended application.

Finally, to enact our learning out, we also need to bridge the crucial link, i.e., layout. For each topology there are different traces and components that are "critical." Failure to understand this will never get us the desired results. So at the end of the day, we must also pay close heed to Chap. 12 dealing with PCB layout.

## 2.2   What is *Ground*?

We need to start with some basics. In a dc-dc converter there are two input rails (connected to the dc source) and two output rails (connected to the load). Of these, one rail is always shared between the input and the output. By convention, this common rail is designated the (system) *ground*. It is the ground for the power stage. But there is another ground too. This is the IC/control ground. This may or may not always be connected to the system ground, especially if the IC is being used in a manner other than its primary intended application. In that case, there could be certain associated problems. For example, the feedback path may not be *direct* anymore, since the feedback voltage will always need to be finally referenced to the IC ground. Since the system ground may be at a different voltage level than the IC ground, sensing of the output would now need to be *differential* in nature. The sensed difference voltage would then need to be level-shifted or "translated" to the IC ground. There are several standard ways of doing this—all involving external components. These will be discussed later.

Note:  The feedback pin has absolute maximum voltage ratings that should not be exceeded. Though this pin is inherently self-stabilizing (at least in steady state), the designer may need to watch out for transients, during start-up or shutdown or under short circuits on the output, that could damage the feedback pin. This is usually of concern only when the IC is used "unconventionally," i.e., in a configuration other than its basic intended application.

Since the many different configurations can appear bewildering at first sight, it will help to point out their common threads. This will lead to a better visualization in terms of their basic building blocks and highlight all the things we can do—without missing any possibilities. Therefore, we will start off with some basic definitions.

## 2.3 The *N-switch* and *P-switch*

For bringing the myriad possibilities under smaller and smaller umbrellas, we need to make some rather unconventional definitions in this chapter. The reader should bear with us, as he or she will see that it really does help in seeing the bigger picture.

In Fig. 2.1 we have indicated that a voltage of magnitude $v$ needs to be applied with respect to the source terminal of both N-channel FETs (field effect transistors) and P-channel FETs to turn the FET *on* (assuming enhancement type FETs only). The dotted triangles (alongside the label $v$) indicate the direction of *increasing* voltage (in terms of magnitude). We have also indicated that a voltage of magnitude $v$ needs to be applied with respect to the emitter terminals of both NPN and PNP BJTs (bipolar junction transistor) to turn the BJT *on*.

Throughout this chapter we will usually attempt to keep the lower voltage input rail on the bottom side of the figure and the higher voltage input rail on the top side. Further, the input is usually on the left side and the output on the right. These steps will help in keeping all the schematics visually appealing, easy to follow, and mutually consistent.

In Fig. 2.1 we have also shown that the easiest way to turn the FET or BJT *off* is to connect the gate/base to the source/emitter.

Because of the drive similarities, in this chapter we will generally talk in terms of an *N-switch* (being either an *N*-channel FET or an NPN BJT) or a *P-switch* (either a P-channel FET or a PNP BJT). There are some differences between the circuit requirements of BJTs and FETs, but we will discuss these later.

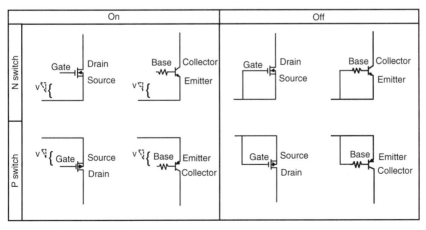

**Figure 2.1**  N-switches and P-switches.

## 2.4  The LSD Cell

Coming to the basic structure of power conversion circuits, in all cases we have an inductor (L), a switch (S), and a diode (D) connected to each other. This is hereafter generically nicknamed an *LSD cell*. The LSD cell as introduced here goes a long way in understanding which hidden applications are a "natural possibility" for a given IC and which are not. It will be shown that if we identify the LSD cell type occurring in the primary intended application of a switcher IC, we can easily apply it to any other topology in which the *same LSD cell type occurs* (irrespective of the topology or configuration). We will also see that in some cases, by a special technique, we can even make an IC perform in an LSD cell type *other than its originally intended LSD cell*.

The designer should preferably start thinking in terms of HI (high) and LO (low) rails rather than positive or negative rails, or even ground, because the designations of the rails may change as we change topologies and configurations but what is HI will still remain a high voltage rail with respect to the LO rail.

From the LSD cell we can generate the "− cell" which is so called because as we can see from Fig. 2.2, this requires the switch to be *below* the switching node. The "+ cell" has the switch positioned above the switching node. Each class leads to two more types—one with a P-channel FET (or PNP BJT) and one with an N-channel FET (or NPN BJT). So we get the N+, P+, and the N− and P− cells. Note that in the figure the body diode for each FET is shown in gray. Clearly, it cannot

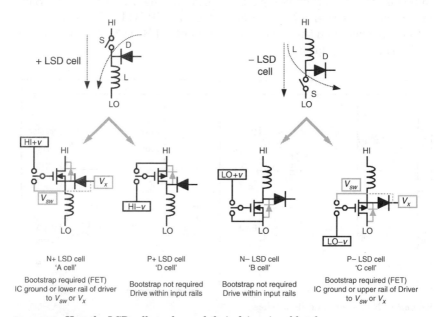

**Figure 2.2**  How the LSD cells evolve and their drive signal levels.

be the other way round or the input supply will see a short. Note that the *LSD node* is the same as the switching node of the converter.

We also see that *two of the four cases require a drive voltage outside the input rails* (the first and the fourth) whereas two don't. This is actually true only for FETs, and not BJTs, and has a great bearing on the usefulness and applicability of the device. If a drive signal is required to be outside the available input rails, we would need a *bootstrap circuit*. But any bootstrap capacitor needs to be "refreshed" every cycle. This can happen only when the switch turns *off* because that is when the internal circuitry releases a current source for charging up the capacitor (close to the HI). Therefore, any IC requiring a bootstrap is not allowed to have a duty cycle of 100 percent.

We note that a duty cycle of 100 percent is often considered quite useful for a buck topology because as the input falls below the set output level, the output will track the input very closely. The input to output differential is just one switch forward voltage drop. The device is then effectively just functioning as an LDO (linear dropout regulator) of sorts. In portable battery-operated devices this tracking can greatly extend the maximum operating time of the unit (as the batteries slowly discharge). So for the buck, the P+ cell is getting increasingly popular, especially for portable devices like MP3 players. However, we also note that *100 percent duty cycle is not acceptable for either the boost or the buck-boost*. These topologies are fundamentally different from the buck in that energy is delivered to the output only when the switch turns *off*. So we can end up with a curious situation at start-up where the control maximizes the duty cycle (because the output voltage is low and it is trying to get it to rise), but since the duty cycle is 100 percent the output just cannot rise (because there is no off-time). This is a Catch-22 situation at a circuit level! Surprisingly, there have been cases in the industry where such a poorly defined IC was released. Look around!

Note also that when the diode conducts, i.e., the switch turns *off*, $Vx$ and $Vsw$ in Fig. 2.2 become almost the same. $Vsw$ here refers to the voltage at the switching node (it is not the forward drop across the switch, which is also called $Vsw$ elsewhere in this book), and $Vx$ is the voltage on the *other side* of the diode. So in two of the cases we could have turned the switch *off* by connecting to $Vx$ instead of the more obvious node $Vsw$. In some applications we actually have to do exactly this.

We see that we can have in all four LSD cell types. Two correspond to N-switches and two to P-switches. We have called them A, B, C, D cells in the order shown in the figure. Therefore

- If the cathode of the diode connects to the LSD node, we call it a "+" LSD cell

- If the anode of the diode connects to the LSD node, we call it a "−" LSD cell

And further we have:

- Type A (<u>N+cell</u>): cathode is at LSD node, with N-channel FET or NPN BJT

- Type B (<u>N−cell</u>): anode is at LSD node, with N-channel FET or NPN BJT

- Type C (<u>P−cell</u>): anode is at LSD node, with P-channel FET or PNP BJT

- Type D (<u>P+cell</u>): cathode is at LSD node, with P-channel FET or PNP BJT

## 2.5    Configurations of Switching Regulator Topologies

We note that the words "boost," or "buck," or "buck-boost" always refer only to the *magnitudes* of the input and output voltages. So now we see the need for qualifiers like "negative to negative," "negative to positive," and the like, to fully describe the actual *configurations*. The negative (or inverted) form of the popular positive to positive boost converter is in full a negative to negative boost converter. It would convert say −12 V to −48 V, relative to the (common) ground rail.

We have four possible configurations for each of the standard topologies. In Figs. 2.3, 2.4, and 2.5 we show the possibilities for the buck, the

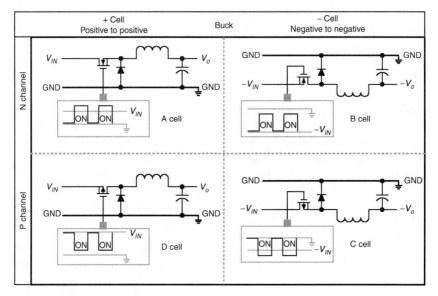

**Figure 2.3**  Buck configurations (with FETs).

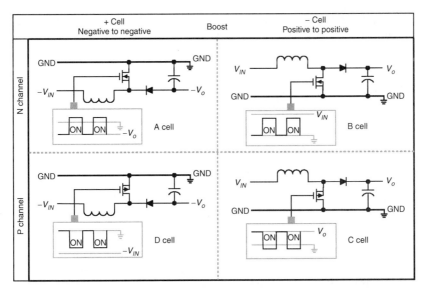

**Figure 2.4**  Boost configurations (with FETs).

boost, and the buck-boost converters respectively. The gate drive levels are also shown. Note the following

1. By "inversion" we can convert an entire schematic (including switch-drive signals) involving an N+ cell to a P− cell and vice versa, that is, $A \Leftrightarrow C$

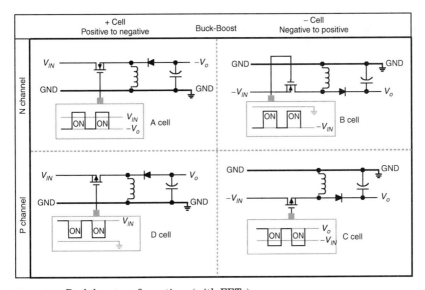

**Figure 2.5**  Buck-boost configurations (with FETs).

2. Similarly we can invert a schematic (with its drive signals) using an N–cell to a P+ cell, that is, $\boldsymbol{B} \Leftrightarrow \boldsymbol{D}$.

3. *Inversion* means we change the polarity of all voltages (e.g., $V$ to $-V$), we also change polarities of the capacitors, we also reverse the diode direction, and we change an N-channel FET to a P-channel FET (and vice versa), or an NPN BJT to a PNP BJT. Remember, however, that for transistors we do not change the *name* (label) of the pin. For example, source remains source, drain too remains the same. Note also that in the process of inversion, HI changes and becomes a LO, and vice versa. Therefore, after inversion, we may like to flip the schematic vertically to comply with our preferred convention of keeping HI on the top and LO below. The reader should try out this procedure here and see how it works.

**Example 2.1**  If we need to create a schematic for a negative to positive buck-boost using a P-channel FET (Type C from Fig. 2.5), we can start by generating a positive to negative buck-boost schematic using an N-channel FET (Type A from Fig. 2.5) and then invert it.

Keep in mind that N-switches are generally preferred as they reduce die size for a given $Rds/V_{ce\_sat}$ as compared to P-switches. Here Rds is the drain to source on-resistance and $V_{ce\_sat}$ the collector to emitter saturation voltage drop. Therefore we will now focus almost exclusively on ICs using N-switches only. But the rules we will uncover in the process will apply universally, even to P-switches.

## 2.6  Basic Types of Switcher ICs

In the previous figures, the details of the IC and control were not shown. Let us now study some typical integrated switcher ICs first to see how they are internally configured.

Commonly, there are four basic switcher IC types available (all use N-switches) as shown in Fig. 2.6. On closer examination we see that these fall into two basic categories, hereby designated Type 1 and Type 2 ICs. Note *the bold trace shown in the figures is the connection between the switch and the control section*. And this is what makes the two types really different. Type 2 ICs are generally considered to be flyback/buck-boost/boost ICs, and Type 1 ICs are considered buck ICs. We will soon see that *Type 1 ICs are generally more versatile.*

Here are some related comments:

Type 1 connects the source/emitter (lower voltage switch pin) to the $-$pin of the control block. This is sometimes also called an IC with *low-side drive*.

Type 2 connects the drain/collector (higher voltage switch pin) to the $+$pin of the control block. This is called an IC with a *floating drive* or *high-side drive*.

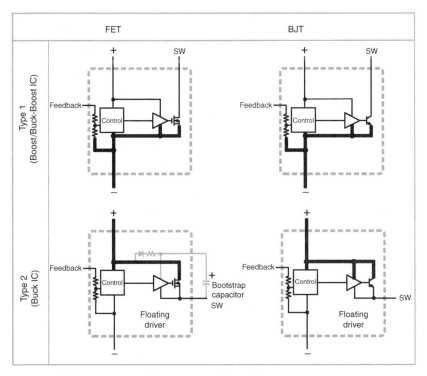

**Figure 2.6**  The two common IC types using N-switches.

We can see that a Type 2 IC needs a bootstrap circuit if a FET is being used. The bootstrap capacitor will get refreshed by the resistor and diode (as shown) when the switch turns *off*. Therefore 100 percent duty cycle will not be possible nor allowed for this type of IC.

The primary intended application for Type 1 IC is the positive to positive boost and we can see from Fig. 2.4 that this uses the Type B cell.

The Type 2 IC is primarily intended for positive to positive buck applications, and we can see from Fig. 2.3 that it corresponds to Type A cell. Note that for N-switches we have only two possible types of cells— Type A or Type B.

In all cases in Fig. 2.3 through Fig. 2.5 we see that for a Type A cell the drive signal is outside the available rails and therefore bootstrap is required (when using a FET).

Note that NPN BJTs are generally easier to drive than FETs since the base has to be taken only slightly higher than the emitter to turn the switch *on* (in fact even the small existing collector-to-base drop can be used for this purpose as in Darlington/$\beta$-multiplier drive arrangements). But for this we cannot drive the transistor into hard saturation. The advantage is that we do not need a bootstrap (or floating driver) since we have now effectively forced the required drive signal levels to be within the available rails. With this convenience comes the realization that BJTs will usually always have a higher forward drop than FETs and therefore they will typically have higher conduction losses. They will also usually be slower and will thus have higher switching losses too.

Note that the actual SW pin labeling in Fig. 2.6 depends on the *perceived* application for the part, not necessarily on how it is *actually* used. It should not be assumed that the SW pin is always the switching node of the converter power stage. In Fig. 2.6 the SW pin is simply the uncommitted pin of the transistor, i.e., the one *not* directly connected to either the "+" or "−" pins (going to the supply rails). Under "normal" expectations, it is expected to be the switching node (as it then has the required degree of freedom to "swing"). But as we will see later, this pin can be connected to a fixed rail, and in fact either the "+" or "−" pins may be forced to be the swinging/switching node! We had also seen earlier that the "−" pin (IC ground for negative ground schematics) may not be the system ground either. Therefore in all cases the designer needs to take the labeling of the IC pins with a pinch of salt, never forgetting what they really are in terms of the internal construction of the IC.

The feedback pin of the IC is usually referred by the control section to the lower (−) rail of the IC. But in reality, how the output voltage of the converter is actually sensed and the voltage "translated" so as to reference it correctly to the IC, depends on the actual application. This will become clearer later.

Type 1 ICs usually have two voltage ratings: one for the control (+pins in Fig. 2.6) and a higher rating for the switch (drain/collector), both measured with respect to the IC ground (the −pins in Fig. 2.6).

Type 2 ICs, almost invariably by internal design, do not permit the SW pin to be taken more than about 1 V below IC ground. This limits some possible applications, particularly some clever ideas using tapped inductors. Switcher ICs that are built to handle tapped inductor topologies are extremely rare. We will now take up Type 1 and Type 2 ICs in more detail below.

## 2.7 Flyback/Buck-Boost/Boost ICs Compared

These are generically referred to as Type 1 ICs in this chapter. We will now see why there is no essential difference between an IC intended primarily for a boost application and one, say, for a flyback/buck-boost

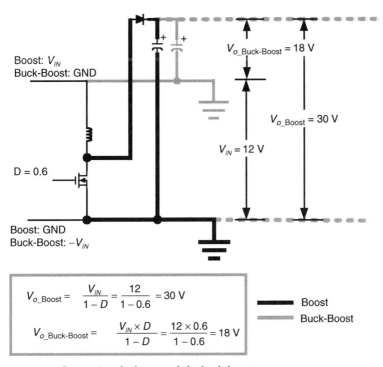

**Figure 2.7**  Comparing the boost and the buck-boost.

application. We should first be aware of the basic topological difference between a boost and a buck-boost power stage.

In Fig. 2.7 we can see that the physical change from a "positive to positive" boost to a "negative to positive" buck-boost is actually very simple—it involves just redirecting the connection of the negative terminal of the output capacitor from the *lower rail* to the *upper rail*. The two topologies are clearly not all that different! In fact as *far as the switch and its drive are concerned, there is no difference at all*, because basically only the designation (or labeling) of the rails has changed. In the example we also observe that in either case, for the same applied duty cycle, the output voltage rail is exactly 30 V as measured from the *IC ground* (not with respect to the system ground, since that designation changes depending on the topology). So, from the viewpoint of the IC, it just won't "know" if we go from a boost to buck-boost topology.

The main remaining difference between the topologies is in the method of implementing *feedback*. Since for a boost, the IC control is typically always connected to the lower rail, a simple resistive divider across the output capacitor can be used to connect (directly) to the feedback pin of the IC. But for the buck-boost, the output voltage is required to be regulated with respect to the system ground (the "upper rail"), whereas the IC control (and its feedback circuitry) is still referenced to the lower rail. Therefore a more elaborate solution is required. This usually takes the form of a differential amplifier stage to sense the output voltage of the buck-boost and then to "translate" it to the lower rail. However, we can see that the requirements, specifications, and ratings of such a differential stage can be extremely diverse depending on the input/output levels. Therefore this extra stage is rarely (if ever) integrated into the switcher IC itself. That means that a true negative to positive buck-boost integrated switcher (with integrated feedback) may be virtually nonexistent. Therefore, since the feedback implementation block is external to the IC, there is no longer any remaining architectural difference between an IC meant for a boost topology and one for a buck-boost topology. They are one and the same IC. The bottom-line is that a switcher IC meant for a flyback/buck-boost application can usually always be successfully used for a boost application and vice versa.

Note that in the above discussion we are talking about a negative to positive buck-boost. For a positive to negative buck-boost application we can see from the various schematics that follow that direct feedback using a simple divider is again possible. But for this we have to float the IC on the switching node (for a Type 1 IC) or on the negative output rail (for a Type 2 IC).

In this chapter we will use the word "flyback" to refer exclusively to a buck-boost stage with inherent primary to secondary isolation. Obviously this requires a transformer. But we could also have a transformer-based buck-boost with *no* isolation present because in that case the primary and secondary windings are deliberately connected together for easier implementation of feedback (without an optocoupler). We can also, in principle, thus create various buck-boost configurations (e.g., negative to positive, positive to positive, and the like).

## 2.8   Other Possible Applications of Buck and Buck-Boost ICs

We know that for the buck-boost and boost topologies the average inductor current goes as $I_o/(1 - D)$. So with a current ripple ratio of around 0.4, we can add 20 percent to the average value of the inductor current to get the peak inductor (and switch) current. The current limit *CLIM* must therefore be set at least that high. But this clearly depends on duty cycle, i.e., input and output voltages. Therefore for a Type 1 IC when we say it is a *3-A device* we are talking about current limit, not load current, as is the case for Type 2 ICs.

For a Type 2 IC, which is basically intended for a buck topology, the average inductor current is the load current $I_o$. We can add about

20 percent to this to get the peak current. We can see that the peak value is almost independent of the input voltage *for this topology*. Therefore when we talk of a Type 2 IC as being a 3-A device we are talking about the load current possible (irrespective of application conditions), but only *provided we are still using it for a buck application*. The current limit of a Type 2 IC is typically set 20 to 40 percent higher than its declared load current capability. If a 3-A Type 2 IC, for example, is used in a boost or buck-boost application, then it *cannot* necessarily deliver 3 A of load current (just like any other 3-A Type 1 IC in a similar application). Therefore, in general, we always have to explicitly calculate how much load current we can actually get from a certain IC, and that may depend on the specific application conditions and topology. In Table 2.1 we have provided all the equations we require to go about the process of formally qualifying a given device for any given application.

**Note:** A word of design caution here: unlike off-line power supplies where we conventionally size the inductor/transformer according to the upper limit ("max") of the current limit tolerance, for low voltage applications (up to about 40 V) we usually size the inductor not according to current limit, but according to the load current. The inductor is thus *expected* to saturate momentarily under start-up or step load conditions. That is usually not a problem since integrated switchers have such fast acting current limits and second-level protection circuitry (often not even declared in the datasheet), that the switch (especially when it is a FET) easily survives almost any abnormal condition. But when the voltages are high (e.g., greater than about 40 V), *we may need to change our inductor selection criterion* for dc-dc converters, from one being based on load current to being sized as per current limit of the device. Sometimes, we may also be operating in a gray area where there is no such hard and fast rule. In that case, bench testing and careful verification on a case-to-case basis will likely be required.

We expect that some of the qualifying conditions/equations may depend on the minimum and/or maximum input voltages, $V_{inmin}$ and $V_{inmax}$ respectively. In addition, every controller is designed with a certain maximum possible duty cycle limit $D_{max}$. Clearly, if the input and output voltages demand more than $D_{max}$ the converter cannot work. Therefore the equations to check for this possible limit are also provided. The feedback scheme is indicated, and the equations to set the resistor values are also provided. $V_{fb}$ is the voltage on the feedback pin of the IC under regulation. Ignoring the small dc offset error, this is equal to the internal reference voltage applied at the other pin of the error amplifier for an *adjustable output part*. Integrated switcher ICs generally come either as *adjustable*, requiring an external resistive divider to set output, or *fixed* voltage parts in which the divider is internal to the IC.

**TABLE 2.1  IC Selection Table for the Various Topologies (Part 1).**

**With inductor:** (all are magnitudes only)

| Topology | Configuration | IC Type* | Figure | | Equations |
|---|---|---|---|---|---|
| Buck | +ve to +ve | 1 | 13 | $V_{sw\,max} \geq V_{in\,max}$<br>$V_{IC\,max} \geq V_{in\,max}$<br>$V_{IC\,min} \leq V_{in\,min}$ | $I_o \leq 0.8 \bullet ICLIM$<br>$R_2 \approx R_1 \bullet \left[\dfrac{V_o}{V_{fb}} - 1\right]$<br>$D_{max} \geq \dfrac{V_o}{V_{in\,min}}$ |
| | | 2 | 14 | $V_{IC\,max} \geq V_{in\,max}$<br>$V_{IC\,min} \leq V_{in\,min}$ | $I_o \leq 0.8 \bullet ICLIM$<br>$R_2 = R_1 \bullet \left[\dfrac{V_o}{V_{fb}} - 1\right]$<br>$D_{max} \geq \dfrac{V_o}{V_{in\,min}}$ |
| | −ve to −ve | 1 | 10** | $V_{sw\,max} \geq V_{in\,max}$<br>$V_{IC\,max} \geq V_{in\,max}$<br>$V_{IC\,min} \leq V_{in\,min}$ | $I_o \leq 0.8 \bullet ICLIM$<br>$R_2 \approx R_1 \bullet \left[\dfrac{V_o - 0.6}{V_{fb}} - 1\right]$<br>$D_{max} \geq \dfrac{V_o}{V_{in\,min}}$ |
| | | 2 | Does not exist (opposite cell) | | |

| | | | | | |
|---|---|---|---|---|---|
| Boost | +ve to +ve | 1 | 8 | $V_{sw\,max} \geq V_o$ <br> $V_{IC\,max} \geq V_{in\,max}$ <br> $V_{IC\,min} \leq V_{in\,min}$ | $I_o \leq 0.8 \bullet ICLIM \bullet \dfrac{V_{in\,min}}{V_o}$ <br> $R_2 = R_1 \bullet \left[ \dfrac{V_o}{V_{fb}} - 1 \right]$ <br> $D_{max} \geq \dfrac{V_o - V_{in\,min}}{V_o}$ |
| | | 2 | Does not exist (opposite cell) | | |
| Boost | −ve to −ve | 1 | 11 | $V_{sw\,max} \geq V_o$ <br> $V_{IC\,max} \geq V_o$ <br> $V_{IC\,min} \leq V_{in\,min}$ | $I_o \leq 0.8 \bullet ICLIM \dfrac{V_{in\,min}}{V_o}$ <br> $R_2 \approx R_1 \bullet \left[ \dfrac{V_o}{V_{fb}} - 1 \right]$ <br> $D_{max} \geq \dfrac{V_o - V_{in\,min}}{V_o}$ |
| | | 2 | 16 | $V_{IC\,max} \geq V_o$ <br> $V_{IC\,min} \leq V_{in\,min}$ | $I_o \leq 0.8 \bullet ICLIM \bullet \dfrac{V_{in\,min}}{V_o}$ <br> $R_2 = R_1 \bullet \left[ \dfrac{V_o}{V_{fb}} - 1 \right]$ <br> $D_{max} \geq \dfrac{V_o - V_{in\,min}}{V_o}$ |

**Note:** By convention, $R_2$ is always connected to the higher voltage rail of output and $R_1$ to the lower.

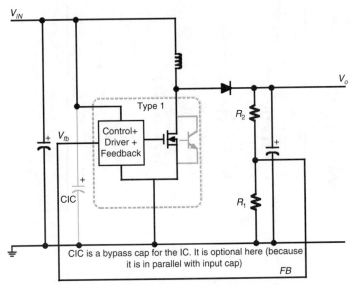

Figure 2.8  Positive-to-positive boost using a boost IC (Cell B).

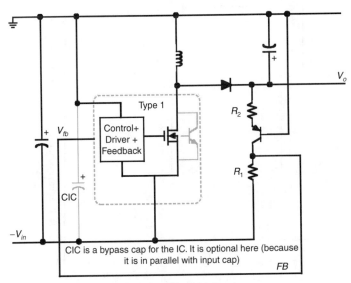

Figure 2.9  Negative-to-positive buck-boost using a boost IC (Cell B).

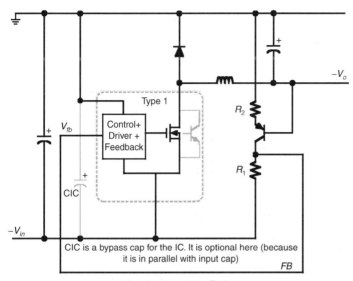

Negative-to-negative Buck

**Figure 2.10**  Negative-to-negative buck using a boost IC (Cell B).

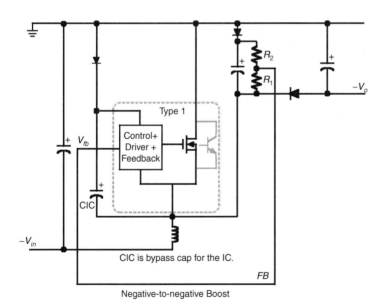

Negative-to-negative Boost

**Figure 2.11**  Negative-to-negative boost using a boost IC (Cell A).

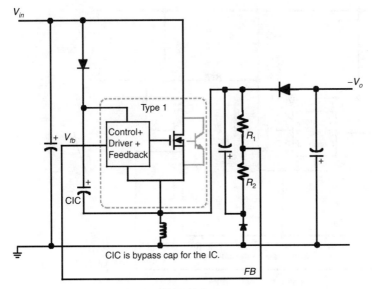

Positive-to-negative Buck-Boost

**Figure 2.12** Positive-to-negative buck-boost using a boost IC (Cell A).

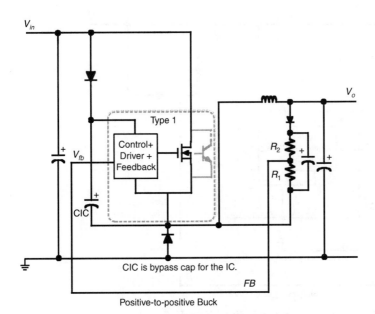

Positive-to-positive Buck

**Figure 2.13** Positive-to-positive buck using a boost IC (Cell A).

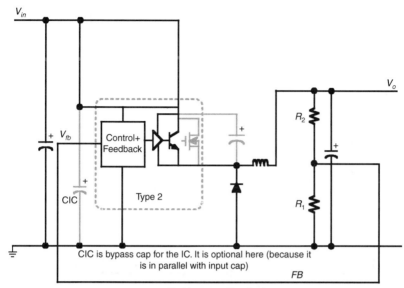

Positive-to-positive Buck

**Figure 2.14** Positive-to-positive buck using a buck IC (Cell A).

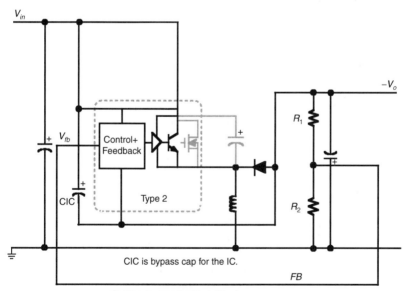

Positive-to-negative Buck-Boost

**Figure 2.15** Positive-to-negative buck-boost using a buck IC (Cell A).

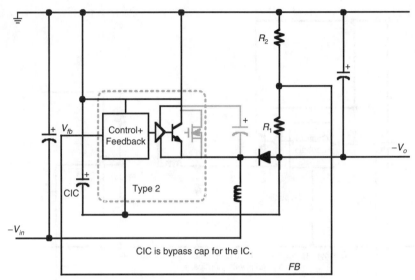

**Figure 2.16**  Negative-to-negative boost using a buck IC (Cell A).

In all of the equations that follow in this chapter, the switch and diode forward drops are generally assumed to be negligible. So a little additional guardbanding and discretion may be necessary to account for these parasitics.

Now we come to the crucial chain of logic behind identifying the various possible applications of switcher ICs:

The primary intended application for a Type 1 IC is the positive to positive boost. We can confirm that this configuration involves an N–cell (Type B cell). Therefore we conclude that this IC is "comfortable" with any topology/configuration, provided that involves a (similar) Type B cell. This cell is thus considered a "natural choice" for this IC. Note that we also now start seeing the advantage of talking in terms of LSD cells rather than in terms of topologies/ configurations. See Figures 2.8–2.16.

The primary intended application for a Type 2 IC is the positive to positive buck. We can confirm that this configuration involves a Type A cell. Therefore we conclude that this IC is most "comfortable" with any topology/configuration provided that involves a (similar) Type B cell. This cell is thus considered a "natural choice" for this IC. See Figures 2.14–2.16.

However, because the Type 1 IC has an *additional degree of freedom* due to the fact that the rail going to the switch is *not* connected to the supply rail of the control, a certain technique has come about by which we can use a Type 1 IC in an application that involves the *opposite* cell i.e., a Type A cell in this case. But in the process, the accuracy of the output regulation does suffer a great deal. Further, some ICs just may not "like" to be operated in this manner, because it requires the IC (control) ground to be a *swinging* node. So these extra possibilities which involve using a Type 1 IC in a cell configuration other than its natural choice should be actually tried out on the bench before confirming. In addition, besides layout differences, there are also some other subtle nuances based on topologies that might come in the way of using an IC in different configurations, but let us say that in principle all these possibilities exist.

If we use a transformer-based buck-boost converter, we can connect the primary and secondary winding terminations together in various ways to get any combination of positive or negative input and outputs. We can also leave the two windings isolated, in which case we need an error amplifier on the secondary side and an optocoupler. In this case $V_{fb}$ would refer to the reference voltage of the secondary-side error amplifier, not of the primary-side. Any primary-side error amplifier is in any case usually disabled because two error amplifiers would be just too much to handle and stabilize. The required equations are provided in Table 2.2.

## 2.9  Some Practical Cases

We now present some typical examples to clarify the selection procedure further.

In Tables 2.1 and 2.2 we have provided the equations which should be used to verify that the chosen IC suits the application on hand, mainly in terms of voltage and current. Let us use these in the examples below.

**Example 2.2**  The LM2585 is a 3-A flyback regulator. The minimum value of its internal current limit (see its datasheet table of electrical characteristics) is 3 A. Its input operating voltage range is 4 V to 40 V. Its switch can withstand 65 V. Can it be used in a boost topology? And for what applications?

The following steps are required in this analysis.

1. We identify that the LM2585 is a Type 1 IC by our nomenclature.
2. Referring to Table 2.1 we see that it can be used as, say, a positive to positive boost.
3. From the equations we see that the input voltage must be below 40 V and the output voltage must be below 65 V (since $V_{swmax} > V_o$ and

**TABLE 2.2  IC Selection Table for the Various Topologies (Part 2).**

| Topology | Configuration | IC Type* | Figure | With inductor: (all are magnitudes only) | Equations |
|---|---|---|---|---|---|
| Buck-Boost | +ve to –ve | 1 | 12 | $V_{sw\,max} \geq V_{in\,max} + V_o$ <br> $V_{IC\,max} \geq V_{in\,max} + V_o$ <br> $V_{IC\,min} \leq V_{in\,min}$ | $I_o \leq 0.8 \bullet ICLIM \bullet \dfrac{V_{in\,min}}{V_{in\,min} + V_o}$ <br> $R_2 = R_1 \bullet \left[\dfrac{V_o}{V_{fb}} - 1\right]$ <br> $D_{max} \geq \dfrac{V_o}{V_{in\,min} + V_o}$ |
|  |  | 2 | 15 | $V_{IC\,max} \geq V_{in\,max} + V_o$ <br> $V_{IC\,min} \leq V_{in\,min}$ | $I_o \leq 0.8 \bullet ICLIM \bullet \dfrac{V_{in\,min}}{V_{in\,min} + V_o}$ <br> $R_2 = R_1 \bullet \left[\dfrac{V_o}{V_{fb}} - 1\right]$ <br> $D_{max} \geq \dfrac{V_o}{V_{in\,min} + V_o}$ |
|  | –ve to +ve | 1 | 9 ** | $V_{sw\,max} \geq V_{in\,max} + V_o$ <br> $V_{IC\,max} \geq V_{in\,max}$ <br> $V_{IC\,min} \leq V_{in\,min}$ | $I_o \leq 0.8 \bullet ICLIM \bullet \dfrac{V_{in\,min}}{V_{in\,min} + V_o}$ <br> $R_2 \approx R_1 \bullet \left[\dfrac{V_o - 0.6}{V_{fb}} - 1\right]$ <br> $D_{max} \geq \dfrac{V_o}{V_{in\,min} + V_o}$ |
|  |  | 2 |  | Does not exist (opposite cell) |  |

## With Transformer: (all are magnitudes only)

| | | | | |
|---|---|---|---|---|
| Buck-Boost/Flyback | All | 1 | $V_{sw\,max} \geq V_{in\,max} + V_z$ <br> $V_{IC\,max} \geq V_{in\,max}$ <br> $V_{IC\,min} \leq V_{in\,min}$ <br> $n \equiv Np/N_s$ <br> $V_r \equiv V_o \bullet n$ <br> $V_z > V_r$ | $I_o \leq \left[ 0.8 \bullet ICLIM \bullet \dfrac{V_{in\,min}}{V_{in\,min}+V_r} \right] \bullet n$ <br> $R_2 = R_1 \bullet \left[ \dfrac{V_o}{V_{fb}} - 1 \right]$ <br> $D_{max} \geq \dfrac{V_r}{V_{in\,min} + V_r}$ |
| Flyback | | 1 | $V_{sw\,max} \geq V_{in\,max} + V_z$ <br> $V_{IC\,max} \geq V_{in\,max}$ <br> $V_{IC\,min} \leq V_{in\,min}$ <br> $n \equiv Np/Ns$ <br> $V_r \equiv V_o \bullet n$ <br> $V_z > V_r$ | $I_o \leq \left[ 0.8 \bullet ICLIM \bullet \dfrac{V_{in\,min}}{V_{in\,min}+V_r} \right] \bullet n$ <br> $R_2 = R_1 \bullet \left[ \dfrac{V_o}{V_{fb}} - 1 \right]$ <br> $D_{max} \geq \dfrac{V_r}{V_{in\,min} + V_r}$ |

**Note:** By convention, $R_2$ is always connected to the higher voltage rail of output and $R_1$ to the lower.

$V_{ICmax} > V_{inmax}$). These define the input/output voltage conditions for any suitable application.

4. The maximum load current is

$$I_o = 0.8 \times ICLIM \times \left[\frac{V_{in\ min}}{V_o}\right]$$

So if the output is set to 60 V and the input ranges from say 20 V to 40 V, the maximum load (with a suitably designed practical inductor) is

$$I_o = 0.8 \times 3 \times \left[\frac{20}{60}\right] = 0.8\ \text{A}$$

**Example 2.3**  The required application conditions are $V_{in}$ ranging from 4.5 V to 5.5 V. The output requirement is −5 V at 0.5 A. Can the LM2651 be used?

LM2651 is a 1.5-A buck regulator. Note firstly that this IC can deliver 1.5 A in a buck configuration but not so in any other configuration/topology. The load rating must therefore be recalculated here. The following steps are performed.

1. We identify the LM2651 as a Type 2 IC according to our nomenclature.

2. We refer to the table and we can see that a positive to negative buck-boost is possible with this IC.

3. Referring to the datasheet of this device we get

   - $V_{ICmin} = 4$ V
   - $V_{ICmax} = 14$ V
   - ICLIM $= 1.55$ A (minimum of tolerance band)
   - $D_{max} = 92$ percent (minimum of tolerance band)

4. Therefore we now check sequentially for these conditions:

   - $V_{ICmax} > V_{inmax} + V_o$
     14 V > 5.5 V + 5 V = 10.5 V   (**OK**)
   - $V_{ICmin} < V_{inmin}$
     4 V < 4.5 V   (**OK**)
   - $I_o < 0.8 \times ICLIM \times (V_{inmin}/(V_{inmin} + V_o)$
     0.5 < 0.8 × 1.55 × {4.5/(4.5 + 5)} = 0.587   (**OK**)
   - $D_{max} > V_o/(V_o + V_{inmin})$
     0.92 > 5/(5 + 4.5) = 0.53   (**OK**)

Therefore, the LM2651 is acceptable for the intended application.

## 2.10   Differential Voltage Sensing

In Figs. 2.9 and 2.10 we used a crude differential sense stage to reference the feedback to the IC ground. A more accurate sensing scheme can be implemented by using an op-amp (like the LM324), as shown in Fig. 2.17. There are two ways of setting up such a differential amplifier. The lower schematic block has a higher gain. Note that the inputs to the op-amp are labeled $V_{o\_hi}$ and $V_{o\_lo}$. This means that irrespective of how the schematic actually labels the output rails, the inputs to the op-amp connect to the HI rail or the LO rail respectively. Some of the relevant aspects of op-amps must be kept in mind. For example, note that an op-amp has a specified input voltage common-mode range. For the LM324 series this number is specified to be 1.5 V below the upper supply rail and this parameter is hereby called $v'$. We require that the voltage on both the input pins of the op-amp stays within this allowed range or the op-amp cannot be considered fully functional. Since the voltages on these pins are fixed by virtue of the resistors, if the resistors are also considered fixed, the only way to ensure that the

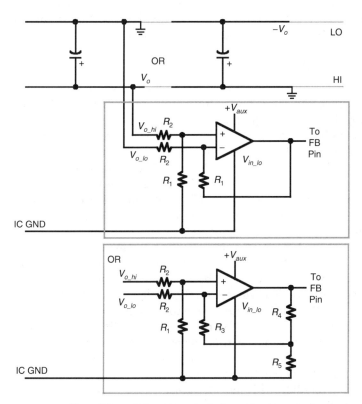

**Figure 2.17**  Two differential sensing techniques for improved output regulation.

TABLE 2.3  **Design Table for Differential Stages.**

| Op-amp | Equation set |
|---|---|
| Standard differential amp. | $R_2 = R_1 \bullet V_o / V_{fb}$ |
| | $V_{aux} \geq v' + \left[ \dfrac{R_1}{R_1 + R_2} \bullet (V_{inmax} + V_o) \right]$ |
| Hi-gain differential amp. | $R_2 = R_x \bullet V_o / V_{fb}$ |
| | $R_x \equiv R_3 + R_4 + \dfrac{R_3 \bullet R_4}{R_5}$ |
| | $R_2 = R_3 + \dfrac{R_4 \bullet R_5}{R_4 + R_5}$ |
| | $V_{aux} \geq v' + \left[ \dfrac{R_1}{R_1 + R_2} \bullet (V_{inmax} + V_o) \right]$ |

common-mode condition is met is to set the op-amp supply rail $+V_{aux}$ sufficiently higher. This limit equation is therefore also provided in Table 2.3. Note that if the required minimum $+V_{aux}$ value is still low enough it may be possible to connect it to an available dc rail. If not, an additional external rail will need to be created to run the op-amp stage.

## 2.11   Some Topology Nuances

Some of the concerns when we traverse topologies have to do with nuances of the topologies themselves. In particular, we must remember that a buck topology has no right half plane (RHP) zero but the boost and the flyback/buck-boost do (in continuous conduction mode). Therefore when we try to take a buck IC (with internal fixed compensation) and make it perform in a boost or buck-boost application, we may not have the desired ability to tailor the crossover frequency to less than 1/4th of the RHP zero frequency (as is generally recommended for avoiding this particular mode of instability). So how do we successfully take a Type 2 IC and apply it to other topologies?

To answer that we first must remember the "intuitive explanation" behind the RHP zero. This is said to occur as follows. If we suddenly increase the load on the output of a boost or buck-boost regulator the output dips momentarily. The voltage on the feedback pin therefore falls slightly and this commands the duty cycle to increase to try and correct for this. But both the boost and the buck-boost are different from the buck in that during the switch on-time, no energy flows into the output, since we are basically just building up energy in the inductor during that time. So if the duty cycle increases in response to the load increase, in fact there is a smaller off-time. Therefore we get less, rather than more, current to flow into the output. This causes the output to decrease further, as it gets momentarily starved. Eventually, after a few cycles, the average inductor current does ramp up progressively and the output dip will get corrected. But before that happens, we see a situation where the load disturbance is in effect reinforcing itself. In severe cases this may lead to sustained oscillations. In the Bode plot, the RHP zero shows up as a normal (LHP) zero in the gain plot, causing the slope to increase (upward) by 20 dB/decade, but it behaves as a pole in the phase plot, causing a $-90°$ shift (downward).The location of the RHP zero is

$$f_{RHPZ\_BOOST} = \frac{R_L \times (1-D)^2}{2\pi L}$$

$$f_{RHPZ\_BUCK\_BOOST} = \frac{R_L \times (1-D)^2}{2\pi L \times D}$$

where $R_L$ is the load resistance. As mentioned, we should generally ensure that we roll-off the gain fast enough to be well clear of this frequency. In effect, this means that we are just not allowing the control to react too fast to the sudden change in load.

Two well-known practical RHP zero suppression techniques are used when using a buck IC to generate other topologies. We show them as applied to a positive to negative configuration. Both are shown together in Fig. 2.18, but one or both can be used. The one on the right senses when the duty cycle increases suddenly and pushes up the feedback pin slightly so that it doesn't dip too low in response to the sudden load demand. On the left we see that a diode has been inserted and also the IC bypass cap CIC needs to be sized much bigger now. It is no wonder that the schematic then looks much closer to a buck rather

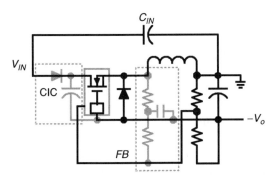

**Figure 2.18** Practical techniques to alleviate RHP zero.

than a buck-boost. And, in fact, during a load transient it does behave temporarily as a buck, because now some energy can be transferred to the output even during the switch on-time. Note that the input diode is for reverse protection and can often be omitted if sudden input steps are not expected. Also, if we follow the current on initial application of power we will see that a reverse current will flow through the output capacitor. So we may need to put in a reverse diode at that point too, whenever the second method is used.

## 2.12   Composite Topologies

We will find that possibly the most frequently discussed composite topology is a series combination of a boost and a buck. This has understandably always been of interest as it would provide a step-up or step-down (buck/boost) response leading to a virtually flexible peg to fit almost any hole. But there are many ways of actually combining the boost and the buck. For instance, "cascaded" (series) or "interleaved" or various parallel combinations have been amply discussed in literature. Within each of these categories, subcategories are spawned by the way the switches are actually driven. In fact, it is almost as important as the schematic combinations themselves to understand the actual drive mechanism further.

An elementary implementation of a cascaded boost-buck is one in which one switch could be PWM-controlled with the other fully *on* or *off* (as applicable). This would require the boost switch to be PWM-controlled, with the buck switch turned fully *on* whenever a step-up operation is required. And whenever a step-down operation is required, the buck switch would be PWM-controlled and the boost switch fully *off*. This is strictly speaking not a composite topology, since only one topology works at any given time. It is more like flipping channels, or "topology-on-demand." But we could also have a case where both the switches have independent PWM control (in which they could also be run in-phase or out of phase). However, these more complex dual control types are not going to be discussed further here.

Our focus here is on the more cost-effective two-stage combinations of the basic three topologies in which *both stages are driven with the same duty cycle*. This calls for only one PWM controller ("master"), the other serving as a "slave," though the designations are virtually interchangeable. Within this category of interest we also have versions which use just one active switching element. This is what was once rather grandiosely, but aptly, called the "topological reduction of a switch" during the development of the Cuk converter. We can build a cascaded two-switch boost-buck combination as shown in the upper left side of Fig. 2.19.

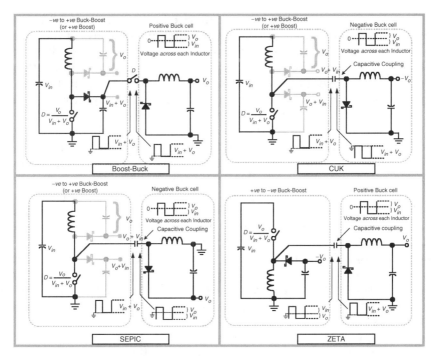

**Figure 2.19**  Understanding composite topologies.

We will find that surprisingly, this boost-buck composite has the same input to output transfer function as the standard buck-boost topology. So are the two really just the same topology? Consider the shape of the current waveform of the input and output capacitors. In a buck, the input current is pulsating, and the output current is relatively smooth (as it comes through an inductor). For a boost the input current is smooth as it is in series with the inductor, but the output current pulsates. In a buck-boost, both the input and output capacitor currents are pulsating. So, in a boost-buck two-switch composite, we essentially combine the best of the smooth boost input with the smooth buck output, though at the cost of an extra inductor (and possibly an extra switch). Therefore despite mimicking known dc transfer functions, the composites in general behave very differently.

In the last circuit in Chap. 16 we have explained an interesting *transfer function coincidence*. We also now know that the buck-boost and the boost topologies are not all that different, it's just a question how we referenced our output rail. On the basis of these facts, we can show that we have a hidden but available auxiliary regulated rail (grayed-out) across the first inductor for the boost-buck composite in Fig. 2.19.

Now this involved two switches. Can we eliminate one switch? Yes, if we use capacitive coupling to inject the switching waveform at the node of the following *buck cell*. For all practical purposes we would still have a boost-buck composite. In fact we can do this in at least three different ways, as shown in the remaining schematics of Fig. 2.19. The Cuk gives an *inverted* step-up or step-down response (i.e., a positive to negative buck/boost configuration). But by re-referencing the ground, we can correct this and we get the SEPIC (single ended primary inductor converter), which is a noninverting step-up or step-down configuration (i.e., positive to positive buck/boost). Or we can also get two independently regulated rails as in the Zeta converter. The latter is a clear illustration of the transfer function coincidence we mentioned above.

Note that in the Cuk converter, for example, we could have drawn power from the grayed-out (unused) intermediate rails. That rail would have been automatically regulated too by the transfer function coincidence (though for that we have to ensure that both inductors are in CCM).

As mentioned, the dc transfer function is the same for all four schematics in the figure. And it happens to be the same as for the standard buck-boost topology. That is easy to understand because the duty cycle is determined purely by voltseconds balance and in each case we have the same voltage *across* each inductor, and for all topologies (see insets in Fig. 2.19). But note that the *current waveforms are different in each case*. It is just that they do not affect the duty cycle calculation, since duty cycle is determined only by the voltages, just so long as the current is high enough to keep the inductors operating in CCM. Note also that, historically, in the Cuk converter development it was observed that the voltage waveforms across the two inductors were identical, so it was decided to "save" a core and wind them together. But in the process a remarkable phenomenon called *ripple steering* was observed. By very careful design of the coupled inductor it is possible to get virtually zero input and output ripple. In the commercial arena, however, such finely tuned coupled inductors are not practical.

For some more interesting composite topologies refer again to Chap. 16.

# 3

# Reference Equations and Graphs for Converter Design

## 3.1 The Defining Difference between the Topologies

The essential difference between the topologies is

$$\boxed{I_{L\_AVG} \equiv I_{DC} = I_O} \text{ (Buck)}$$

$$\boxed{I_{L\_AVG} \equiv I_{DC} = \frac{I_O}{1-D}} \text{ (Boost/buck-boost)}$$

This follows from

$$\boxed{I_{L\_AVG} = I_O} \text{ (Buck)}$$

$$\boxed{I_{D\_AVG} = I_O} \text{ (Boost/buck-boost)}$$

## 3.2 Definition of Current Ripple Ratio

We define the *current ripple ratio* of the inductor current waveform as

$$\boxed{r = \frac{\Delta I}{I_{L\_AVG}} \equiv \frac{\Delta I}{I_{DC}}}$$

We should set this parameter to about 0.4 at maximum load condition (for all the topologies).

## 3.3   Graphical Tools for Inductor Selection

In Figs. 3.1, 3.2, and 3.3 we have defined the *conversion ratio* $x = V_o/V_{in}$ (magnitudes of voltages only) and this forms the abscissa (horizontal axis).

The curves are all for a normalized load current of 1 A.

The designer can extend these values to any application by remembering that inductance will vary inversely with load current and the energy rating of the inductor (i.e., $1/2 \times L \times I_{PK}^2$) varies as square of load current (for a given $r$).

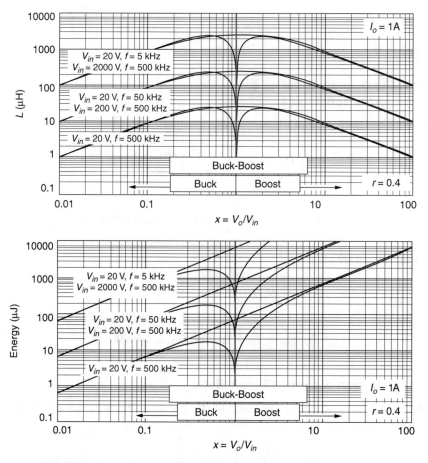

**Figure 3.1**   Inductance and energy curves if input voltage is considered fixed.

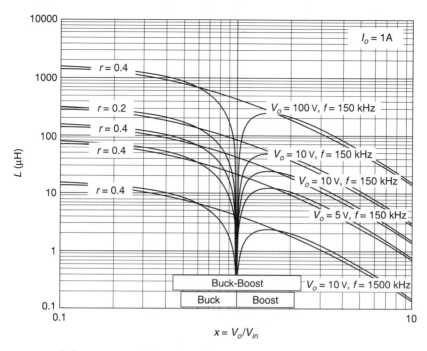

**Figure 3.2** Inductance curves if output voltage is considered fixed.

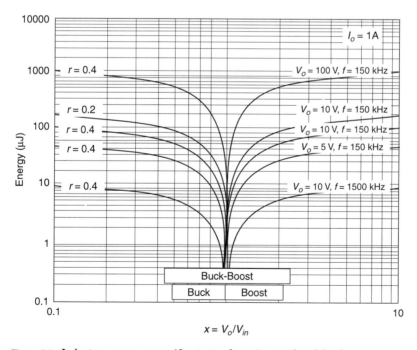

**Figure 3.3** Inductor energy curves if output voltage is considered fixed.

**TABLE 3.1  Reference Design Table for Continuous Conduction Mode.**

| | Buck | Boost | Buck-Boost |
|---|---|---|---|
| Duty cycle | $\dfrac{V_O + V_D}{V_{IN} - V_{SW} + V_D}$ | $\dfrac{V_O - V_{IN} + V_D}{V_O - V_{SW} + V_D}$ | $\dfrac{V_O + V_D}{V_{IN} + V_O - V_{SW} + V_D}$ |
| $V_{IN\_50}$ (V) | $(2 \bullet V_O) + V_{SW} + V_D$ $\approx 2 \bullet V_O$ | $\dfrac{1}{2} \bullet [V_O + V_{SW} + V_D]$ $\approx V_O/2$ | $V_O + V_{SW} + V_D$ $\approx V_O$ |
| Output voltage, $V_O$(V) | $V_{IN} \bullet D - V_{SW} \bullet D - V_D \bullet (1 - D)$ | $\dfrac{V_{IN} - V_{SW} \bullet D - V_D \bullet (1 - D)}{1 - D}$ | $\dfrac{V_{IN} \bullet D - V_{SW} \bullet D - V_D \bullet (1 - D)}{1 - D}$ |
| Volt$\mu$seconds ($V\mu s$) | $\dfrac{V_O + V_D}{f} \bullet (1 - D) \bullet 10^6$ | $\dfrac{V_O - V_{SW} + V_D}{f} \bullet D \bullet (1 - D) \bullet 10^6$ | $\dfrac{V_O + V_D}{f} \bullet (1 - D) \bullet 10^6$ |
| $L$ ($\mu$H) | $\dfrac{V_O + V_D}{I_O \bullet r \bullet f} \bullet (1 - D) \bullet 10^6$ | $\dfrac{V_O - V_{SW} + V_D}{I_O \bullet r \bullet f} \bullet D \bullet (1 - D)^2 \bullet 10^6$ | $\dfrac{V_O + V_D}{I_O \bullet r \bullet f} \bullet (1 - D)^2 \bullet 10^6$ |
| Inductor current ripple ratio $r$ | $\dfrac{V_O + V_D}{I_O \bullet L \bullet f} \bullet (1 - D) \bullet 10^6$ | $\dfrac{V_O - V_{SW} + V_D}{I_O \bullet L \bullet f} \bullet D \bullet (1 - D)^2 \bullet 10^6$ | $\dfrac{V_O + V_D}{I_O \bullet L \bullet f} \bullet (1 - D)^2 \bullet 10^6$ |
| $\Delta I$(A) | $\dfrac{V_O + V_D}{L \bullet f} \bullet (1 - D) \bullet 10^6$ | $\dfrac{V_O - V_{SW} + V_D}{L \bullet f} \bullet D \bullet (1 - D) \bullet 10^6$ | $\dfrac{V_O + V_D}{L \bullet f} \bullet (1 - D) \bullet 10^6$ |
| rms current in input cap (A) | $I_O \sqrt{D \bullet \left[1 - D + \dfrac{r^2}{12}\right]}$ | $\dfrac{I_O}{1 - D} \bullet \dfrac{r}{\sqrt{12}}$ | $\dfrac{I_O}{1 - D} \bullet \sqrt{D \bullet \left[1 - D + \dfrac{r^2}{12}\right]}$ |
| $I_{PP}$ in input capacitor (A) | $I_O \bullet \left[1 + \dfrac{r}{2}\right]$ | $\dfrac{I_O \bullet r}{1 - D}$ | $\dfrac{I_O}{1 - D} \bullet \left[1 + \dfrac{r}{2}\right]$ |

| | | | |
|---|---|---|---|
| rms current in output cap (A) | $I_O \bullet \dfrac{r}{\sqrt{12}}$ | $I_O \bullet \sqrt{\dfrac{D + \frac{r^2}{12}}{1 - D}}$ | $I_O \bullet \sqrt{\dfrac{D + \frac{r^2}{12}}{1 - D}}$ |
| $I_{PP}$ in output capacitor (A) | $I_O \bullet r$ | $\dfrac{I_O}{1-D} \bullet \left[ 1 + \dfrac{r}{2} \right]$ | $\dfrac{I_O}{1-D} \bullet \left[ 1 + \dfrac{r}{2} \right]$ |
| Energy handling capability (μJ) | $\dfrac{I_O \bullet V \mu s}{8} \bullet \left[ r \bullet \left( \dfrac{2}{r} + 1 \right)^2 \right]$ | $\dfrac{I_O \bullet V \mu s}{8 \bullet (1-D)} \bullet \left[ r \bullet \left( \dfrac{2}{r} + 1 \right)^2 \right]$ | $\dfrac{I_O \bullet V \mu s}{8 \bullet (1-D)} \bullet \left[ r \bullet \left( \dfrac{2}{r} + 1 \right)^2 \right]$ |
| rms current in inductor (A) | $I_O \bullet \sqrt{1 + \dfrac{r^2}{12}}$ | $\dfrac{I_O}{1-D} \bullet \sqrt{1 + \dfrac{r^2}{12}}$ | $\dfrac{I_O}{1-D} \bullet \sqrt{1 + \dfrac{r^2}{12}}$ |
| Average current in inductor (A) | $I_O$ | $\dfrac{I_O}{1-D}$ | $\dfrac{I_O}{1-D}$ |
| rms current in switch (A) | $I_O \bullet \sqrt{D \bullet \left[ 1 + \dfrac{r^2}{12} \right]}$ | $\dfrac{I_O}{1-D} \bullet \sqrt{D \bullet \left[ 1 + \dfrac{r^2}{12} \right]}$ | $\dfrac{I_O}{1-D} \bullet \sqrt{D \bullet \left[ 1 + \dfrac{r^2}{12} \right]}$ |
| Peak current switch/diode/ inductor (A) | $I_O \bullet \left[ 1 + \dfrac{r}{2} \right]$ | $\dfrac{I_O}{1-D} \bullet \left[ 1 + \dfrac{r}{2} \right]$ | $\dfrac{I_O}{1-D} \bullet \left[ 1 + \dfrac{r}{2} \right]$ |
| Average current in switch (A) | $I_O \bullet D$ | $I_O \bullet \dfrac{D}{1-D}$ | $I_O \bullet \dfrac{D}{1-D}$ |
| Average current in Diode (A) | $I_O \bullet (1 - D)$ | $I_O$ | $I_O$ |
| Average input current (A) | $I_O \bullet D$ | $I_O \bullet \dfrac{D}{1-D}$ | $I_O \bullet \dfrac{D}{1-D}$ |
| Output voltage ripple ($\pm$mV)* | $1/2 \bullet I_O \bullet r \bullet ESR\,(m\Omega)$ | $1/2 \bullet \dfrac{I_O}{1-D} \bullet \left[ 1 + \dfrac{r}{2} \right] \bullet ESR\,(m\Omega)$ | $1/2 \bullet \dfrac{I_O}{1-D} \bullet \left[ 1 + \dfrac{r}{2} \right] \bullet ESR\,(m\Omega)$ |

$r = \Delta I / I_{DC}$; $L$ in $\mu H$, $f$ in Hz, All voltages and currents are magnitudes,* ESL ignored

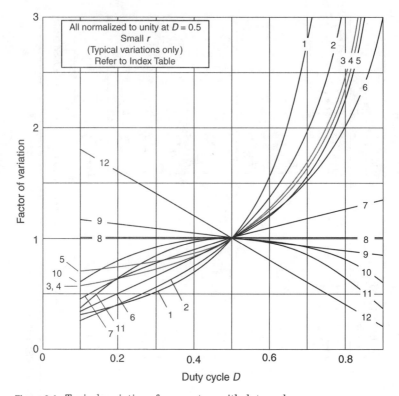

Figure 3.4 Typical variation of parameters with duty cycle.

We can easily see from the curves all the dependencies on frequency, input voltage, output voltage, and $r$. This should help in quickly picking the right inductor for the job for any application.

Thereafter, the reader can refer to Table 3.1 which has all the key design equations for the three topologies in terms of $r$. $V_{SW}$ is the drop across the switch when it is *on*, and $V_D$ is the diode forward drop.

For design optimization and power supply test guidance study Fig. 3.4 and Table 3.2 to see how the parameters typically vary with input voltage (duty cycle).

### 3.4  The Design Equations Table

As mentioned above, for optimum results set $r \approx 0.4$. For a wide input range, set this value at the minimum input voltage for boost and buck-boost, but at maximum input voltage for the buck.

The inductance and energy rating of the core should also be decided at minimum input voltage for boost and buck-boost, but at maximum input voltage for the buck.

TABLE 3.2  Worst-Case Input Condition for Each Parameter (Index Table for Fig. 3.4).

| Parameters | Buck | Boost | Buck-Boost |
|---|---|---|---|
| Inductor current swing $\Delta I$ ($2 \times I_{AC}$) | $V_{IN\_MAX}$ 12 | $V_{IN\_50}$ 11 | $V_{IN\_MAX}$ 12 |
| Core loss | $V_{IN\_MAX}$ | $V_{IN\_50}$ | $V_{IN\_MAX}$ |
| Inductor energy/ core saturation | $V_{IN\_MAX}/V_{IN}$ 8 | $V_{IN\_MIN}$ 1 | $V_{IN\_MIN}$ 1 |
| Average current in inductor | $V_{IN}$ 8 | $V_{IN\_MIN}$ 3 | $V_{IN\_MIN}$ 3 |
| rms current in inductor | $V_{IN\_MAX}/V_{IN}$ 8 | $V_{IN\_MIN}$ 3 | $V_{IN\_MIN}$ 3 |
| Copper loss/temperature of inductor | $V_{IN\_MAX}/V_{IN}$ | $V_{IN\_MIN}$ | $V_{IN\_MIN}$ |
| rms current in input capacitor | $V_{IN\_50}$ 10 | $V_{IN\_50}$ 11 | $V_{IN\_MIN}$ 6 |
| Input voltage ripple | $V_{IN\_MAX}/V_{IN}$ 8 | $V_{IN\_MAX}$ 12 | $V_{IN\_MIN}$ 3 |
| rms current in output capacitor | $V_{IN\_MAX}$ 12 | $V_{IN\_MIN}$ 6 | $V_{IN\_MIN}$ 6 |
| Output voltage ripple | $V_{IN\_MAX}$ 12 | $V_{IN\_MIN}$ 3 | $V_{IN\_MIN}$ 3 |
| rms current in switch | $V_{IN\_MIN}$ 7 | $V_{IN\_MIN}$ 2 | $V_{IN\_MIN}$ 2 |
| Average current in switch | $V_{IN\_MIN}$ | $V_{IN\_MIN}$ | $V_{IN\_MIN}$ |
| Peak current in switch/diode/inductor | $V_{IN\_MAX}$ 9 | $V_{IN\_MIN}$ 4 | $V_{IN\_MIN}$ 5 |
| Average current in diode | $V_{IN\_MAX}$ 12 | $V_{IN}$ 8 | $V_{IN}$ 8 |
| Temperature of diode | $V_{IN\_MAX}$ 12 | $V_{IN}$ 8 | $V_{IN}$ 8 |
| Worst-case efficiency (typical) | $V_{IN\_MAX}$ | $V_{IN\_MIN}$ | $V_{IN\_MIN}$ |

Numbers in the columns refer to corresponding numbered curves in Fig. 3.2
$V_{IN}$ means any input voltage is appropriate
$V_{IN\_50}$ is input voltage at which D = 0.5

Note that all these equations assume CCM (continuous conduction mode). If $L$ is so chosen that $r$ as calculated from Table 3.1 works out to be more than 2, we have transited into DCM (discontinuous conduction mode). Even if we are in CCM, to start with, by decreasing the load and/or increasing the input voltage (if applicable), the current ripple ratio can become greater than 2. Under that condition the equations in this chapter are not valid, and therefore Chap. 4 which deals with DCM should be referred to instead. Note that both sets of equations (for CCM and DCM) must yield exactly the same results at the boundary of transition (critical boundary).

# Discontinuous Conduction Mode Equations

## 4.1 Introduction

In *continuous conduction mode* (CCM), the current ripple ratio $r = \Delta I/I_{DC}$ of the inductor is always less than 2. This ratio is, by our definition, applicable only to CCM. If $r = 2$, we are at the *critical boundary* between CCM and *discontinuous conduction mode* (DCM). If $r$ is calculated using the CCM equations of Chap. 3 and its value turns out to be greater than 2, though that has no physical meaning, it does imply that the system is now in DCM. Once we set $r$ to a certain optimum value within its valid range of 0 to 2 (usually around 0.4), the current ripple ratio can increase and thus the converter could enter DCM due to several reasons:

*Decreasing the inductance.* The (magnitude of the) slopes of the current up-ramp and down-ramp will now increase. But since the average current has a direct relationship to the load current, it will remain fixed. Therefore $\Delta I$ will increase though $I_{DC}$ will remain constant. So $r$ will increase. Note that some inductors are by design "swinging chokes," and their inductance may change substantially (usually decrease) as the load increases.

*Increasing/decreasing the input voltage.* In any converter, as we increase the input voltage, $D$ decreases. So, the available time for the current to ramp up decreases. This tends to reduce $\Delta I$. However, during this time, the voltage across the inductor has increased, and so $\Delta I$ tends to increase on this account. The net result on $\Delta I$ is thus not obvious, and it depends on the topology. For a buck and a buck-boost, in fact, $\Delta I$ increases with input voltage, but for the boost it has a maximum at around $D = 0.5$. Similarly, the center of the inductor ramp is a constant for a buck, but it decreases as D decreases (input increases) for the boost and the buck-boost. $r$ being the ratio $\Delta I/I_{DC}$, the overall effect is that $r$ increases as input voltage increases for the buck and the buck-boost. For a boost, however, $r$ has a maximum value (most likely point to be in DCM) at around $D = 0.33$. Ignoring parasitics, this amounts to an input of about two-third of the output.

*Decreasing the load.*    The dc value of the inductor current always has a direct relationship to the load current. So if we decrease $I_O$, we will decrease $I_{DC}$ and so $r$ will increase.

Eventually, at $r = 2$ the converter will start passing into discontinuous conduction mode. This happens provided there is a diode structure in the freewheeling path (nonsynchronous operation). The diode will prevent the current from going negative and so the current will stay at zero, indicating DCM.

One exception is the synchronous buck in which there is no DCM. Here the inductor current can reverse direction through the (lower) synchronous FET (provided this FET is always *on* whenever the upper FET is *off*). This reverse current continues for a while through the body diode of the upper FET when the lower FET turns *off* and the upper FET turns *on*. This maintains a quasi-CCM mode, often called *forced PWM* or simply CCM. Note that in this case *all the equations for CCM as presented in Chap. 3 are still valid, despite the fact that r can now be greater than 2.*

The parameter tables for CCM given in Chap. 3 can be quickly solved for (a) the inductance at which DCM occurs, (b) the input voltage at which DCM occurs, and (c) the load at which DCM occurs. This is found out by simply equating $r$ to 2 in the corresponding equations.

DCM has certain advantages and disadvantages. The current components in almost all the power related components are more "peaky" and so rms currents are higher (with a typical accompanying fall in efficiency). Since in DCM the duty cycle decreases as we reduce load, we may face minimum load restrictions or pulse-skipping. But DCM is still worth considering for several reasons:

The efficiency may not be as bad as expected. It may even be better. This is attributable mainly to the fact that in DCM the switch turns *on* when the freewheeling diode has fully recovered. So there is no reverse recovery spike through it. Especially in boost converters, this spike has a greatly deleterious effect on efficiency, and therefore its suppression is a key design issue.

There is almost no visible leading edge spike in the switch current waveform when operating in DCM. Such converters tend to run smoothly, with very little jitter, and there is also no need for blanking time in the current sense or for delay in the current limit circuit.

The magnetics are smaller. See Chap. 10, dealing with flyback transformer design, for some related discussions.

DCM converters are also easier to stabilize. A Type 2 (or simpler) compensation should suffice. Also, the phenomenon of subharmonic instability is absent (see Chap. 14, dealing with slope compensation). Note that there is still some debate on the significance of the right half plane zero in DCM. But it is clear that for the buck and its derived topologies (e.g., forward converter, push-pull, and the like) the RHP zero just cannot occur.

In integrated switchers the current limit is usually internal and fixed. So if we are operating the converter with a load much less than the rated maximum load, we have to be concerned with the "overload margin." This is the available headroom between the peak operating current and the set current limit. If the headroom is too small, we will not get a good transient response to step changes in line or load, since the converter will "hit the stops." Note that the converter may hit the current limit even under a normal power-up (if there is no soft-start present). Holdup time, if required, will also be affected if we have inadequate overload margin. On the other hand, excessive overload margins can cause reliability problems. For example, the inductor may saturate. And if the input voltage is high enough, the current may start rising at a rate faster than the response time of the current limit circuit, and the switch may thus get destroyed. DCM gives us the option that for a given load, we can actually *decrease* the headroom between the peak and the current limit, leading to enhanced reliability.

## 4.2  How DCM Equations Are Calculated

We all know that in CCM the duty cycle does not depend on load current or inductance. This makes CCM "easy." In DCM the picture looks rather complicated at first sight. The duty cycle can still be easily calculated from basics as we have shown in 4.3 below, but the rest of the equations seem quite imposing. This is possibly why DCM equations are shunned by most designers and in return rarely even seen in related literature. But realizing (and hoping) that more experienced designers are not yet completely through with DCM, in Table 4.1 we have presented the generalized forms of the key stress components. This table suffices completely, provided we carry out the calculation from "top to bottom." Such a table is also a good learning tool as it illustrates several of the principles and techniques we introduced in Chap. 1. However, to

TABLE 4.1  Generalized Top to Bottom Equation Set for DCM.

| | Buck | Boost | Buck-Boost |
|---|---|---|---|
| $V_{ON}$ | $V_{IN} - V_O - V_{SW}$ | $V_{IN} - V_{SW}$ | $V_{IN} - V_{SW}$ |
| $V_{OFF}$ | $V_O + V_D$ | $V_O - V_{IN} + V_D$ | $V_O + V_D$ |
| Duty cycle $D$ | $\sqrt{\dfrac{2 \times I_O \times L \times f \times V_{OFF}}{V_{ON} \times (V_{ON} + V_{OFF})}}$ | $\sqrt{\dfrac{2 \times I_O \times L \times f \times V_{OFF}}{V_{ON}^2}}$ | $\sqrt{\dfrac{2 \times I_O \times L \times f \times V_{OFF}}{V_{ON}^2}}$ |
| Peak current $I_{PK}$ (A) | $\dfrac{V_{ON} \times D}{L \times f}$ | $\dfrac{V_{ON} \times D}{L \times f}$ | $\dfrac{V_{ON} \times D}{L \times f}$ |
| Switch average (A) | $\dfrac{I_{PK}}{2} D$ | $\dfrac{I_{PK}}{2} D$ | $\dfrac{I_{PK}}{2} D$ |
| Switch rms $I_{SW\_RMS}$ (A) | $I_{PK} \sqrt{\dfrac{D}{3}}$ | $I_{PK} \sqrt{\dfrac{D}{3}}$ | $I_{PK} \sqrt{\dfrac{D}{3}}$ |
| Diode duty cycle $D'$ | $D' = \dfrac{2 \times I_O}{I_{PK}} - D$ | $D' = \dfrac{2 \times I_O}{I_{PK}}$ | $D' = \dfrac{2 \times I_O}{I_{PK}}$ |
| Inductor average $I_{L\_AVG}$ (A) | $\dfrac{I_{PK}}{2} D' + \dfrac{I_{PK}}{2} D$ | $\dfrac{I_{PK}}{2} D' + \dfrac{I_{PK}}{2} D$ | $\dfrac{I_{PK}}{2} D' + \dfrac{I_{PK}}{2} D$ |
| Inductor rms $I_{L\_RMS}$ (A) | $I_{PK} \sqrt{\dfrac{D'}{3} + \dfrac{D}{3}}$ | $I_{PK} \sqrt{\dfrac{D'}{3} + \dfrac{D}{3}}$ | $I_{PK} \sqrt{\dfrac{D'}{3} + \dfrac{D}{3}}$ |
| Switch rms $I_{SW\_RMS}$ (A) | $I_{PK} \sqrt{\dfrac{D}{3}}$ | $I_{PK} \sqrt{\dfrac{D}{3}}$ | $I_{PK} \sqrt{\dfrac{D}{3}}$ |
| Switch average $I_{SW\_AVG}$ (A) | $\dfrac{I_{PK}}{2} D$ | $\dfrac{I_{PK}}{2} D$ | $\dfrac{I_{PK}}{2} D$ |
| Diode average $I_{D\_AVG}$ (A) | $\dfrac{I_{PK}}{2} D'$ | $\dfrac{I_{PK}}{2} D'$ | $\dfrac{I_{PK}}{2} D'$ |
| Diode rms $I_{D\_RMS}$ (A) | $I_{PK} \sqrt{\dfrac{D'}{3}}$ | $I_{PK} \sqrt{\dfrac{D'}{3}}$ | $I_{PK} \sqrt{\dfrac{D'}{3}}$ |
| Input cap rms $I_{CIN\_RMS}$ (A) | $\sqrt{I_{SW\_RMS}^2 - I_{SW\_AVG}^2}$ | $\sqrt{I_{L\_RMS}^2 - I_{L\_AVG}^2}$ | $\sqrt{I_{SW\_RMS}^2 - I_{SW\_AVG}^2}$ |
| Output cap rms $I_{COUT\_RMS}$ (A) | $\sqrt{I_{L\_RMS}^2 - I_{L\_AVG}^2}$ | $\sqrt{I_{D\_RMS}^2 - I_{D\_AVG}^2}$ | $\sqrt{I_{D\_RMS}^2 - I_{D\_AVG}^2}$ |
| DC input current $I_{IN}$ (A) | $I_{SW\_AVG}$ | $I_{L\_AVG}$ | $I_{SW\_AVG}$ |

go even further, in Table 4.2 we have introduced a simplifying factor called $\alpha$. This is the ratio of the load current to the peak current. With this factor we can see that even the DCM equations start looking elegant, manageable, and thus better suited for optimization.

To calculate the duty cycle from first principles, we must consider the following:

In DCM the diode conducts for a period defined by $D'/f$. So $D'$ is the diode conduction duty cycle. We do the same for CCM, but the difference

**TABLE 4.2 Alternative Form of DCM Equations (in Terms of $I_O$ and $\alpha$).**

$\alpha = I_O / I_{PK}$

$I_{PK} = \dfrac{V_{ON}D}{Lf}$

| | Buck | Boost | Buck-Boost |
|---|---|---|---|
| $V_{ON}$ | $V_{IN} - V_O - V_{SW}$ | $V_{IN} - V_{SW}$ | $V_{IN} - V_{SW}$ |
| $V_{OFF}$ | $V_O + V_D$ | $V_O - V_{IN} + V_D$ | $V_O + V_D$ |
| Duty cycle $D$ | $\sqrt{\dfrac{2 \times I_O \times L \times f \times V_{OFF}}{V_{ON} \times (V_{ON} + V_{OFF})}}$ | $\sqrt{\dfrac{2 \times I_O \times L \times f \times V_{OFF}}{V_{ON}^2}}$ | $\sqrt{\dfrac{2 \times I_O \times L \times f \times V_{OFF}}{V_{ON}^2}}$ |
| Peak current $I_{PK}$ (A) | $\equiv I_O/\alpha$ | $\equiv I_O/\alpha$ | $\equiv I_O/\alpha$ |
| Switch average (A) | $= \dfrac{I_O D}{2\alpha}$ | $= \dfrac{I_O D}{2\alpha}$ | $= \dfrac{I_O D}{2\alpha}$ |
| Switch rms $I_{SW\_RMS}$ (A) | $= \dfrac{I_O}{\alpha}\sqrt{\dfrac{D}{3}}$ | $= \dfrac{I_O}{\alpha}\sqrt{\dfrac{D}{3}}$ | $= \dfrac{I_O}{\alpha}\sqrt{\dfrac{D}{3}}$ |
| Diode duty cycle $D'$ | $= 2\alpha - D$ | $= 2\alpha$ | $= 2\alpha$ |
| Inductor average $I_{L\_AVG}$ (A) | $= I_O$ | $= \dfrac{I_O(2\alpha + D)}{2\alpha}$ | $= \dfrac{I_O(2\alpha + D)}{2\alpha}$ |
| Inductor rms $I_{L\_RMS}$ (A) | $= \dfrac{I_O}{\sqrt{1.5 \times \alpha}}$ | $= \dfrac{I_O\sqrt{2\alpha + D}}{\alpha\sqrt{3}}$ | $= \dfrac{I_O\sqrt{2\alpha + D}}{\alpha\sqrt{3}}$ |

*(Continued)*

**TABLE 4.2 Alternative Form of DCM Equations (in Terms of $I_O$ and $\alpha$) (Continued).**

$\alpha = I_O / I_{PK}$

$I_{PK} = \dfrac{V_{ON} D}{Lf}$

| | Buck | Boost | Buck-Boost |
|---|---|---|---|
| Switch rms $I_{SW\_RMS}$ (A) | $= \dfrac{I_O\sqrt{D}}{\alpha\sqrt{3}}$ | $= \dfrac{I_O\sqrt{D}}{\alpha\sqrt{3}}$ | $= \dfrac{I_O\sqrt{D}}{\alpha\sqrt{3}}$ |
| Switch average $I_{SW\_AVG}$ (A) | $= \dfrac{I_O D}{2\alpha}$ | $= \dfrac{I_O D}{2\alpha}$ | $= \dfrac{I_O D}{2\alpha}$ |
| Diode average $I_{D\_AVG}$ (A) | $= I_O\left(1 - \dfrac{D}{2\alpha}\right)$ | $= I_O$ | $= I_O$ |
| Diode rms $I_{D\_RMS}$ (A) | $= \dfrac{I_O\sqrt{2\alpha - D}}{\alpha\sqrt{3}}$ | $= \dfrac{I_O\sqrt{2}}{\sqrt{3}\times\alpha}$ | $= \dfrac{I_O\sqrt{2}}{\sqrt{3}\times\alpha}$ |
| Input cap rms $I_{CIN\_RMS}$ (A) | $= \dfrac{I_O\sqrt{D(4 - 3D)}}{2\alpha\sqrt{3}}$ | $= \dfrac{I_O\sqrt{4\alpha(2 - 3\alpha - 3D) + D(4 - 3D)}}{2\alpha\sqrt{3}}$ | $= \dfrac{I_O\sqrt{D(4 - 3D)}}{2\alpha\sqrt{3}}$ |
| Output cap rms $I_{COUT\_RMS}$ (A) | $= \dfrac{I_O\sqrt{1 - 1.5\alpha}}{\sqrt{1.5}\times\alpha}$ | $= \dfrac{I_O\sqrt{1 - 1.5\alpha}}{\sqrt{1.5}\times\alpha}$ | $= \dfrac{I_O\sqrt{1 - 1.5\alpha}}{\sqrt{1.5}\times\alpha}$ |
| DC input current $I_{IN}$ (A) | $= \dfrac{I_O D}{2\alpha}$ | $= \dfrac{I_O(2\alpha + D)}{2\alpha}$ | $= \dfrac{I_O D}{2\alpha}$ |

in DCM is that $D' \neq 1 - D$. So the calculation is a little tricky. Now we have to work out the PWM duty cycle from the voltseconds law combined with the fact that we need a certain load current $I_O$ at the output. See the derivation below.

We need to remember that the buck is different in that to get the required load current, we have to average over the current flowing during both the switch and the diode conduction times (since energy flows to the output during both these intervals). But for the boost and buck-boost we have to average only over the diode conduction time.

As mentioned in Chap. 2, the boost and buck-boost are very similar to each other. Thus the duty cycle calculations below are the same for either topology, as shown below. The only difference between them is that the applied voltages across the inductor during the on-time ($V_{ON}$) and the off-time ($V_{OFF}$) are different. For these values we can consult Table 4.1 and 4.2.

### 4.3 The Duty Cycle Equations

The calculation is fairly self-explanatory.

| **Boost/buck-boost** | **Buck** |
|---|---|

$$I_{PK} = \frac{V_{ON} \times D}{L \times f} = \frac{V_{OFF} \times D'}{L \times f} \qquad\qquad I_{PK} = \frac{V_{ON} \times D}{L \times f} = \frac{V_{OFF} \times D'}{L \times f}$$

$$\frac{I_{PK}}{2} \times D' = I_O \qquad\qquad\qquad\qquad \frac{I_{PK}}{2} \times D' + \frac{I_{PK}}{2} \times D = I_O$$

$$D' = \frac{2 \times I_O}{I_{PK}} \qquad\qquad\qquad\qquad D' = \frac{2 \times I_O}{I_{PK}} - D$$

$$\frac{V_{ON} \times D}{L \times f} = \frac{V_{OFF}}{L \times f} \times \frac{2 \times I_O}{I_{PK}} \qquad \frac{V_{ON} \times D}{L \times f} = \frac{V_{OFF}}{L \times f} \times \left(\frac{2 \times I_O}{I_{PK}} - D\right)$$

$$D = \frac{2 \times I_O \times V_{OFF}}{I_{PK} \times V_{ON}} \qquad\qquad D = \frac{2 \times I_O \times V_{OFF}}{I_{PK} \times (V_{ON} + V_{OFF})}$$

$$D = \frac{2 \times I_O \times V_{OFF}}{V_{ON}} \times \frac{L \times f}{V_{ON} \times D} \qquad D = \frac{2 \times I_O \times V_{OFF}}{(V_{ON} + V_{OFF})} \times \frac{L \times f}{V_{ON} \times D}$$

$$\boxed{D = \sqrt{\frac{2 \times I_O \times L \times f \times V_{OFF}}{V_{ON}^2}}} \qquad \boxed{D = \sqrt{\frac{2 \times I_O \times L \times f \times V_{OFF}}{V_{ON} \times (V_{ON} + V_{OFF})}}}$$

# 5

# Front-End of Off-Line Power Supplies

## 5.1 Conventional Front-End Design

### 5.1.1 Input voltage shape

The voltage across the input bulk capacitor is effectively the supply rail for the converter. The input capacitor is in fact indispensable, not only because of input voltage ripple considerations, but because in a typical off-line power supply the diode bridge conducts only for a small part of the ac half-cycle. During the remaining period, all the required input energy can come only from the bulk capacitor. Therefore, when the appropriate diode pair of the input bridge rectifier gets forward biased, an input current surge lasting for a few milliseconds pushes its way in. Its main effect is to recharge ("top up") the bulk capacitor. An energy balance condition occurs automatically under *steady state* in which the amount of energy pushed into the capacitor is just enough for it to continue providing input power to the converter, right up to the moment where the whole sequence can repeat itself exactly. In Fig. 5.1 we show several half-cycles of the ac input *voltage*. Later in this chapter we will also discuss the shape of the input surge *current* waveform.

As we see from the figure, there is a normal voltage ripple occurring between the limits $V_{INMIN\_HI}$ and $V_{INMIN\_LO}$, with an average value of $V_{INMIN\_AVG}$. However, there could be a *power line disturbance* (PLD) in which the next ac half-cycle may "sag." If the sag is severe enough, for all practical purposes that ac half-cycle can be considered missing altogether, since it is too low to be able to forward bias either of the diode pairs of the input bridge anyway. Under these conditions the voltage on the bulk capacitor follows a definite trajectory (locus of constant

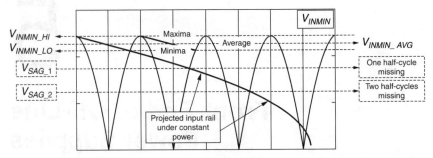

**Figure 5.1**  The rectified ac input and the voltage to the converter.

power in our case) as seen in the figure. It may get restored if the next half-cycle is normal, in which case the lowest point would be $V_{SAG\_1}$. But the disturbance may last longer and it could miss that particular half-cycle too. The lowest point is then $V_{SAG\_2}$. It is important for the designer to know what these trough levels are, but unfortunately, a closed form equation that is accurate enough probably does not exist, and the solution invariably requires mathematical simulation or some iterations. If the designer ends up doing unsubstantiated guesswork on these very important numbers, the design is going to be poor from the get-go. The importance of relating a power supply design to its front-end is often not clearly understood. In Fig. 5.2 through Fig. 5.5 we

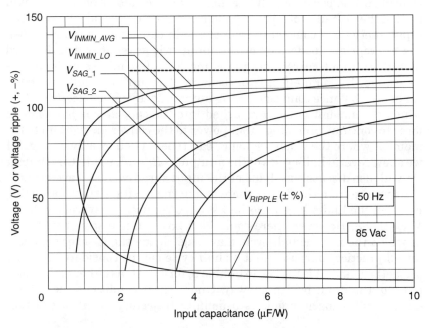

**Figure 5.2**  Design curves for input ripple (85 Vac/50 Hz).

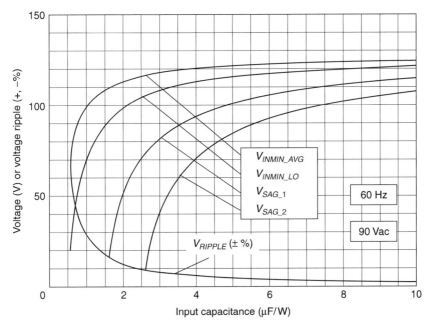

**Figure 5.3**  Design curves for input ripple (90 Vac/60 Hz).

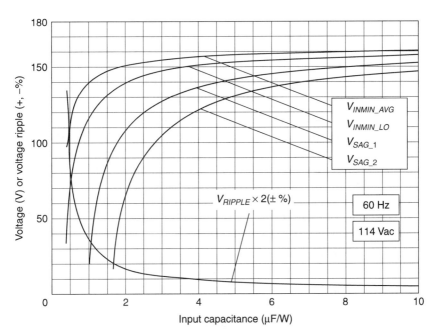

**Figure 5.4**  Design curves for input ripple (114 Vac/60 Hz).

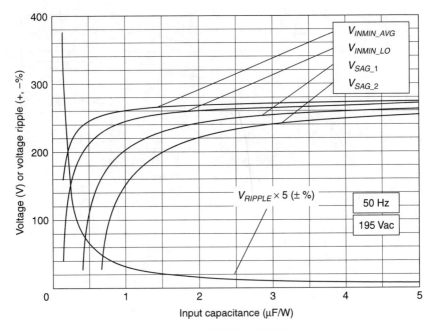

**Figure 5.5**  Design curves for input ripple (195 Vac/50 Hz).

have therefore used mathematical simulations to provide several plots for easy reference. They are based on the most prevalent or commonly designed-for line conditions.

The designer may still demand to know why this information is vital to his or her design, so here are the main points:

Under normal operation the effective average voltage input to the converter is $V_{INMIN\_AVG}$. The dissipation in the switch, for example, is based on this average voltage, *not the peak* value (which is $V_{INMIN\_HI}$). So if we use our waveform analysis of the switch current trapezoid using the duty cycle at $V_{INMIN\_HI}$, we will just be approximating. We could be very detailed and carry out integration over the full ac ripple cycle, but the results actually match up very closely to simply assuming a steady zero-ripple dc input of $V_{INMIN\_AVG}$.

Efficiency calculations or predictions must therefore also be done at $V_{INMIN\_AVG}$ (not $V_{INMIN\_HI}$) for better accuracy.

If one or more ac half-cycles are missed as the input voltage falls, our ability to keep the converter running depends upon a) not hitting current limit and b) not hitting duty cycle limit. In effect, we are looking to have a certain *holdup time*. The amount of holdup time we get is not only dependent on our selection of the bulk capacitance but also on how we set the limits of our controller. In boost and buck-boost (or flyback) topologies the center of the current ramp climbs steeply as duty cycle increases, so eventually we will hit either or both these limits. If we know the exact instantaneous voltage level just before the ac resumes ($V_{SAG\_1}$ or $V_{SAG\_2}$ as applicable) then we can set the current limit by analyzing the switch waveform at this input level. For duty cycle we can set the $D_{MAX}$ of the controller (if we have that flexibility), otherwise we should set the normal operating duty cycle sufficiently below the maximum, so as to leave enough "headroom" for the purpose.

Therefore, if the ac input is removed (or just temporarily missing), the converter continues drawing current from the bulk capacitor even as the voltage on the capacitor keeps slewing down. Since the current goes up steeply as the voltage decreases, we have to design the size of the relevant magnetic component of the converter according to the maximum current it will see under these conditions, not the current under normal operation. Sure, *we do choose the inductance based on normal operation, but not its physical size.* That is determined by the holdup time (or the brownout voltage level) the power supply is expected to tolerate. Clearly, though most components may just get briefly hot during a PLD event, we cannot afford to saturate our magnetics even for a moment, for that can lead to immediate destruction of the switch. Note that the same extreme situation may be encountered even under a normal power-down event.

Which brings us to the question—what if our current limit and duty cycle limit "headrooms" are too generous? Now the converter will keep trying to function down to lower and lower voltages if the ac does not resume (as during a normal removal of ac power). The escalating currents may end up by being limited only by various parasitics. Such an undesirable situation can easily arise if we had made a wrong guess about the input voltage shape, and/or set our current limit and duty cycle limit incorrectly.

Conversely, if we set current limit and $D_{MAX}$ too low (inadequate headroom), we will not meet the required holdup time. We already know that if we set these limits too high we could destroy the switch due to saturation. But note that either way the required setting is wholly *dependent on the selected input capacitance*.

Any current limit or $D_{MAX}$ setting will always have a certain tolerance range. We should confirm the power throughput capability (and holdup time) at the *lower* tolerance end of the set current limit and $D_{MAX}$ (using also the lower tolerance limit of the bulk capacitance). But we should always rule out the possibility of inductor saturation at the *upper* limit of the current limit and duty cycle tolerance. Therefore "good" controllers/switchers must either have an accurately set current limit and $D_{MAX}$, or should allow us to set these externally with similar accuracy.

The designer should also be watchful for the much-advertised *power capability* of certain integrated switchers available for off-line flyback applications. Between the lines, and somewhere in the fine print, you may see that there is probably no holdup time guaranteed under the stated peak power. Holdup time is something that one rightfully expects of any equipment connected to the vagaries of the input mains lines, but clearly not everyone agrees!

Also, peak power capability is not necessarily continuous power capability. It may be so only if we have perfect thermal management. Usually that is not so. So the designer should also request the vendor of the integrated off-line switcher device to guide him or her on the *size of the heatsink* required to achieve the claimed maximum power capability. This will ensure a more realistic estimation of the power actually possible in a real application. If the power capability figure is just inflated by marketing, it is better that the engineers understand this earlier rather than later.

Notice that the curves presented in this chapter are drawn up in terms of $\mu F/W$ of input capacitance. As we said, everything is ultimately related to the input capacitance. Now we can use the curves to also pick the optimum input capacitance to do the job. A typical thumb-rule for universal input (worldwide) off-line power supplies is said to be 3 $\mu F/W$. From the low-line curves we can see that this is close to the "knee," thus justifying that choice in general. But the designer must remember that this is the input stage and it cares only about *input* power. It has no way of knowing how much load power we are getting out of the converter. For example, if we have a 60 W flyback with any abysmal 50 percent efficiency then it is actually drawing 120 W at its input, and the input capacitor needs to be able to process that power. So we would choose a (theoretical) value of $3 \times 120 = 360 \mu F$ of input capacitance and not $3 \times 60 = 180 \mu F$. In addition, we should also account for initial $\pm 20$ percent tolerance and another 20 percent fall in capacitance (ageing) over time when using aluminum electrolytics. So we actually need to start with 56 percent more capacitance than the theoretically calculated value of $360 \mu F$. Often designers tend to ignore some of these considerations. But we certainly cannot compound all the errors by choosing a capacitance based on output power and also ignoring tolerance and ageing. Yes, if we are anyway ignoring holdup time altogether, we can probably do whatever we want! Finally we must also remember to verify the ripple current rating of the capacitor and its life expectancy as discussed in Chap. 17. The capacitor we finally choose may in fact end up being based on ripple current capability rather than holdup considerations. Some vendors of switcher ICs in fact even ignore the ripple current (and life) considerations when making their nifty evaluation boards. They "look better" they say, and if pressed, their "justification" is that if an adapter ever fails, most customers throw away the adapter anyway (of course, now they don't even have a choice, and they may even have to abandon the connected equipment too).

We should also keep in mind that if we have too high an input voltage ripple, the gain of the system might not be enough to reject this sufficiently. For example, if we desire ±50 mV of output ripple, out of this typically at least half will be the line frequency component (the rest being high-frequency ripple). So if the input ripple is ±10 percent at say 85 Vac and $3\mu$F/W, then our average input voltage is about 110 Vdc, and the ripple is ±11 V (see Fig. 5.1). We therefore need to reduce this ripple by a factor of $11/0.025 = 440$. So the dc gain of the loop must be greater than $20 \times \log(440) = 53$ dB. That is rather too high a gain anyway, so we may prefer to add post LC filters instead. That is also cheaper than trying to increase the input capacitance injudiciously.

Note that in the figures, the ripple $\Delta V/V$ is also presented, though in some cases it may have been multiplied by a certain integer, just to bring it into the range of display of the other curves. This factor is indicated too. The vertical axis represents the absolute value of the voltages but is also used for the ripple expressed in terms of ± percent.

We also present the basic equation commonly used in literature for the lowest point of input voltage

$$V_{SAG} = \sqrt{V_{INMIN\_HI}^2 - \frac{2 \times 10^6 \times t_{holdup}}{\mu F/W}}$$

This equation is based purely on energy considerations and gives results slightly lower than the curves presented so far in this chapter. The difference is that earlier in this chapter we were looking at the exact number of missing ac half-cycles rather than some "blind" or arbitrary holdup time figure, which is what the above equation uses. Holdup time is stated rather blandly in most original equipment manufacturer (OEM) specifications, though in reality we know that the spirit behind the holdup time figure is not just an arbitrary duration but the number of missing ac half-cycles. We may like to discuss this issue further with the equipment manufacturer, since a holdup specification in terms of missing cycles (rather than time) may lead to smaller input capacitance.

Here is a practical example of how the curves impact the design of a converter.

**Example 5.1** We are designing a 5 V output forward converter for the range 85 Vac to 265 Vac. We want a holdup time of 20 ms at 50 Hz. What should be the turns ratio of the transformer?

A holdup of 20 ms is two ac half-cycles, corresponding to $V_{SAG\_1}$. From Fig. 5.2 we can see that for an "economical" choice of 3 $\mu$F/W we will get 60 Vdc at the end of the holdup time. At this point we can assume the duty cycle has gone to its maximum of 48 percent (typical forward converter controller limit). Since the output must still be in regulation, we have

$$V_o = V_{IN} \times \frac{D}{n}$$

where $n$ is the turns ratio $n_P/n_S$. So

$$n = \frac{60 \times 0.48}{5} = 5.8$$

The calculated duty cycle under steady operation at 85 Vac is

$$D = \frac{5 \times 5.8}{110} = 0.26$$

Note that we have not considered the forward voltage drops across the switch and the freewheeling diode. That will push up the duty cycle by a few percent. Also, the turns ratio may have to change slightly since the number of primary and secondary turns must both be integers. So, if for example, $n_S$ is 3 then the closest integral value for $n_S$ is 17, which makes the turns ratio 5.67. Some iterations are therefore always required.

We also need to consider whether $D = 0.26$ is a good choice from the viewpoint of switch and capacitor dissipations. Usually we try to set $D \approx 0.3$ to 0.35 for a forward converter. So in our case we will probably need to go back and increase the input capacitance somewhat.

### 5.1.2 The input current shape

The input bridge conducts only for a part of the ac half-cycle. The input (bulk) capacitor discharges slowly during the remaining period at a rate that depends on the power it is delivering to the converter (which is $P_{IN} = P_{OUT}/$Efficiency). If we put larger and larger bulk capacitance we can end up with extremely high peak and rms currents through the input bridge and the electromagnetic interference (EMI) filter chokes (along with a lot of low-frequency harmonic content too). This is attributable to the fact that the bridge conduction time becomes shorter as we increase the bulk capacitance. Since a certain amount of average input power is being demanded by the converter, the current bursts

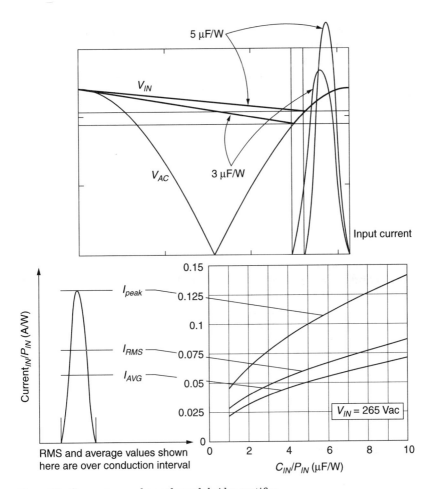

**Figure 5.6**  Current waveshape through bridge rectifier.

must increase in amplitude to compensate for the shorter gating inter-val (see Fig. 5.6).

Knowledge of the input rms current is necessary to correctly esti-mate the copper losses in both the common-mode EMI chokes and the differential-mode EMI chokes, whereas knowledge of the peak current is necessary to correctly estimate the core volume of the differential-mode choke (its energy handling capability). In the common-mode choke, the input ac line current effectively cancels out, so saturation is not nor-mally a consideration.

The shape of the input current into the power supply is usually de-scribed as a *haversine*. This is simply a sine waveform offset vertically so as to make its lowest point coincide with the horizontal axis. The current waveform shown in the figure is a half-cycle haversine (with a period equal to the diode conduction time).

We can show that the following equations apply here:

The time for which each diode conducts is

$$t_C = \frac{\cos^{-1}[A]}{2 \times \pi \times f_{LINE}} \quad \text{(seconds)}$$

where

$$A = \sqrt{1 - \frac{P_{IN}}{2 \times C \times f_{LINE} \times V_{AC}^2}}$$

C is in Farads above.

The rms and average values of the current waveform (calculated only over the conduction time) are

$$I_{AVG} = \frac{\sqrt{2} \times \pi \times P_{IN}}{V_{AC} \times [1 + A] \times \cos^{-1}[A]} \quad \text{(amperes)}$$

$$I_{RMS} = I_{PEAK} \times \frac{\sqrt{1.5}}{2} \equiv 0.612 \times I_{PEAK} \quad \text{(amperes)}$$

The peak value of the current is

$$I_{PEAK} = 2 \times I_{AVG}$$

Let us illustrate the use of the graphical aids provided in Fig. 5.6.

**Example 5.2**   A power supply delivering 5 A@14 A at 70 percent efficiency has an input capacitor of 330 $\mu$F. What are the rms and peak input currents at 265 Vac/50 Hz?

$$P_{IN} = \frac{5 \times 14}{0.7} = 100 \text{ W}$$

So $A = 0.978$, and $t_C = 0.67$ ms. We calculate: $I_{AVG} = 4.05$ A, $I_{PEAK} = 8.1$ A, $I_{RMS} = 8.1 \times 0.612 = 5.03$ A.

From the figure, the input capacitance per (input) watt is $330/100 = 3.3$ $\mu$F/W. We locate this value on the horizontal axis and then we can see that this gives us about 0.05 A for the rms current on the vertical axis. But the vertical axis is the current per watt of input power. So for our case, the rms current is $100 \times 0.05 = 5$ A. This agrees with our numerical calculation above.

Since we have two conduction intervals in each ac cycle, the input average and rms currents now calculated over the entire cycle are (with $T = 1/f_{LINE}$)

$$I_{IN\_AVG} = 4.05 \times \frac{2 \times t_C}{T} = 0.27 \text{ A}$$

and

$$I_{IN\_RMS} = \left[ 5.03^2 \times \frac{2 \times t_C}{T} \right]^{1/2} = 1.3 \text{ A}$$

So we have to ensure that our differential-mode input filter choke does not saturate with a peak instantaneous current of 8 A. All the input chokes must have copper thick enough to handle 1.3 A rms. Some engineers pick the choke based on certain personal thumb rules. It can be shown that they typically underestimate the rms input current by almost a factor of 2 (and the heating in the EMI chokes by about four times). In the next section we will take up *power factor correction* (PFC). Here we just point out that if we had a power supply with PFC (PF $\approx$ 1), we would have got at 265 Vac input

$$I_{IN\_RMS} = \frac{P_{IN}}{V_{IN\_RMS}} = \frac{100}{265} = 0.38 \text{ A}$$

and the peak of this is

$$I_{IN\_PEAK} = I_{IN\_RMS} \times \sqrt{2} = 0.534 \text{ A}$$

Note this is much less than the non-PFC case, which partly explains why the EMI filter chokes in a power-factor-corrected power supply are usually much smaller.

Note that we also want to repeat the calculations at low-line conditions. At low-line, despite the fact that the conduction time increases significantly, the required average input current is much higher too. So the peak/rms currents actually increase.

## 5.2   Front-End with PFC

### 5.2.1   Regulatory issues

We will briefly cover the key terminology first. *Apparent power* is defined as $V_{AC} \times I_{AC}$ where we have used the rms values of the line voltage and the line current. This may or may not be equal to the useful power available, which is called *real power*. The ratio of real power to apparent power is called *power factor* (PF). The usual goal is to make PF close to unity. We do that by implementing *power factor correction* (PFC). This additional PFC stage is interposed between the input mains and the usual PWM (pulse width modulated) power converter stage.

Note that real power is mathematically determined by dividing the time axis into a very large number of small segments and multiplying the voltage in each time segment by the current flowing at that moment.

Finally, we take the average of the result. Therefore, we are in effect using the *instantaneous* values of current and voltage in calculating the real power, and that is what we need to do by the basic definition of work (or energy). But if we take *independent* or separate measurements (of the rms values of the voltage and current) and take their product, we may not get the right answer. We may, for example, be ignoring any *phase relationship* between the voltage and current. The (wrong) result of such a calculation is the apparent power. Apparent power may be further interpreted as being the vector sum of the real power and a term called the *imaginary power*.

With near perfect power factor correction, the real power which is $V_{AC} \times I_{AC} \times PF$ becomes almost equal to the apparent power which is $V_{AC} \times I_{AC}$. This means that the power factor which is also expanded as $PF = \mathrm{Cos}\,\phi \times \mathrm{Cos}\,\Theta$ becomes almost unity. Here $\phi$ is the *phase angle* and $\Theta$ is the *distortion angle*. In non-PFC power supplies, $\phi \approx 0$, but $\mathrm{Cos}\,\Theta$ is typically around 0.6 to 0.7. Too high a bulk capacitor (without PFC) will cause the power factor to decrease further. It will also significantly increase the EMI and associated filtering costs.

The first thing we have to keep in mind is that though electricity meters luckily don't charge us for apparent power, but only for real power (or we would have acted much faster in introducing PFC!), the amount of power we can pass through the wiring *does depend on the apparent power*. If the PF is low, then for a given *load power*, the apparent power is much higher. That is because the input current pulses drawn are now narrower and more "peaky," thus increasing their rms value and causing more heating in the wiring. This effectively limits the amount of useful power we are permitted to draw from the line.

Here is an example based on the standard 120 V/15 A outlet circuit commonly found in offices and homes in the United States. This line circuit should not be used to handle anything more than $120 \times 15 \times 0.8 = 1440$ W of apparent power. Note that here we have included a 20 percent mandated safety margin (which also prevents nuisance tripping of the line circuit breakers). Assuming that the overall efficiency (PFC + PWM stages) is 75 percent, our power supply can be used to deliver a maximum of 120 V $\times$ (15 A $\times$ 0.80) $\times$ 0.75 = 1080 W of load power, but *provided PF = 1*. If we had not implemented PFC, and assuming a typical PF of 0.6, the maximum output power from our (75 percent efficient) power supply is now only 120 V $\times$ (15 A $\times$ 0.80) $\times$ 0.75 $\times$ 0.6 = 648 W. We cannot connect a similar power supply with any higher output power than this, without first improving the efficiency and/or increasing the power factor.

The relevant standard most power supply designers historically referred to for low frequency line harmonic reduction was called IEC555.

This was later renamed IEC-61000-3-2 and is now called the European norm EN61000-3-2. This norm currently applies to most equipment between 75 W to 1000 W. Each of these power thresholds refers to the power we are attempting to draw *from* the mains, and is *not* the output (useful) power coming out of our power supply (or any attached system). So, for example, a 70 W flyback with 70 percent efficiency is a $70/0.7 = 100$ W load as far as the mains line is concerned. And since 100 W clearly falls between 75 W and 1000 W, this power supply certainly needs to incorporate power factor correction (particularly for use in Europe). Note that the responsible European agency CENELEC (which when translated from French becomes the *European Committee for Electrotechnical Standardization*) has been contemplating further reducing the lower threshold from 75 W to 50 W.

The most common way of dealing with low-frequency line harmonic reduction is the ubiquitous *average current-mode controlled, continuous conduction mode, fixed frequency boost power factor correction*, which provides a typical PF of about 0.98. Other ways of improving PF may not always be adequate because we have to remember that the line harmonic standard referred to above *does not* actually specify any minimum "acceptable" figure for PF. What it really demands is that we reduce the harmonic content of the input current waveform to within certain specified limits. Compliance is required up to the *39th harmonic* of the line frequency. Therefore, a high PF is not really mandatory, but is just a better indication that we are close to achieving a pure sine waveshape (no harmonic content). On the other hand we could hypothetically encounter an odd shaped waveform with a high PF (say >0.9), but with one stubborn harmonic that would make us fail the entire line harmonic test. So, most people have discovered one way or the other, that the "cleanest and surest" way to meet the line harmonic standard is to use a dedicated PFC controller IC, preferably implementing an average current-mode control boost topology.

### 5.2.2  Boost PFC

We are going to go past most of the better known aspects of this "common cure," and into its optimization and some vital control aspects. As a starting point, we have tabulated the key equations concerning component selection in Table 5.1 and plotted out their variations with respect to line voltage. These equations can be occasionally found in related literature, but surprisingly they are usually lacking because they assume a *resistive* load connected at the output of the PFC stage. In fact, a key to possible cost reduction is in understanding how the PFC and the succeeding PWM stages mutually interact, and how they therefore may be properly synchronized to our advantage.

**TABLE 5.1    Basic Design Table for (unsynchronized) PFC.**

| | |
|---|---|
| Input voltage before rectification | $\sqrt{2} \times V_{AC} \times \sin(2\pi f_{AC} t)$ |
| Input voltage after rectification | $\left\| \sqrt{2} \times V_{AC} \times \sin(2\pi f_{AC} t) \right\|$ |
| Input current before rectification | $\sqrt{2} \times \dfrac{P_{IN}}{V_{AC}} \times \sin(2\pi f_{AC} t)$ |
| Input current after rectification | $\left\| \sqrt{2} \times \dfrac{P_{IN}}{V_{AC}} \times \sin(2\pi f_{AC} t) \right\|$ |
| Input power | $V_{AC} \times I_{AC} = \dfrac{V_{IPK} I_{IPK}}{2}$ |
| Average input current | $\dfrac{2}{\pi} \times I_{IPK} \equiv 0.637 \times I_{IPK}$ |
| Average input voltage | $\dfrac{2}{\pi} \times V_{IPK} \equiv 0.637 \times V_{IPK}$ |
| Duty cycle of PFC | $1 - \left\| \sqrt{2} \times \dfrac{V_{AC}}{V_o} \times \sin(2\pi f_{AC} t) \right\|$ |
| | *This is (almost) identical to the large signal duty cycle equation for a conventional boost topology, that is, $D = (V_o - V_{in})/V_o$. A slight imbalance is required to cause the current to progressively ramp up or down with the ac cycle.* |
| Conduction duty cycle of diode | $\left\| \sqrt{2} \times \dfrac{V_{AC}}{V_o} \times \sin(2\pi f_{AC} t) \right\|$ |
| Switch rms current | $\dfrac{P_{IN}}{V_{IPK}} \sqrt{2 - \dfrac{16 \times V_{IPK}}{3 \times \pi \times V_o}}$ |
| | *Proportional to $P_{IN}$* |
| Inductor rms current | $\sqrt{2} \times \dfrac{P_{IN}}{V_{IPK}}$ |
| | *Proportional to $P_{IN}$, inversely proportional to input voltage* |
| Capacitor rms current (total)    *(unsynchronized)* | $I_o \sqrt{\dfrac{16 \times V_o}{3 \times \pi \times V_{IPK}} + \dfrac{1}{D_{PWM}} - 2}$ |
| | *Proportional to $P_{IN}$* |

*(Continued)*

**TABLE 5.1  Basic Design Table for (unsynchronized) PFC (*Continued*).**

| | |
|---|---|
| Capacitor rms current (high frequency) (*unsynchronized*) | $I_o\sqrt{\dfrac{16 \times V_o}{3 \times \pi \times V_{IPK}} + \dfrac{1}{D_{PWM}} - 2.5}$ |
| | *Proportional to $P_{IN}$* |
| Capacitor rms current (low frequency) (*unsynchronized*) | $\dfrac{I_o}{\sqrt{2}}$ |
| | *Proportional to $P_{IN}$, does NOT depend on input voltage* |
| Diode rms current | $\dfrac{P_{IN}}{V_{IPK}}\sqrt{\dfrac{16 \times V_{IPK}}{3 \times \pi \times V_o}}$ |
| | *Proportional to $P_{IN}$* |
| Peak current (*diode, switch, and inductor*) | $2\dfrac{P_{IN}}{V_{IPK}}$ |
| | *Proportional to $P_{IN}$, inversely proportional to input voltage* |
| Diode average current | $I_o$ |
| | *Proportional to $P_{IN}$, does NOT depend on input voltage or Duty Cycle* |
| *where* $\quad P_{IN} = V_{AC} \times I_{AC} = \dfrac{V_{IPK} \times I_{IPK}}{2}, \quad V_{IPK} = \sqrt{2} \times V_{AC}, \quad I_o = P_{IN}/V_o$ <br> *"flat-top approximation used" (high inductance)* | |

Since the conversion efficiency of an off-line boost PFC stage is typically very high (over 90 percent with possible help from turn-on snubbers), the input and output power of the boost stage are almost equal, i.e., $P_{IN} \cong P_o$. Note that $P_o$, which is the output power for the PFC stage forms the input power for the succeeding PWM stage. The bulk capacitor which provides the high voltage dc rail (HVDC) for the PWM is the output capacitor of the boost stage. The duty cycle of the PWM is called $D_{PWM}$ in Table 5.1. This table is actually two tables rolled into one. The lower half provides the relevant rms and average currents. Note that all these current components *are proportional to $P_{IN}$*. Equivalently, they are also proportional to the load current flowing out of the PFC stage $I_o$. In this chapter we have at various places provided some useful numbers for quick reference. Some of these are arbitrarily benchmarked to $P_{IN} = 1000$ W, or we may have set $I_o$ to some stated value. Therefore, in either case the designer can easily *scale* these numbers for the power level in his or her application.

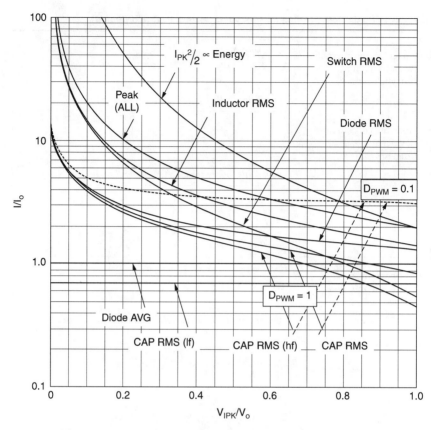

**Figure 5.7**  The current components as they vary with line voltage.

In Fig. 5.7 we have presented the actual variations as we change the line voltage. On the $y$ axis we have the currents corresponding to 1 A load current (flowing out of the bulk capacitor). For example, at a typical setting of 385 V for the high voltage dc rail (HVDC), this corresponds to power going into the PWM stage of 385 V/1 A = 385 W. On the $x$ axis we have the ratio of conversion $V_{IPK}/V_o$. For example, if the ac input is 90 Vac, the rectified peak value of the input rail $V_{IPK}$ is $90 \times (2)^{0.5} = 127$ V. Note that if the output is set between 385 V and 400 V, $V_{IPK}/V_o$ is about 0.33 at 90 Vac. Thus looking at the figure we can calculate what the current components are in the PFC circuit for any particular load and line condition.

### 5.2.3  Capacitor selection

This is a key design issue for a PFC stage. There are actually two capacitors to consider.

There is always a low capacitance film or ceramic input capacitor next to the input bridge rectifier (usually after the rectifier). This forms the input capacitor to the PFC stage. The thumb-rule value for this is $3 \, \mu F/kW$ and this value is considered a good compromise. If we use too large an input capacitance it would remain somewhat charged up as the instantaneous ac input voltage goes toward zero, and this would cause the input current waveform to get deformed from a pure sine wave (lowering the power factor). Too low of an input capacitance may affect the overall EMI filtering efficacy, but further, since the capacitor will discharge to very low values, the PFC stage will have to "struggle" to boost the voltage to the required HVDC level, thus increasing the stresses. Note that if this capacitor is chosen to be a fairly large value, but is erroneously placed *before* rather than *after* the bridge, the voltage into the PFC stage will pass through zero (as it tries to follow the unrectified ac input). This too will therefore create much the same effect as a very low capacitance placed after the bridge. So, the correct position of this component is obvious. Note that this capacitor, being the input to a boost topology, has a smoothing choke between it and the switching transistor, so it is not required to provide any high-frequency pulses as is the case for an input capacitor of a buck or buck-boost topology. This capacitor therefore does not necessarily need to be an expensive high-frequency metallized polypropylene (*MKP*) type. Cheaper metallized polyester (*MKT*) capacitors often work fine. However, the designer is cautioned that in this position the capacitor does need to be rugged enough to withstand the high inrush charging current, especially under high-line conditions. A metallized capacitor can degrade significantly over time in such a front-end position, because though it does tend to "self-heal," it keeps losing a little capacitance each time that happens. The process is cumulative, and the eventual failure of this component may not have been anticipated in initial bench testing. We should remember that *MKP capacitors have a typical* $dV/dt$ *rating three to four times better than MKT capacitors*, and are more suited in this application for this reason itself. From the viewpoint of safety, any line-to-line capacitor placed *before* the bridge has to be a *safety approved X-capacitor*, thus adding to the cost. After the bridge, we can use a cheaper capacitor. But some engineers prefer to use a safety approved capacitor even *after* the bridge for its excellent $dV/dt$ capability (which can be traced to its rather special construction). In brief, this input capacitor can become a key *reliability* issue and it must be very carefully selected.

The output capacitor of the boost stage is the input capacitor to the PWM stage. It provides holdup time to keep the PWM stage working during a momentary line sag condition. Most engineers know how to do a simple capacitance versus holdup time calculation, based on the well-known energy balance equation

$$\frac{1}{2}C \times \left(V_{initial}^2 - V_{final}^2\right) = P_o \times t_{holdup}$$

But we should remember that the value of the required capacitance is intricately linked to the design of the PWM stage. This stage must be able to continue to operate with normal output regulation down to $V_{final}$. As explained in the previous sections of this chapter, its maximum duty cycle and current limit must be set accordingly.

Some power supply manufacturers prefer to use a 450 V bulk capacitor, but these are much more costly than 400 V capacitors. With very careful design, including the use of a PFC control IC with a tight tolerance on the feedback/reference voltage (e.g., the UC3854), we can often get away with a 400 V capacitor, even for a worldwide 90 to 270 Vac power supply. The derating margins may not look very good on paper, but then there are also little statistical data to prove that derating really helps improve the reliability of a modern aluminum electrolytic capacitor (but check with your specific vendor on this).

The second design issue with the output capacitor is that the *life of the capacitor* is determined by the rms value of the current through it. At no time should we exceed the stated ripple current capability of the selected capacitor. We may even need to pick a larger capacitance (preferably same voltage rating) just to get a larger size which can better handle the ripple current that it is going to encounter. See Chap. 17 for a deeper understanding on how to calculate the life expectancy of aluminum electrolytic capacitors. Incidentally, forced air cooling is *not* usually used by capacitor manufacturers when they test and characterize life. So, we have to be really careful in all our assumptions and estimates. For example, we should not use any *temperature multipliers* to estimate life, even if they have been provided in the capacitor datasheet itself. Finally, what we *can and should do* when performing a life calculation is to treat the high-frequency rms current component differently from the low-frequency component, or we would likely end up with an overdesigned capacitor. The *equivalent series resistance* (ESR) of such a capacitor goes *down* with frequency, thus permitting higher ripple currents at high frequencies (for which there are frequency multipliers provided in the datasheet). Therefore, we must *normalize* the current components—either to an equivalent high frequency net value, or an equivalent low-frequency one. *A typical frequency multiplier is about 1.43 for high frequencies (around 100 kHz)*, and we can increase the ripple current by this factor above the low-frequency ripple current rating. For example, if the rated (low frequency) ripple current is $I_{RATED\_LO\_FREQ}$ and the high-frequency multiplier is $f_{MULT}$, then the effective low-frequency ripple current is actually

$$I_{RMS\_EFFECTIVE\_LO} = \sqrt{I_{RMS\_LO}^2 + \left(\frac{I_{RMS\_HI}}{f_{MULT}}\right)^2}$$

The criterion for correctly selecting the bulk capacitor is thus

$$I_{RMS\_EFFECTIVE\_LO} \leq I_{RATED\_LO}$$

Note that *only when this condition is satisfied are we allowed to use the doubling rule*, i.e., that the life of an aluminum electrolytic doubles every $10°C$ fall in temperature. Note that *hf* stands for high frequency, and *lf* for low frequency. The low-frequency components are all input line related. The high-frequency components come from two sources—the high-frequency switching of the PFC stage and the high-frequency switching of the succeeding PWM stage.

The question also arises: Can we somehow synchronize the two sources of the high-frequency rms component so as to *reduce* the ripple current through the capacitor? This "trick" has in fact been implemented rather quietly for quite some time now by several different manufacturers of commercial power supplies. But unfortunately no equations for this technique seem to be available in literature. Therefore we have used a powerful mathematical simulation program to generate the results presented in the next section, and these results have been verified on production models.

### 5.2.4   Synchronizing the PFC and PWM stages

Some power factor control ICs offer *synchronization capability* but that phrase is usually meant to mean that the PFC is *in-phase* with the PWM. So whenever the PFC switch turns *on*, so does the PWM switch. But in general this synchronization scheme is not necessarily conducive to lowering EMI. Admittedly, it may make the EMI spectrum more predictable, but in fact it can severely *increase* the required EMI filtering cost. So, many engineers feel that it is actually better to run the PFC and PWM independently (no synchronization), just taking care to keep their respective switching frequencies spaced somewhat apart to avoid beat frequencies.

In the remaining part of this chapter we will henceforth ignore the in-phase synchronization in favor of the out-of-phase synchronization discussed next.

This synchronization scheme was introduced as a combo IC a few years ago and billed as the "industry's first leading edge/trailing edge modulation scheme" PFC/PWM controller IC. What this essentially tried to accomplish was based on the following intuitive thought process: we know that the PWM draws current out from the bulk capacitor whereas the PFC dumps current into it. What if we could make these opposing currents cancel? Maybe what we really want to do is to turn *on* the PWM switch at the very same moment as the PFC starts to turn *off*. This is an *out-of-phase* synchronization in that sense. So we would then expect that the freewheeling current (coming out from the PFC diode) would head straight into the PWM stage (most of the time) *without having to recycle through the bulk capacitor*. Of course, the PFC duty cycle varies from very low to very high values, and the PWM duty cycle is fixed. So this "cancellation" would work but by different amounts over the ac line cycle (see Fig. 5.8). It is therefore hard to provide any easy closed-form equation for calculating the net reduction in the rms current through the capacitor. Not surprisingly, there was, in fact, no such detailed application information provided even for the combo IC mentioned.

This combo IC had a single clock of fixed frequency, and whereas the PWM switch would turn *on* at the clock edge, the PFC would switch *off* at the same edge. In most power converters regulation is carried out by varying the moment at which the switch turns *off*. This is called *trailing edge* modulation. Therefore, in this IC the PWM worked the same way, but for the PFC we use *leading edge* modulation. So now we have to regulate *knowing beforehand* that the switch must turn *off* at the clock edge and then determining the moment at which we need to turn the switch *on*.

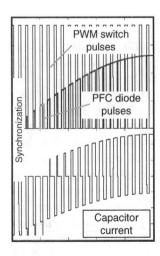

**Figure 5.8**  How out-of-phase synchronization reduces capacitor current.

It turns out that this is in reality an unnecessarily elaborate way of implementing out-of-phase synchronization. An easier way uses more standard parts—the popular UC3854, and the UC3844. We can synchronize the two ICs much the way we normally do for such ICs (with a small resistor in series with the timing capacitor), except that now we use the *off* edge of the 3844 output signal to reset the clock of the UC3854. In principle we could also reverse the order and use the UC3854 as the "master." We may then intuitively expect a slight inherent line frequency modulation in this process, but in practice it was found to be transparent at the output of the PWM.

In Fig. 5.9 and Fig. 5.10 we have plotted the ripple currents for 90 Vac and 270 Vac respectively, versus the PWM stage duty cycle. We see that using synchronization we get the highest percentage improvement at low-line (only the total rms is shown). The corresponding numbers are provided in Table 5.2 for quick lookup. We also see that at high-line, the rms capacitor currents go up steeply again, because the much narrower PWM pulse widths can subtract much less from the diode current pulses. Note that the total rms current through the capacitor is by definition

$$I_{RMS} = \sqrt{I_{RMS\_LO}^2 + I_{RMS\_HI}^2}$$

(here this is not yet *frequency normalized* as we showed earlier).

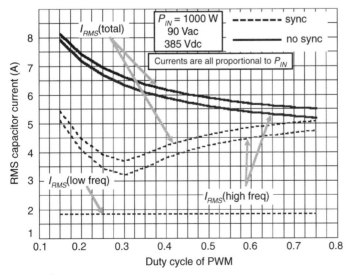

**Figure 5.9** Current components as a function of duty cycle at 90 Vac, unsynchronized and synchronized compared.

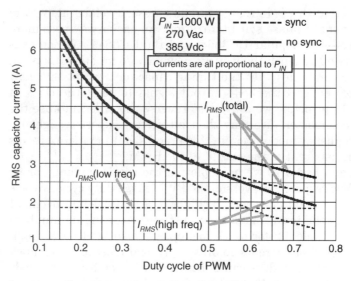

**Figure 5.10** Current components as a function of duty cycle at 270 Vac, unsynchronized and synchronized compared.

### 5.2.5  Synchronization over a wide input range

We must not forget that a capacitor is chosen not on the basis of a perceived ripple current "improvement," or percentage thereof, but on the *actual ripple current*. We may see a big "improvement" as compared to what the current would have been *at that particular input voltage* without synchronization, but we also need to know its absolute value. We should actually look at *both extremes* of input voltage and *pick the higher* of the synchronized rms currents so reported. And that would be the number to use for confirming the capacitor rating. We can in fact glean that information line-by-line from Table 5.2 but the process would be slow and cumbersome. To speed things up, what we did is as follows. We looked at both the input voltage extremes (with synchronization present) and picked the higher of the two net rms readings (note that in general, with synchronization present, this could be occurring *at high-line or at low-line*). We then compared this higher reading with the maximum rms current without synchronization (that we know, from Fig. 5.7, always occurs at low-line). Thus we calculated the improvement *over the entire input range*. This is in the form $\Delta I / I_{NO\_SYNC}$. We then displayed these numbers in Table 5.3 for quick reference.

The conclusions from Table 5.3 are:

1. The maximum improvement (*and* the lowest absolute value) for the rms current is at around $D_{PWM} = 0.325$. With a conventional forward

TABLE 5.2  Current Components as a Function of Duty Cycle at 90 Vac and 270 Vac, Unsynchronized and Synchronized.

| | No Synchronization | | | Synchronization | | | |
|---|---|---|---|---|---|---|---|
| | $I_{RMS}(A)$ | $I_{RMS}(A)$ | | $I_{RMS}(A)$ | $I_{RMS}(A)$ | | |
| Duty cycle | Low freq. | High freq. | $I_{RMS}(A)$ | Low freq. | High freq. | $I_{RMS}(A)$ | Improvement (%) |
| **1000 W/90 Vac/385 Vdc** | | | | | | | |
| 0.15 | 1.837 | 7.922 | **8.132** | 1.837 | 5.112 | **5.432** | 33.2 |
| 0.2 | 1.837 | 7.177 | **7.408** | 1.837 | 4.052 | **4.449** | 39.9 |
| 0.25 | 1.837 | 6.691 | **6.938** | 1.837 | 3.439 | **3.898** | 43.8 |
| 0.3 | 1.837 | 6.346 | **6.606** | 1.837 | 3.204 | **3.693** | 44.1 |
| 0.35 | 1.837 | 6.087 | **6.358** | 1.837 | 3.465 | **3.92 1** | 38.3 |
| 0.4 | 1.837 | 5.886 | **6.166** | 1.837 | 3.796 | **4.217** | 31.6 |
| 0.45 | 1.837 | 5.725 | **6.012** | 1.837 | 4.035 | **4.433** | 26.3 |
| 0.5 | 1.837 | 5.592 | **5.886** | 1.837 | 4.217 | **4.6** | 21.9 |
| 0.55 | 1.837 | 5.481 | **5.781** | 1.837 | 4.363 | **4.734** | 18.1 |
| 0.6 | 1.837 | 5.387 | **5.692** | 1.837 | 4.478 | **4.84** | 15 |
| 0.65 | 1.837 | 5.306 | **5.615** | 1.837 | 4.574 | **4.929** | 12.2 |
| 0.7 | 1.837 | 5.236 | **5.549** | 1.837 | 4.654 | **5.003** | 9.8 |
| 0.75 | 1.837 | 5.174 | **5.491** | 1.837 | 4.723 | **5.067** | 7.7 |
| **1000 W/270 Vac/385 Vdc** | | | | | | | |
| 0.15 | 1.837 | 6.297 | **6.56** | 1.837 | 5.997 | **6.272** | 4.4 |
| 0.2 | 1.837 | 5.331 | **5.638** | 1.837 | 4.973 | **5.301** | 6 |
| 0.25 | 1.837 | 4.655 | **5.004** | 1.837 | 4.247 | **4.627** | 7.5 |
| 0.3 | 1.837 | 4.144 | **4.532** | 1.837 | 3.693 | **4.125** | 9 |
| 0.35 | 1.837 | 3.736 | **4.163** | 1.837 | 3.24 | **3.724** | 10.5 |
| 0.4 | 1.837 | 3.398 | **3.863** | 1.837 | 2.863 | **3.402** | 11.9 |
| 0.45 | 1.837 | 3.11 | **3.612** | 1.837 | 2.545 | **3.139** | 13.1 |
| 0.5 | 1.837 | 2.859 | **3.398** | 1.837 | 2.259 | **2.911** | 14.3 |
| 0.55 | 1.837 | 2.636 | **3.213** | 1.837 | 2.015 | **2.726** | 15.1 |
| 0.6 | 1.837 | 2.434 | **3.049** | 1.837 | 1.797 | **2.57** | 15.7 |
| 0.65 | 1.837 | 2.25 | **2.904** | 1.837 | 1.601 | **2.436** | 16.1 |
| 0.7 | 1.837 | 2.078 | **2.774** | 1.837 | 1.438 | **2.333** | 15.9 |
| 0.75 | 1.837 | 1.918 | **2.655** | 1.837 | 1.299 | **2.249** | 15.3 |

converter with a duty cycle set to about 0.3 to 0.35, the improvement due to synchronization is less than around 40 percent.

2. In fact, over the duty cycle range of 0.23 to 0.4 we can expect more than 32 percent improvement.

TABLE 5.3  **Overall Capacitor RMS Current Improvement due to Synchronization, Over the Entire Input ac Range.**

| $D_{PWM}$ | Total cap rms current (A) with synchronization | Total cap rms current (A) without synchronization | Net improvement % |
|---|---|---|---|
| 0.1 | 7.876 | 9.414 | 16.339 |
| 0.15 | 6.273 | 8.132 | 22.863 |
| 0.175 | 5.742 | 7.727 | 25.687 |
| 0.2 | 5.301 | 7.408 | 28.445 |
| 0.225 | 4.94 | 7.151 | 30.923 |
| 0.25 | 4.626 | 6.938 | 33.324 |
| 0.275 | 4.359 | 6.759 | 35.511 |
| 0.3 | 4.124 | 6.606 | 37.575 |
| 0.325 | 3.912 | 6.474 | 39.564 |
| 0.35 | 3.921 | 6.358 | 38.328 |
| 0.375 | 4.079 | 6.256 | 34.808 |
| 0.4 | 4.217 | 6.166 | 31.609 |
| 0.425 | 4.335 | 6.085 | 28.758 |
| 0.45 | 4.433 | 6.012 | 26.255 |
| 0.475 | 4.523 | 5.946 | 23.927 |
| 0.5 | 4.6 | 5.886 | 21.855 |
| 0.6 | 4.84 | 5.692 | 14.96 |
| 0.7 | 5.003 | 5.549 | 9.829 |
| 0.8 | 5.124 | 5.439 | 5.803 |
| 0.9 | 5.214 | 5.352 | 2.582 |

385 Vdc, $P_{IN} = 1000$ W, 90-270 Vac

## 5.2.6  Calculating the high- and low-frequency rms components

For the purpose of estimating the life of the aluminum capacitor we need to know the breakup as explained earlier. Here is a sample calculation.

**Example 5.3**  We have a worldwide input 70 W flyback running off the PFC stage. Its efficiency is 70 percent. What are the components of the capacitor rms current at a set duty cycle of 50 percent at 90 Vac?

Let us first calculate the results assuming an input power of 1000 W. At an HVDC of 385 V, the load current is

$$I_o = \frac{1000}{385} = 2.597 \text{ A}$$

We know from Table 5.1 that the total unsynchronized rms current at

90 Vac is

$$I_o \sqrt{\frac{16 \times V_o}{3 \times \pi \times V_{IPK}} + \frac{1}{D_{PWM}} - 2}$$

So,

$$2.597 \times \sqrt{\frac{16 \times 385}{3 \times \pi \times 127} + \frac{1}{0.50} - 2} = 5.891 \text{ A}$$

Note that this agrees very closely with the numerical results provided in Table 5.2 (which we could have used directly here). This would be the value based on which we would have picked the capacitor had there been no synchronization.

The improvement from Table 5.3 is 21.855 percent. So using synchronization the rms current requirement must be

$$I_{RMS\_SYNC} = 5.891 \times \left(1 - \frac{21.855}{100}\right) = 4.604 \text{ A}$$

We also know that the low-frequency component of this is

$$I_{RMS\_SYNC\_LO} = \frac{I_o}{\sqrt{2}} = \frac{2.597}{\sqrt{2}} = 1.836 \text{ A}$$

Therefore the high-frequency component is

$$I_{RMS\_SYNC\_HI} = \sqrt{I_{RMS\_SYNC}^2 - I_{RMS\_SYNC\_LO}^2}$$

$$I_{RMS\_SYNC\_HI} = \sqrt{4.604^2 - 1.836^2} = 4.222 \text{ A}$$

But these numbers assumed a 1000 W converter. In our case we have $70W/0.7 = 100$ W. Therefore, all we need to do here is to reduce the numbers calculated above by a factor of 10. The high-frequency and low-frequency components of the rms current are therefore 0.42 A and 0.18 A, respectively. Knowing the frequency multiplier for the chosen capacitor family, we can now normalize these values to be an equivalent low-frequency current as explained earlier and thus select our capacitor.

**Note:** We should remember that after the above wide input comparison-based calculation, we don't know (and don't really need to know) whether the (worst case) synchronized rms calculated above occurs at high-line or at low-line. For selection of the capacitor, the information is sufficient. But as a matter of fact, if we want to know, this worst case is at low-line. This can be easily figured out by looking at Table 5.2 for $D_{PWM} = 0.5$.

### 5.2.7 Sequencing, protection, and some related observations

A sequence of external logic is necessary to maintain integrity of sequencing during fault conditions. This should not be underestimated. A recommended control scheme is as follows:

1. A small low-power auxiliary flyback switcher must be present. This should work all the way from very low-line voltages (∼ 30 Vac) to maximum line voltage.

2. This switcher provides power from its primary side to the PFC and PWM ICs, as also for the primary-side sequencing and protection circuitry. From its secondary side it provides an auxiliary rail which powers the OCP (overcurrent protection) and the output OVP (overvoltage protection) circuitry.

3. Two diodes *before* the bridge rectifier from the L (live) and N (neutral) lines should be OR-ed to sense the presence of the unrectified ac line, and thus detect any missing cycles. Note that after the bridge, since the bulk capacitor keeps the voltage help high, we cannot know quickly enough from that node whether a dropout is occurring or not. And if there is a momentary line dropout (say 10 to 20 ms), this should be sensed rapidly and the output of the PFC IC should be disabled (no switching, but the IC stays powered up) for the duration of the dropout. This helps avoid "overworking" the PFC choke, which could otherwise cause it to saturate. The IC should recover promptly to restore the output voltage as soon as the dropout is over. Note that the PWM stage must continue to function normally during this time.

4. Another technique to simplify the sequencing logic and also to save on the size of the boost choke is to *simply reprogram the HVDC rail to say 250 V (rather than 385 V) under severe line sags (or brownout conditions)*. We can do this because no holdup time is usually demanded under a severe brownout condition. It is enough just to keep the outputs in regulation during this duration. With this reprogramming, the PFC stage needs to struggle much less to maintain its regulation. This can help reduce the size of the PFC choke too. In effect, we are shifting some of the burden from the PFC stage to the PWM stage for a short duration, while keeping the PFC section alive (still switching).

5. The UC3854A/B featured an improvement over the previous generation UC3854 in that by limiting its maximum multiplier output current, it provides foldback under line brownout and extreme line conditions. This may help reduce the complexity of the external logic.

6. The UC3854B has a lower startup threshold voltage of 10.5 V as compared to 16 V for the UC3854A, and is usually considered a better choice for this reason. Also note that the drive signal from the PFC controller to the FET is almost equal to the upper supply rail of the IC, and for reliability reasons we want to keep the gate only high enough to ensure the FET turns *on* fully.

7. A reference voltage for the additional external comparators (typically 2.5 V) is therefore required on the primary side. This may, for example, come from an LM431 or a TL431, which in turn derives its power from a primary side auxiliary rail.

8. We should also use this 2.5 V reference to set up an accurate OVP (overvoltage protection) for the HVDC rail. Normally this circuit should only act to drag down the soft-start pin of the PFC controller. The intention should be only to pinch off the duty cycle, not to reset the PFC IC completely. This helps avoid nuisance tripping under power-up or line/load transients. But in more severe cases we should be able to cause complete shutdown of the entire power supply and PFC. The last thing we want to see is a high voltage bulk capacitor explode!

**Note:** If a 400 V or 450 V electrolytic capacitor is overcharged there is danger of a catastrophic and hazardous failure. But we can remember that the failure in such a condition is purely a heat buildup issue. Typically we may actually be allowed to exceed the voltage rating of an electrolytic capacitor 1.2 to 1.4 times, provided the duration is less than about 1 s (but consult the vendor). However, a sustained overvoltage just cannot be tolerated. There is a subtler safety issue involved too. Some control ICs may simply sense the voltage on the feedback pin to initiate OVP protection. While this is OK in low voltage applications, IEC (International Electrotechnical Commission) does not allow any *single point* failure to cause a hazardous condition. So, effectively this means that for either the PFC IC or the PWM IC in any off-line power supply, *the pin being used for normal regulation cannot be used for fault protection* (either internal to the IC or via external logic). Because in doing so we are actually trying to use an object to sense its own malfunction. For example, what if that pin just had a bad solder joint? This single pin issue was actually overlooked inadvertently by safety test agencies themselves till some years ago.

9. We should include fault detection for the auxiliary switcher stage, and disable the PFC stage if, for example, the auxiliary rail is not being able to get up close to its set regulation level.

10. We should have proper sequencing logic under normal power-up, or if the user uses an *on/off* logic switch to turn the unit *on*. A recommended sequence for power-up is as follows. The inrush current protection circuit must first start working. At the same time, the auxiliary switcher turns *on*. Then provided the voltage on the bulk capacitor has increased beyond a certain level, the PFC IC gets enabled. When the reference voltage of the PFC IC comes up, it enables the PWM IC.

11. Many companies do not wish to use a single-source part. While PFC ICs from different vendors are certainly not pin compatible, it may be helpful to know that the UC3854 is manufactured under license by Toko (see *www.toko.com*). It is then called the TK83854D.

12. For practical examples of inrush protection circuits, turn-on snubbers, and an interesting technique to get two UC3844's to mutually synchronize, see Chap. 16.

13. There is an *automatic* PFC control scheme used sometimes, which requires a flyback PFC stage operating in critical conduction mode (boundary between continuous conduction and discontinuous conduction modes). Since $V = LdI/dt$, in critical conduction we get $V = LI_P/t_{ON}$. So if we keep a constant *on* time, $I_P$ becomes proportional to $V$. But the average of the *on* pulse is $I_P/2$; thus the average is also proportional to $V$. If the input is a pure sine wave, then so is the average input current. This scheme however requires differential sensing of the output voltage for regulation since the ground of the control IC is no longer the same as the power ground. We can use the differential sensing techniques provided in Chap. 2.

14. The biggest impact on efficiency in a boost stage may come from the severe shoot-through current from the bulk capacitor that passes through the yet unrecovered 600 V PFC diode when the switch turns *on*. Since this current passes through the FET while there is still a high voltage present across it, we can get large $V \times I$ crossover losses. Besides the problem of the fall in efficiency, the required heatsinking of the FET is also affected significantly. Therefore, in this position it is critical to use an extremely fast diode (sometimes called "hyperfast"). Any diode with more than 20 to 30 ns recovery time is unacceptable except in very low power and noncritical applications.

**Hint:** Check not just the "typical" value of the recovery time as given in the vendor's datasheet but its maximum stated limit. Remember that only the limits (minimum and maximum) are guaranteed by component vendors and that typical values don't really mean a thing.

15. Therefore for low-power applications engineers often prefer to use boost PFC ICs that operate in critical conduction mode. In this case, since the current through the diode has fallen to zero when the switch turns *on*, the diode has fully recovered and there are no reverse recovery issues and no shoot-through. But such ICs necessarily operate with variable frequency and their EMI filtering often poses a major problem.

16. In low to medium power PFC stages, the turn-on snubber, which is virtually indispensable in high-power stages, may seem like a luxury (for a practical turn-on snubber circuit see Chap. 16). In that case some engineers like to reduce the reverse recovery current spike by replacing the single 600 V PFC diode with, say, two 300 V diodes in series. Here they are relying on the fact that low-voltage diodes recover much faster than high-voltage diodes and so, despite their higher combined forward drop and consequent increased conduction losses, we actually improve efficiency by reducing the $V \times I$ crossover losses. But note, we cannot allow the full voltage to appear across either one of the diodes at any moment, however brief. So they must be well matched, especially

in terms of their dynamic characteristics. That is not easy. Some engineers try to achieve this by placing ballasting resistors across each diode, much as we do for capacitors in series. But under dynamic conditions (during transitions) this cannot help even in principle because the lead inductances will prevent any immediate equalizing response. The other option is to use two series diodes *in one package*, on the assumption that since they pass through exactly the same fabrication steps, they will automatically be well matched. Well, they don't always. Some manufacturers are forced to take separate chips and try to combine them on one substrate. Part of the problem is that this diode combination requires an odd internal connection—the anode of one is to be connected to the cathode of the other.

17. Recently some vendors (e.g., ST Microelectronics at *www.st.com*) have come up with *Tandem diodes* based on this principle. They are rated (and look) like a single 600 V diode and have two leads, but they offer typical recovery times of about 12 ns.

18. A critical component of a commercial PFC implementation is a surge diode placed from the + terminal of the bridge rectifier to the cathode of the PFC diode. At initial power-up, this allows the bulk capacitor to charge while keeping the huge inrush current away from the PFC section's choke and ultrafast diode. But this surge component is also a key reliability issue for the entire power supply. *It is, in fact, the component most likely to fail under repeated application of ac input voltage*. It need not be a fast diode as it goes completely out of the picture as soon as the PFC switch turns *on*. It does not therefore get hot either, and can be an axial component. But its nonrepetitive surge rating must be very high. Some designers still prefer to make this diode a fast (expensive) diode possibly because of EMI concerns. They use, for example, a free-standing 16 A/600 V diode (with active *inrush current protection circuitry present*). But a slow (and cheaper) diode like the 1N5408 (from a quality vendor) should also be tried out since these have higher surge ratings (on paper at least). Note that in this position two such diodes in series can also be tried out if we want to shift some of the surge current back into the PFC choke and diode. But the relative sharing then depends on various parasitics like the dc resistance of the choke. Therefore we can do this, but only with great care.

Note: One general problem of using cheap multiple-source parts is that all vendors do not have the same fabrication process or quality even for the *same part number*. So if we specify a 1N5408 for the surge diode, and are careful enough to pick a quality source initially, there is no guarantee that tomorrow a "smart," cost-sensitive purchase officer doesn't start buying the "same part" from an alternative source, with consequent reliability issues.

19. For the boost inductor of a dc-dc converter stage we normally set the inductance so that we get an inductor current ripple ratio $\Delta I/I_{AVG}$ of about 0.4, on the grounds that this represents a good compromise between the size of the inductor and the capacitors. But in a PFC boost stage this holds little meaning since the size of the capacitor is being determined by several other criteria, like the holdup time. Besides, the rms current through the capacitor is also being determined by several factors, not just the current ripple ratio of the inductor. Therefore, for a boost PFC, the inductance of the choke is typically chosen such that at low instantaneous line voltages the inductor is operating in discontinuous conduction mode. Closer to the peak instantaneous voltage at 90 Vac it should be transiting into continuous conduction mode. This design criterion helps minimize the size of the inductor. The required energy rating of the inductor can be looked up from the Energy $\propto I_{PK}^2/2$ curve in Fig. 5.7. From Chap. 10 we can then see how to select a core volume on the basis of the required energy storage capability. The only pending concern here is the calculation of core losses for PFC. We analyze this next.

### 5.2.8 Core losses in the PFC choke

We can refer to Chap. 11 to see how to calculate the core loss for a steady dc input. In particular, we need to know the peak-to-peak current swing, which translates into a proportional flux density swing. But the problem with PFC is that the instantaneous input voltage is varying; thus either we need a summation or integration over the entire ac half-cycle, or we should establish an *equivalent dc level* i.e., a point on the input voltage curve that correctly reflects the average value over the entire ac half-cycle. The situation is further compounded by the fact that the applied input rms voltage can also vary (usually between 90 and 270 Vac).

In Fig. 5.11 we have shown a mathematical simulation of the inductor current. Note that the frequency is scaled down for visual clarity, but since the inductance used in the simulation was also correspondingly scaled up, the peaks, troughs, and the peak-to-peak values of the currents shown are in fact the actual values we will see under the conditions indicated in the figure.

We see that as we increase the line voltage, the peak-to-peak values are much higher somewhere in the middle of the instantaneous input range.

In Fig. 5.12 we have plotted the peak-to-peak values for several input ac voltages. Tabulating the *average* peak-to-peak values in Table 5.4, we see that the worst case occurs at 180 Vac. If we vary inductance, this varies only slightly (between 175 Vac and 185 Vac), unless the inductance is unusually low so as to cause discontinuous mode even at

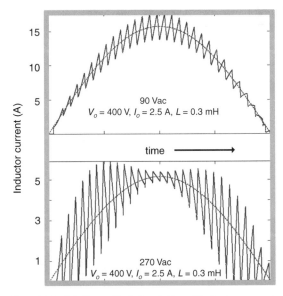

**Figure 5.11** Mathematical simulation of inductor current in boost choke.

90 Vac. Further, the values in the table are *inversely proportional to inductance*. So, if we have an inductance of 155 $\mu$H, we need to double the reported peak-to-peak average current. Note that peak-to-peak values depend on applied voltages and not on load current. So, the values reported in the table are virtually *independent of load current* or power level. But they are inversely proportional to the switching frequency.

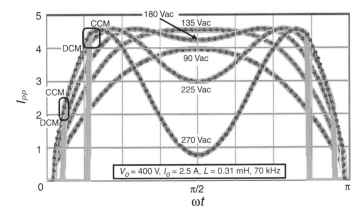

**Figure 5.12** Peak-to-peak currents at various line conditions.

TABLE 5.4   The Peak-to-Peak Current Averaged Over an AC Half-Cycle.

| Ac input voltage (V) | Average peak-to-peak current (A) |
|:---:|:---:|
| 90 | 2.771 |
| 135 | 3.465 |
| 140 | 3.513 |
| 145 | 3.556 |
| 150 | 3.593 |
| 155 | 3.625 |
| 160 | 3.65 |
| 165 | 3.66 |
| 170 | 3.685 |
| 175 | 3.693 |
| 180 | 3.696 |
| 185 | 3.693 |
| 190 | 3.685 |
| 195 | 3.671 |
| 200 | 3.651 |
| 225 | 3.442 |
| 270 | 2.683 |

$V_o = 400$ V, $I_o = 2.5$ A, $L = 0.31$ mH, 70 kHz

**Example 5.4**  We have a worldwide input 500 W PFC stage. It uses a choke inductance of 2 mH, and its switching frequency is 100 kHz. What is the average $\Delta I$ we should use in core-loss estimation?

From Table 5.4, we have the worst-case average over line under the stated conditions as 3.696 A. Therefore, under our new conditions

$$\Delta I = 3.696 \times \frac{0.31 \text{ mH}}{2 \text{ mH}} \times \frac{70 \text{ kHz}}{100 \text{ kHz}} = 0.4 \text{ A}$$

The $\Delta B$ can be calculated using Faraday's law with a knowledge of the core characteristics and the number of turns. The core loss so calculated will then accurately reflect the maximum value, over the range 90 to 270 Vac, of the average core loss computed over each ac half-cycle. See Chap.11.

**Note:**  We realize that the core loss is not proportional to the swing of the $B$-field (or current), but varies as $\Delta B^n$, where the exponent depends on the material. However, the worst-case average still occurs very close to 180 Vac.

# 6

# Isolated Topologies for Off-line Applications

## 6.1 The Forward Converter

### 6.1.1 Introduction and overview

Since this converter has two magnetic elements—the transformer and the output choke—it is best discussed from a magnetics viewpoint. Much of the skill involved in making a good forward converter is in understanding how to design its transformer optimally. That involves understanding the concept of *proximity effect* which is explained in detail in Chap. 11. In this chapter, we are just going to give a brief introduction to this popular topology.

In Fig. 6.1 we have a single-ended (or single-switch) forward converter. Besides the primary and secondary windings we also have an energy recovery winding (or *tertiary winding*, subscript T). This is the freewheeling path for the magnetization current $I_M$, shown as a gray area in the current waveforms. Almost all the magnetization energy that was stored in the transformer during the switch on-time freewheels into the input bulk capacitor through the tertiary winding when the switch turns *off*. The tertiary winding usually has the same number of turns as the primary winding and is sometimes even wound bifilar with it for good coupling. The maximum voltage across the switch is then $2 \times V_{IN}$. For a worldwide input power supply we can thus make do with an 800 V switch despite the somewhat reduced voltage derating margin. However, if a front-end PFC preregulator stage is present, we would certainly need at least an 850 V or 900 V switch, because the high voltage input dc rail is now itself around 400 V.

The inductor current can be imagined as having been reflected to the primary side with an average (dc) level of $I_{LR} = I_L/n$, where $n =$

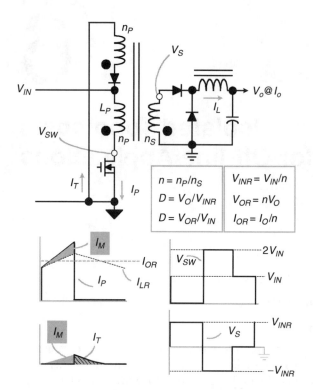

**Figure 6.1** The forward converter.

$n_P/n_S$. As for any buck-derived topology, which this too is, the average (reflected) inductor current is equal to the (reflected) load current. Thus

$$I_L = I_O \quad \text{and} \quad I_{LR} = I_{OR}$$

where we have defined the *reflected output current* as $I_{OR} = I_O/n$.

The voltage across the primary winding when the switch is *on* is $V_{IN}$. When the switch turns *off* this voltage is again $V_{IN}$, though in the opposite direction. Every magnetic element naturally tries to reset itself (if allowed to do so), and it achieves this by reversing the voltage polarity (see the section on the *voltseconds law* in Chap. 1). But the fact that the reverse voltage has the same *magnitude* here, is simply because the tertiary to primary turns ratio is 1:1. The total peak-to-peak voltage across the transformer is, thus, $2 \times V_{IN}$. This is also the voltage the switch is required to hold off, and it thus forms the minimum switch voltage rating. But if, for example, we decrease the number of turns on the tertiary winding, the inductance of this winding would decrease. Then the (magnitude of the) slope of the down-ramp of the current in the tertiary winding, i.e., $V_{IN}/L_T$, would increase, where $L_T$ represents the inductance of this winding (measured separately, all other windings open). So now $L_T \neq L_M$. This will ensure that the transformer "resets" a little earlier during the switch off-time interval. *Reset* implies that the current returns to the same value it had at the start of the cycle, which in this case is *zero*. Thus changing the turns ratio between primary and tertiary windings throws open the possibility of allowing the switch on-time to exceed 50 percent of the time period, i.e., D > 0.5. But nothing comes for free. Now when the switch turns *off*, since we have changed from the simple 1:1 tertiary to primary turns ratio to a *step-up* ratio, we will reflect a voltage equal to the voltage across the tertiary winding (which is clamped to $V_{IN}$), multiplied by the step-up turns ratio. So we will need a switch with a voltage rating in excess of $2 \times V_{IN}$.

But all this is "in principle." Power supplies don't always believe in academics, the way we are wont to do. So in practice even if we could afford higher voltage rating FETs, we need to be able to control the maximum duty cycle very tightly, all tolerances considered. We note that of all the possible values of $D_{max}$ we may want, it is unarguably the easiest to design and guarantee a $D_{max}$ of 50 percent. In the popular UC3844, for example, every alternate clock pulse is simply omitted from reaching the output, giving us an output pulse once every two clock cycles i.e., an automatic duty cycle limit of 50 percent and one that is clearly assured even in principle. This is one reason why this IC is one of the most popular choices for implementing a conventional forward converter (or other isolated buck-derived topologies). With a simple, achievable, and guaranteed $D_{max}$, also comes the sheer convenience of needing the simplest possible primary to tertiary turns ratio of 1:1. All this makes for convenient mass production.

The tertiary winding can in principle be wound bifilar with the primary winding, increasing their mutual coupling in the process and thus minimizing leakage. However, we note that there will then be a large *voltage difference* between adjacent segments of the two windings (study the schematic again). There may also be pinholes in the wire insulation due to manufacturing variations, or from accidental scratches incurred during production. Therefore to avoid flashovers from occurring and destroying the power supply, especially in off-line or similar high-voltage applications, we may prefer *not* to wind the primary and tertiary windings bifilar. Note that then because of the poorer coupling, we will almost certainly need a small snubber/clamp to protect the switch from the resulting primary to tertiary leakage inductance spike.

In Chap. 7 we have explained "transformer action." In particular, we note that if we increase the load, the current through the primary winding increases accordingly, though the magnetization current $I_M$ is unchanged (or so we expect!). Thus $I_M$ should become the only component of the primary-side current $I_P$ if the load is disconnected. Does that mean that we can measure the no-load primary current, and that would automatically give us the magnetization current component present when the load is connected? We are right in thinking that $I_M$ is not supposed to be affected by the load current, but in this specific case we are in for a slight surprise because the duty cycle "pinches off" under light loads. As a result, $I_M$ is not preserved, but changes shape. The root cause of this is that at light loads, the output choke enters discontinuous conduction mode, in which mode we know duty cycle always becomes dependent on load. So, in that sense, $I_M$ actually does depend on load current! Not what we may have expected at first sight. We certainly learn never to take anything for granted in switching power.

If we ignore the magnetization current component, we can mentally visualize the operation of the forward converter to be a buck converter, though with an equivalent secondary-side reflected input dc voltage rail of $V_{INR} = V_{IN}/n$ (created by transformer action). From the transformer we get the additional benefit of isolation. But the major electrical difference between a buck converter and a forward converter is that in a buck we can, in principle, go up to 100 percent duty cycle, but in a forward converter we need to allow not only the output choke to reset, but the transformer too. The latter is in fact harder to achieve unconditionally, because we have no direct control over the state of the transformer or its current. The converter is being controlled by the output voltage and the load current, which in turn are being influenced primarily by the output choke. The transformer is usually "aware" of only the magnetization current component, not the load current. So the only way we can guarantee transformer reset is by simply leaving enough time for reset to happen on its own accord. We observe that the (magnitude of the) slopes of the up-ramp and down-ramp of the magnetization current are the same for a 1:1 tertiary turns ratio $(=V_{IN}/L_P)$. So if the duty cycle manages to reach exactly 50 percent, the transformer will be in critical conduction mode (the boundary between continuous and discontinuous conduction modes). If we exceed this duty cycle even slightly, in every successive cycle there will be a net increase in the magnetization current. Again, since the feedback loop connects only to the output choke, it will have no way of realizing and correcting this build-up of transformer magnetization current. The transformer will end up "flux-walking," and eventually the switch will be destroyed unless we have *effective and adequate primary-side current limiting*. But such a current limit probably does not exist. We note that the purpose of the primary-side current limit is to respond to changes in load current, not any uncontrolled magnetization current build-up. Any increase in load current does *not* increase the flux inside the transformer, but an increase in magnetization current does. So if a certain amount of primary-side limited current is a magnetization current component rather than a load current component, it can cause immediate transformer saturation and switch destruction. Therefore, we always need to set the primary-side current limit by assuming a known and fixed $I_M$. We see that we cannot even in principle safely maintain the transformer of a forward converter in *continuous conduction mode*, as we can for the output choke. However, by ensuring that *the transformer is always operated in discontinuous conduction mode* we can always prevent it from entering a flux-walk or runaway condition. That means we need to restrict the maximum duty cycle to a little less than 50 percent, *come what may*, accompanied by a simple primary to tertiary turns ratio of 1:1.

A variant—the two-switch (or asymmetric) forward converter—has become a workhorse of the medium to high-power segment. Though still restricted to a $D_{max}$ of 50 percent, it has no tertiary winding. It relies on two 400 V to 500 V switches (driven in-phase) and two primary-side ultrafast diodes to route the magnetization current back into the bulk capacitor. This configuration however requires a floating drive for the upper switch. This can come either from a conventional gate-drive transformer, or by using modern solid-state driver ICs like the IRF2110 or IRF2112 from International Rectifier (*www.irf.com*). Since there is no tertiary winding, and the primary winding is itself going to help freewheel the magnetization current into the bulk capacitor, there is no question of "leakage" anymore, and so we don't need primary-side snubbers or clamps here. Further it can be shown that even if one FET is a little sluggish to turn *on* or *off* compared to the other, this mismatch produces no major problems—certainly no attempted shoot-through or cross-conduction related effects.

The other possible design complexity in a forward converter design involves the coupled inductor approach. This serves as a means of replacing all the individual output chokes of a multioutput forward converter with one component. This also improves the cross-regulation and helps in keeping the choke operating in continuous conduction mode under "corner" conditions—for example if the main output is only lightly loaded but one or more of the other outputs are still fully loaded. The only design-related aspect we need to remember here is that the turns ratio we use between the windings in the choke must be as close as possible to their (mutual and respective) ratios within the transformer. Though coupled inductors also give us an opportunity for *ripple steering*, we should not usually count on being able to implement this nice subtlety successfully in a commercial design. Ripple steering forms the basis of many excellent articles, but it relies on parameters that are not, or cannot, be guaranteed in mass production.

Note that sometimes a separate auxiliary winding is thrown across the output choke, and is not derived from a winding on the transformer.

Its polarity is meant to be such that it will conduct only when the current in the main winding is ramping down, i.e., this winding operates essentially in flyback (or freewheeling) mode as far as the output choke is concerned. By coupling it tightly with the main winding and choosing its turns ratio according to the desired auxiliary output voltage, we can get fairly good regulation (about $\pm 5$ to $\pm 10$ percent) on this rail. This technique works well for auxiliary power outputs of up to about 5 to 10 percent of the power flowing through the main winding. If production allows, we can try twisting this winding tightly with the main winding. In that case we may be able to double the auxiliary power output with acceptable regulation.

The forward converted can actually be configured in several ways, depending upon where the choke is placed in the output loop section and also whether we are using the two secondary diodes in common-cathode or in common-anode formation. Refer to Chap. 17 for an interesting discussion related to this aspect.

## 6.2   The Flyback Topology

### 6.2.1   Introduction

Nothing illustrates the finer twists and nuances of a power converter design process than the arguments put forth to either support or refute a certain design choice. In that spirit, we will delve into this rather tricky topology, also known as the flyback. The flyback forms the basis of most low-power off-line converters but is often underestimated in its complexity. It therefore deserves much more than our passing interest here. We all think we know everything about it, but understanding it *well* is arguably one of the most challenging tasks a power supply designer will likely face. We will also take up some subtle issues that go into making a good commercial flyback power supply—like minimum pulse width considerations, voltage feedforward, and some related protection techniques.

To make it more interesting we will present this topology from the angle of an ongoing comparison with the specifications and recommendations of a hypothetical product. We christen this fictitious off-line integrated switcher the *IP-switch* (for integrated power switch). We will try to see what seems right in this product and what could possibly be done better the next time around.

### 6.2.2   The integrated power switch

Delving into the "history" of our benchmark product, we make a curious observation. The older generation of this product family has a maximum duty cycle of 67 percent. In the new generation, this number has mysteriously become 78 percent. The "improved duty cycle feature" has even become an enduring marketing tool, with the front page of the datasheet bulleting "wider duty cycle for more power, and a smaller input capacitor."

As any engineer should, we take the trouble of going through the supplemental reference design kits, but are a little surprised to find that the evaluation boards themselves often don't seem to comply with the published recommendations. We try the accompanying expert system software and ask it to "auto-design" for a given application, first using the older generation and then the newer generation device family. But we are surprised to see that it almost always suggests virtually the same input capacitor. The promised reduction in the size of the input capacitor, or in fact any lower associated cost, doesn't seem to be bearing out. Our curiosity is piqued, and we decide to check this out further.

### 6.2.3   The equivalent buck-boost models

But first we need to brush up on our basic understanding. Let us examine the flyback shown in Fig. 6.2. The pertinent points are as follows:

- The grayed/hatched areas of the voltage waveforms shown correspond to the switch being *on*.

- The voltage *across* the primary winding scales according to the ratio $n_S/n_P$ to become the voltage *across* the secondary winding.

- The voltage across the secondary gets multiplied by $n_P/n_S$ as it is scaled on to the primary side.

- Therefore knowing that the voltage across the secondary during switch turn-off is $V_O$, we get a *reflected output voltage*, designated $V_{OR}$ on the primary side. We have $V_{OR} = V_O \times n_P/n_S$ by the transformer scaling rule.

- We know that during switch turn-on, the voltage across the primary winding is $V_{IN}$, so the corresponding voltage level at that time appearing across the secondary must be $V_{IN} \times n_S/n_P$. This is in a *reflected input voltage* term, though it is not popularly known by any particular name. Here we will call it $V_{INR}$.

$$V_{INR} = V_{IN} \times \frac{n_S}{n_P}$$

- The peak-to-peak voltage across the secondary winding is the sum of the reflected input voltage $V_{INR}$ *and the output voltage* $V_O$. This is also the amount of reverse voltage seen by the diode when it stops

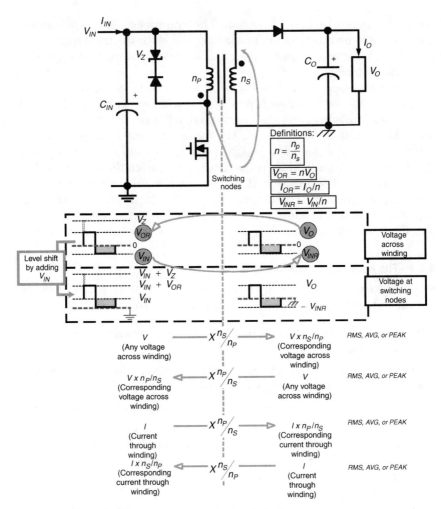

**Figure 6.2** How voltages and currents reflect in a flyback.

conducting. Therefore this forms the required minimum voltage rating of the diode. Note that this rating must be established at the *maximum* input voltage when $V_{INR}$ is highest.

**Note:** In related literature, the second term above is often omitted. So it is generally, though incorrectly, stated that the diode needs to withstand only $V_{IN} \times n_S/n_P$. This could result in a significant error in the component selection procedure, especially for high output voltage applications.

■ Having applied scaling to the voltage levels *across* the windings and thereby having deciphered what they are, we can then deduce the voltage *at* any point (with respect to ground). For this we can apply the required amount of voltage "translation" (i.e., level shifting) to either the primary or secondary voltage waveforms to reference them to a new measurement level. In doing so, we must keep the overall shape and peak-to-peak value unaltered as we shift the waveform vertically up or down. We have to be cautious of the fact that "scaling" must only be applied to the voltage levels appearing *across* any winding, not necessarily to the absolute voltage levels existing at various different points in the primary and secondary loops.

- The currents in the windings scale in inverse manner to the voltage, i.e., in going from primary to secondary, the current *through* the windings goes as $n_P/n_S$. In the other direction we must use the factor $n_S/n_P$. So we can, in effect, apply a sort of "transformer scaling" to the currents involved in the flyback too, even though technically speaking there is no transformer action really involved here. The flyback transformer is actually an inductor with multiple windings. Unlike a forward converter transformer, when current flows in the primary winding in the flyback, it does *not* flow in the secondary and vice versa. However the two current waveforms do "connect" at the moment of every transition (turn-on and turn-off). It is at that moment that a fundamental requirement applies. Since the total ampere-turns of the transformer $\sum NI$ is directly related to the energy residing in the core, ampere-turns cannot change *abruptly*. In our case here, this means that the peak of the primary-side current $I_{PK}$ must be related to the peak of the secondary current $I_{PKS}$ by

$$I_{PK} = I_{PKS} \times \frac{n_S}{n_P}$$

The troughs of the two waveforms are also likewise related.

- In Fig. 6.3, we reveal a useful trick for analyzing the primary or secondary sides of an isolated flyback by drawing the *equivalent primary buck-boost topology or the equivalent secondary buck-boost. Since the equations for a standard dc-dc buck-boost converter are simpler and well known, these equivalent models are also much easier to use and understand*. This helps a great deal in any subsequent visualization and optimization.

- Note that the subscript $R$ stands for reflected. So for example, if any voltage is being reflected from the primary to the secondary, we divide it by $n$, where $n$ is the turns ratio $n_P/n_S$. If any current is being reflected from primary to secondary, it gets multiplied by $n$.

- The switch forward drop, represented by $V_{SW}$ subtracts from the applied input voltage when the switch is *on*, and thus is effectively a component of the voltage *across* the primary winding. We can therefore apply scaling to it too, and so it gets reflected to the secondary side as

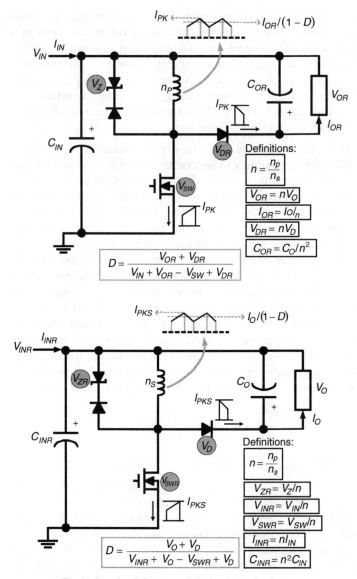

**Figure 6.3** Equivalent buck-boost models for the flyback.

$V_{SW}/n$ as shown in the figure. On similar grounds, the diode voltage drop appears on the primary side as $nV_D$. Notice also how the zener voltage gets reflected.

- We see that voltages and currents scale in a simple *linear* manner. But energy is invariant, and cannot change merely on the basis of any equivalent circuit we come up with. Therefore we can expect that all reactances, being by nature energy storage elements, will reflect in an entirely different manner. For example, the output capacitor of a flyback converter $C_O$ has an energy $1/2(C_O V_O^2)$, and since $V_O^2$ reflects as $n^2 V_O^2$ to the primary, therefore $C_O$ must reflect as $C_O/n^2$. This is shown as $C_{OR}$ in the figure. We can also similarly show that *any leakage inductance on the secondary side will reflect to the primary by getting multiplied by $n^2$* (that happens because $1/2 \times LI^2$ must be invariant). This reflected leakage has severe ramifications on the entire performance and power handling capability of a flyback, as we will see.

- The equivalent primary buck-boost "thinks" that its output voltage is

$$V_{OR} = V_O \frac{n_P}{n_S}$$

$V_{OR}$ actually turns out to be the most fundamental design choice available to the flyback designer. Note that, simplistically speaking, a high $V_{OR}$ leads to a high duty cycle whereas a low $V_{OR}$ produces a low duty cycle. But in fact there are so many other effects of changing $V_{OR}$, that they need a much deeper study.

We note that *as far as the primary side is concerned* (ignoring parasitics for now), it will see no difference between say a 5 V output with a turns ratio of 20 (i.e., a $V_{OR}$ of $5 \times 20 = 100$ V), or a 10 V output with a turns ratio of 10 (i.e., a $V_{OR}$ of $10 \times 10 = 100$ V). *So several different applications, each having the same $V_{OR}$* (turns ratio adjusted according to the output voltage) *are actually the same converter as far as the switch is concerned.* All primary-side currents and voltages are unchanged. The only possible source of difference is the leakage inductance. That is an issue we will take up later.

- The equivalent load current at $V_{OR}$ flowing out of the output terminals of the equivalent primary buck-boost is the reflected output current

$$I_{OR} = I_O \frac{n_S}{n_P}$$

Clearly, if losses are ignored,

$$P_{IN} = V_{IN} I_{IN} = P_O = V_O I_O = V_{OR} I_{OR}$$

- So as far as the primary side is concerned it "thinks" that it is delivering a load current of $I_O/n$ at a voltage $nV_O$. So what is the load power? As expected that is still $V_O \times I_O$. What is the load resistance? The load resistor connected on the secondary side is $R = V_O/I_O$. The primary side however thinks that the load resistor is $nV_O/I_O/n = n^2V_OI_O = n^2R$. So we see that not only reactances, but even resistances reflect from secondary to primary according to the square of the turns ratio. In general we state that since in going from secondary to primary, both $L$ and $R$ reflect according to $n^2$, and $C$ reflects as $n^{-2}$, all *impedances* reflect from primary to secondary *as $n^2$*.

- Knowing that for a standard dc-dc buck-boost converter, the average inductor current (in continuous conduction mode) is $I_O/(1-D)$, the center of the secondary-side current ramp in an actual flyback converter must be

$$I_{CS} = \frac{I_O}{1-D}$$

- The center of the primary-side current must be by the scaling rule

$$I_C = \frac{I_O}{1-D} \times {}^1/_n \equiv \frac{I_{OR}}{1-D}$$

- For a dc-dc buck-boost converter, the duty cycle is (ignoring forward drops across switch and diode)

$$D = \frac{V_O}{V_{IN} + V_O}$$

So for the flyback (or its equivalent buck-boost models)

$$\boxed{D = \frac{V_{OR}}{V_{IN} + V_{OR}} \equiv \frac{V_O}{V_{INR} + V_O}}$$

If we include the switch and diode drops we get for the standard dc-dc buck-boost converter

$$D = \frac{V_O + V_D}{V_{IN} + V_O - V_{SW} + V_D}$$

So for the *equivalent primary buck-boost* we get

$$\boxed{D = \frac{V_{OR} + V_{DR}}{V_{IN} + V_{OR} - V_{SW} + V_{DR}}}$$

and for the *equivalent secondary buck-boost* we get

$$\boxed{D = \frac{V_O + V_D}{V_{INR} + V_O - V_{SWR} + V_D}}$$

both of which are actually identical as we would expect.

### 6.2.4  A worked example

A flyback has 60 turns on the primary and three turns on the secondary. The output voltage is set to 5 V. The load is 20 A. What are the currents in the inductor, switch, diode, and windings at 90 Vac (i.e., $V_{IN}$ is $90 \times \sqrt{2} = 127$ V)?

The $V_{OR}$ is $60/3 = 20$ times the output voltage. Therefore we get $V_{OR} = 100$ V. We know the average load current is 20 A. Reflected to the primary side it appears as a load current $I_{OR}$ of 20 A/20 = 1 A. The duty cycle is $100/(100 + 127) = 0.44$. So the average inductor current in the equivalent primary buck-boost model is $1A/(1 - 0.44) = 1.79$ A. The average value reflects to the secondary side and is therefore 1.79 A $\times 20 = 35.8$ A in the equivalent secondary buck-boost model. Now, usually, for optimum results, we like to set the transformer inductance so that the current ripple ratio $r = \Delta I / I_{AVG}$ in the (equivalent) buck-boost inductor is $\pm 20$ percent. In that case we will get $I_P$ as 1.79 A $\times 1.2 = 2.15$ A. By the scaling rule, the corresponding secondary-side peak current is 2.15 A $\times 20 = 43$ A. We can check that as expected, on the secondary side too, the peak of 43 A is 20 percent higher than the reflected average value of 35.8 A. The average current into the switch is therefore $1.79 \times 0.44 = 0.79$ A. Let us double-check the average current through the diode. This is 35.8 A $\times (1 - 0.44) = 20$ A, which checks out because *we know that for a flyback (or buck-boost), the average diode current must equal the load current*. We have shown the computed waveforms in Fig. 6.4. The *average input current is equal to the average switch current* for this topology. So the input power is 0.79 A $\times 127$ V = 100 W. We see that this implies 100 percent efficiency. But that is what we should have expected would result because we have been ignoring parasitics like the diode and switch forward drops. So the calculations above are in that sense accurate, since they were implicitly based on an initial assumption of zero losses.

### 6.2.5  Dealing with multi-output converters

Here the best way to proceed is to draw out the equivalent primary buck-boost as before. But first we must lump all the output power into an equivalent single output. So if we have the outputs $V_1@I_{O1}$, $V_2@I_{O2}$, $V_3@I_{O3}$, $V_4@I_{O4}$ and so forth, we can lump them to behave as a single output with the following description

$$V_O = V_1$$

$$I_O = I_{O1} + \frac{V_2 I_{O2}}{V_O} + \frac{V_3 I_{O3}}{V_O} + \frac{V_4 I_{O4}}{V_O} \cdots$$

Then for the load in the equivalent primary buck-boost model, we

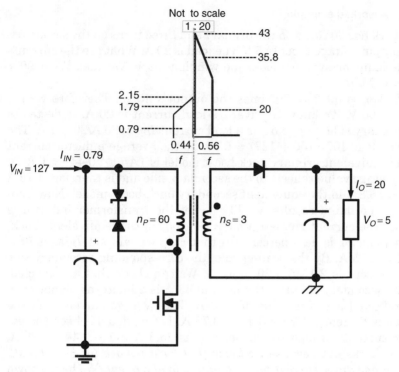

**Figure 6.4** The calculated current components (see example).

have as before

$$I_{OR} = \frac{I_O}{n}$$

and

$$V_{OR} = nV_O$$

This will tell us the primary-side currents, zener dissipation, input capacitor requirements and the like. For calculating secondary-side parameters, we would need to consider each output on an individual basis.

For multioutput converters, the total power is shared between several windings. From the viewpoint of the primary-side they all just appear as a lumped load. Therefore reducing the power output of the converter from the primary-side, though necessary, is not sufficient to preclude a possible overload on any specific output. For example, the entire rated output power of the power supply can hypothetically be derived from only one of its several outputs, but the primary-side would not be able to tell the difference. This could damage the components of the overloaded output, because in all probability (and hopefully so) it wasn't overdesigned to be able to handle much more than its continuous rating. Note that there are also safety considerations in how much energy is allowed from any output for it to qualify as an SELV (safety extra low voltage) output. All told, in multioutput converters it is almost impossible to avoid having separate secondary-side current limiting circuitry for each and every output. Exempted are outputs coming say from a 780x type of series regulator, because these are assumed to be inherently current limited.

We are warned that several design subtleties exist in multioutput converters. For example, how much cross-regulation can we get on the outputs? These issues are very hard to correctly predict on the basis of theoretical models, and so usually a build-and-tweak approach is the fastest way to go.

### 6.2.6   The primary-side leakage term

In Fig. 6.5, we have included a leakage inductance term, which we assume is present on the primary side of the original flyback. At the instant the switch turns *off* we get a voltage spike across the switch, and this is clamped by the zener, as indicated, to a voltage $V_{IN} + V_Z$

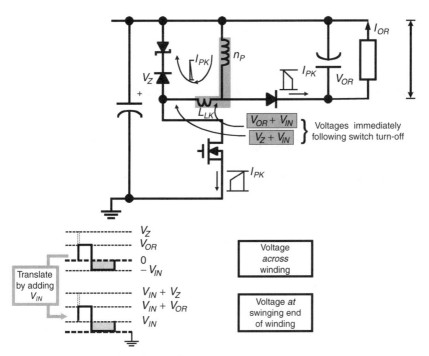

**Figure 6.5**   Primary-side leakage inductance.

(measured with respect to primary ground) and $V_Z$ *across* the winding. During the brief interval that the zener conducts (called $t_Z$ here) the current through the leakage current loop ramps down from $I_{PK}$ to zero at a rate determined by the voltage across the leakage inductor. Let us calculate the energy dissipated in the zener as a result. Since the voltage across the leakage inductor during the time the zener conducts is $V_Z - V_{OR}$, applying $V = LdI/dt$ *to the leakage inductance* we get

$$t_Z = \frac{L_{LK} \times I_{PK}}{V_Z - V_{OR}} \text{ s}$$

During this interval, the energy dissipated in the zener is

$$E_Z = (V \times I \times t) = V_Z \times \frac{I_{PK}}{2} \times t_Z \text{ J}$$

where we have taken the average current through the zener during this interval as $I_{PK}/2$. Simplifying, we get

$$E_Z = \frac{1}{2} L_{LK} I_{PK}^2 \frac{V_Z}{V_Z - V_{OR}} \text{ J}$$

So, the power dissipation in the zener is

$$\boxed{P_Z = \frac{1}{2} L_{LK} I_{PK}^2 \frac{V_Z}{V_Z - V_{OR}} \times f} \text{ W}$$

where $f$ is the switching frequency in Hz.

The energy dissipated in the zener can be expanded as follows

$$E_Z = \frac{1}{2} L_{LK} I_{PK}^2 \frac{V_Z}{V_Z - V_{OR}} = \frac{1}{2} L_{LK} I_{PK}^2 \times \left[ 1 + \frac{V_{OR}}{V_Z - V_{OR}} \right] \text{ J}$$

So, assuming that $1/2 \times L_{LK} \times I_{PK}^2$ is the energy present in the leakage, we seem to have an "unaccounted term"

$$E = \frac{1}{2} L_{LK} I_{PK}^2 \frac{V_{OR}}{V_Z - V_{OR}} \text{ J}$$

The physical explanation for the presence of this term is that during the time $t_Z$, besides the energy residing in the leakage $L_{LK}$, some energy from the primary (magnetizing) inductance $L$ also gets dumped into the zener. This happens because the current flowing in the leakage inductance loop has also to pass through the *voltage gradient* present across $L$, and thus does some work. Let us see how much work is done in the process. Since the voltage across $L$ is $(V_{OR} + V_{IN}) - V_{IN} = V_{OR}$

(see figure), so

$$E_P = (VIt) = (V_{OR}) \times \frac{I_{PK}}{2} \times t_Z \text{ J}$$

Simplifying

$$E_P = \frac{1}{2} L_{LK} I_{PK}^2 \frac{V_{OR}}{V_Z - V_{OR}} \text{ J}$$

which clearly is the "unaccounted term"—now fully accounted for! We can, thus, make the following conclusions:

- We must always set $V_Z$ greater than $V_{OR}$ otherwise the zener will become the *preferred freewheeling path* for the magnetizing current (in preference to the designated output path). If $V_Z$ is set greater than $V_{OR}$, the zener is the preferred (or the only available) path, but during the interval $t_Z$ only.

- To minimize dissipation we also need to set $V_Z$ significantly higher than $V_{OR}$ otherwise the zener dissipation will climb almost exponentially.

- However, the required minimum voltage rating of the switch is $(V_{IN} + V_Z)$. Therefore too high a zener voltage will have an effect on the overall cost and performance, due to the higher-rated FET required. Higher-rated FETs will also tend to have higher switching and conduction losses.

- We must also remember that the voltage across the zener is a function of the current passing through it. A zener is not a perfect device (like every other device we use); thus we may get higher clamping levels across the zener than we theoretically expected. A safety margin must be left for this when selecting the FET voltage rating.

- $V_Z$ is typically chosen to be about 50 to 100 percent higher than $V_{OR}$. For most worldwide input off-line applications, zeners will be found to be rated 150 to 200 V. The turns ratio is usually selected so that $V_{OR}$ is set to about 70 to 140 V, and the switch is then required to be typically rated 600 to 700 V.

- Note that the best-designed commercial off-line flybacks—those which are based on controller ICs and meant for worldwide input voltages— use standard 600 V FETs. Anything more may represent a less than optimum or a cost-ineffective design.

### 6.2.7 The secondary-side leakage term

In Fig. 6.6 we have placed a secondary-side-leakage inductance instead. This could, for example, simply represent the secondary winding lead

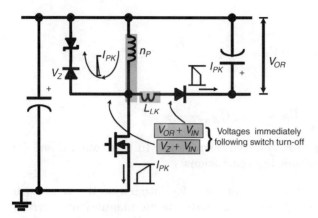

**Figure 6.6**  Secondary-side leakage inductance.

lengths and trace inductances. We note that the voltages across the leakage inductor and at the *drain* of the FET are the same as for Fig. 6.5. *Therefore, for the equivalent primary buck-boost model, there is actually no observable difference in whether a certain leakage is considered primary-side or whether it is secondary-side.* The dissipation in the zener is still represented by the same equation calculated earlier.

But how does a secondary-side leakage play the same role as a primary-side leakage? The reason is that by introducing either a secondary-side or a primary-side leakage inductance, we actually arrive at exactly the same result. In the first case, the primary-side leakage energy has nowhere to go as it is not magnetically coupled to the secondary windings. So it insists on pushing current till the current in the leakage manages to slew down and force a reset (of the leakage inductor). This happens in time $t_Z$. During this interval the secondary output diode just cannot get forward biased, and so the primary magnetization current *is deprived of a freewheeling path* (till $t_Z$ has elapsed). Similarly, for a secondary-side leakage, since there is no current flowing through it at the instant the switch turns *off*, it simply refuses to immediately allow any primary magnetization current to flow (no freewheeling path), till its own current can slew up to the required level (again within time $t_Z$). So, in this case too the primary magnetization current just courses through the zener during $t_Z$, depositing wasted energy in the process. Note that in both cases (slew down or slew up of the leakage current) the governing equation is the same, viz., $V = L dI/dt$. Also, V is the same in either case too. And as soon as the zener stops conducting, in either case the voltage *across* the primary-side winding immediately collapses from its clamped value $V_Z$ to its natural operating value $V_{OR}$.

Returning to the actual flyback schematic, it can be shown that since the current in its secondary loop is $n_P/n_S$ times the reflected primary-side current, and since $1/2 \times LI^2$ is an energy term that is fixed, *any secondary-side leakage inductance present in a flyback (denoted by $L_{SLK}$), must get reflected to the primary side as an equivalent leakage inductance equal to $L_{SLK} \times (n_P/n_S)^2$.* It is this *reflected secondary inductance* that is shown to be exactly equivalent to a primary-side leakage term in the discussion above.

**Example 6.1**  A flyback has 60 turns on the primary and three turns on the secondary. The output voltage is set to 5 V. An estimated 40 nH of secondary-side inductance is present, attributable to the winding terminations and PCB traces *(the thumb rule is 20 nH/in)*. What is the effective leakage inductance as seen by the switch? And what happens if the output was 12 V instead (keeping the same $V_{OR}$)?

Note that

$$V_{OR} = V_O \frac{n_P}{n_S} = 5 \times \frac{60}{3} = 100 \text{ V}$$

The turns ratio is $60/3 = 20$. So, the reflected leakage is

$$L_{LK} = 20^2 \times 40 = 16000 \text{ nH}$$

16 $\mu$H of leakage inductance is certainly not insignificant. Note that if $V_{OR}$ is always set to 100 V irrespective of the set output voltage (and this is a normal design approach), then for a 12 V output we would need to set the following turns ratio

$$\frac{n_P}{n_S} = \frac{V_{OR}}{V_O} = \frac{100}{12} = 8.33$$

In this case, if the number of secondary turns is increased to five, then the calculated number of primary turns is $8.33 \times 5 = 41.67$. Clearly, this is not possible since turns are always integral numbers. So, suppose we choose 42 turns for the primary. Now

$$V_{OR} = V_O \frac{n_P}{n_S} = 12 \times \frac{42}{5} = 100.8 \text{ V}$$

The turn ratio is $42/5 = 8.4$. So, the reflected leakage for a 12 V output is

$$L_{LK} = 8.4^2 \times 40 = 2822 \text{ nH}$$

We can see that the reflected leakage is only 2.8 $\mu$H here, that is, about six times less than what we got for a 5 V output (ignoring any primary-side leakage). The dissipation in the zener being proportional to leakage inductance will also drop by about six times.

**Note:** We stated above that the number of turns must necessarily be an integer. We need to qualify this here. A "half-turn" is often used by power supply designers to get the required output voltage in multioutput converters so as to achieve good "centering." They do this by simply *not* completing a full turn, but passing the winding out from the "wrong" side of the transformer. "Half-turns" don't really exist in nature since current will flow in a complete loop one way or another. By creating a "half-turn" what we are actually doing is only introducing some additional leakage inductance in series with that winding. This is however an acceptable technique, provided a good freewheeling path is still available for the magnetization current. Otherwise the spikes can damage the switch. Therefore, "half-turns" aren't usually placed in the main secondary winding. We should also be careful about safety issues here. By terminating a primary winding on a bobbin pin right next to a secondary winding pin termination (or vice versa), we may be violating mandatory clearance and creepage requirements. See Chap. 17 for more details. We can also keep in mind that there are known textbook techniques on how to create a *genuine* "half-turn." This is done from the point of view of the flux inside the core, and is achieved by winding turns in a special way on the outer limbs of a standard E-core. However, there are even tougher safety issues to overcome here, and the author has never seen this ever implemented in high-volume commercial production.

- In general we can say that in effect, the *secondary-side leakage inductance as seen by the switch is inversely proportional to* $V_O^2$. This assumes a standard (fixed) $V_{OR}$ irrespective of output voltage (turns ratio adjusted accordingly).

- A "good" off-line flyback transformer has a typical primary-side measured effective leakage inductance of less than 1 to 2 percent of its primary inductance. Anything more probably needs a serious redesign of the winding arrangement (like using a split primary winding structure).

- Small lead lengths/PCB traces on the secondary side are crucial for maintaining low effective leakage on a practical board assembly, *especially for low output voltages*.

**Note:** Leakage inductance is a major restriction on the maximum power that a flyback can deliver, efficiently. Actually, even 600 W is easily "doable" with a flyback, provided the output voltage is set high (e.g., 60 V). Admittedly, a flyback may not be the smallest or most cost-effective method for high-power applications, but it certainly can work in many cases.

- Note that when a transformer's leakage is measured in production by putting a short across the transformer's secondary winding pins, we are not accounting for the trace inductances that will get added on to the complete secondary loop in real operation. We cannot always afford to ignore this, especially for low output voltages.

- The overall efficiency of a flyback can actually fall between 5 and 10 percent simply on account of an inch or two of trace length on the secondary side.

- *Flying leads* are occasionally used when we run out of available transformer pins. But these should generally be avoided for the main output, especially if it is of low voltage.

- However, in such cases, as also for secondary windings when they are made out of copper foil, we can get quite a significant reduction in the lead inductance by closely paralleling (or even twisting if possible) the forward and return leads of the secondary winding.

- Secondary-side trace lengths present *after* the output capacitor don't count in this analysis. They have essentially dc current in them, and so their inductance cannot pose any issue from a high-frequency perspective.

  Hint:  A correct measurement strategy would be to take an actual converter board with the transformer in place. Then we should place thick wires across the diode and the output capacitor. We should then cut any one trace leading to the primary winding and measure the inductance across the primary winding with an LCR meter. This will give us the net effective $L_{LK}$ present on the primary side, all PCB trace inductances now included. We may be very surprised to see how much the traces cause the measured leakage inductance to increase, especially for high primary to secondary turns ratios (i.e., low output voltages with a fixed $V_{OR}$).

### 6.2.8  Optimization and analysis

Here we will provide some curves to help in making quick design estimates. If we don't see the larger picture but get bogged down by myriad detailed calculations, we will likely mistake the trees for the wood, and thus never manage proper optimization. We will also see that generally, each set of curves presented has *proportionality statements* along their top edges, which can be used to scale the numbers to our specific application conditions.

The steps to a successful optimization procedure are as follows:

**Step 1: $V_Z/V_{OR}$.**  If the FET is rated 700 V (as in the IP-Switch), then for an ac input voltage of 270 V, the peak rectified value which forms the dc input to the converter is $270 \times 2^{1/2} = 382$ V. The voltage across the FET will be $V_{IN} + V_Z$. Keeping a safety margin of, say, 50 V we are

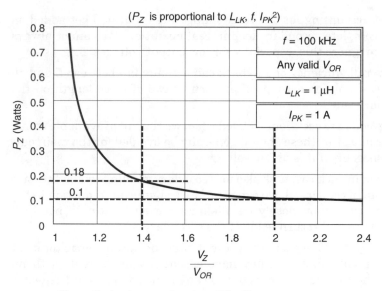

**Figure 6.7** Zener dissipation as a function of $V_Z/V_{OR}$.

trying not to go above 650 V on the switch. Therefore $V_Z$ must be less than $650 - 382 = 268$ V. A typical zener has a certain basic tolerance for its clamping voltage, which is also a function of the current through it. Therefore suppose we choose a 200 V zener here. If we then choose a $V_{OR}$ of 100 V, the ratio $V_Z/V_{OR}$ is 2. For a chosen $V_{OR}$ of 140 V, the $V_Z/V_{OR}$ will be 1.4. The $V_Z/V_{OR}$ ratio is important because it determines the zener dissipation.

**Step 2: Dissipation in zener.**   Now coming to Fig. 6.7 we see that if the switching frequency is 100 kHz and the peak current in the switch just prior to turn-off is 1 A, then for a $V_Z/V_{OR}$ of 2 as compared to a $V_Z/V_{OR}$ of 1.4, the dissipation goes by up almost 80 percent. Realizing that at low-line conditions the zener dissipation may account for 20 to 50 percent of the total energy dissipated inside the switcher, we can expect a steep fall in overall converter efficiency whenever we try to increase $V_{OR}$ *without being able to maintain the ratio* $V_Z/V_{OR}$ *high enough*. This simply means that we can increase $V_{OR}$, but then we must try to use a higher rated switch (which will allow a higher voltage zener).

Note that the zener dissipation formula can be rewritten as

$$P_Z = \frac{1}{2}L_{LK}I_{PK}^2\frac{Vratio_{CLAMP}}{Vratio_{CLAMP} - 1} \times f \quad \text{W}$$

where

$$Vratio_{CLAMP} = \frac{V_{CLAMP}}{V_{OR}} \equiv \frac{V_Z}{V_{OR}}$$

This has been written in a more general form to indicate that the clamp can be a zener or an RCD (resistor-capacitor-diode) type, and the same dissipation equation applies. Clearly, for ensuring even basic operation, $Vratio_{CLAMP}$ must always be at least greater than unity.

We now note that the new generation IP-Switch has a frequency of 130 kHz as compared to 100 kHz earlier. It also has an officially "recommended" $V_Z/V_{OR}$ of about $180V/135V = 1.33$. We see that that the higher $V_Z/V_{OR}$ will significantly increase dissipation. In addition, the higher switching frequency will cause a further 30 percent increase in the zener dissipation, because the zener will now take that many more "hits" every second as compared to 100 kHz.

**Step 3: Duty cycle.**    In Fig. 6.8 we can read off the duty cycle for different input ac line voltages. Therefore at 90 Vac the duty cycle is 0.44 for a $V_{OR}$ of 100 V, but it increases to 0.525 for a $V_{OR}$ of 140 V.

**Step 4: Peak current.**    We use the fact that the average inductor current for a flyback is (ignoring losses)

$$I_L = \frac{I_{OR}}{1 - D}$$

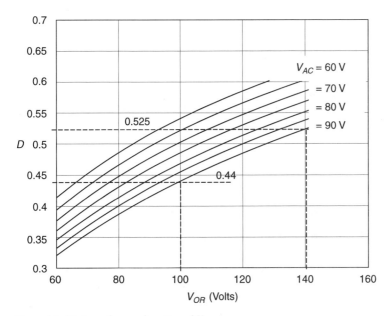

**Figure 6.8**  Duty cycle as a function of $V_{OR}$.

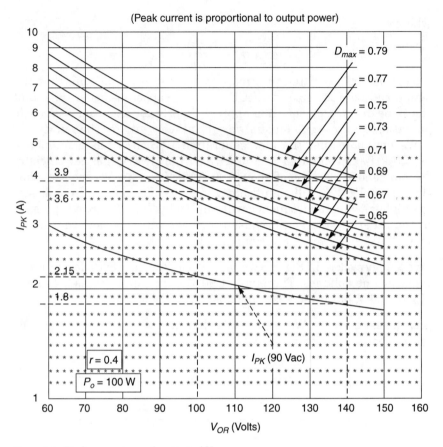

**Figure 6.9** Peak current as a function of $V_{OR}$.

The ramp portion of the inductor current typically adds about 20 percent to its average level to give the peak current (in the switch, inductor, and diode), with an "optimally" designed transformer ($r = 0.4$). So

$$I_{PK} = \frac{1.2 \times I_{OR}}{1 - D}$$

**Note:** To account for the losses inside the converter, we may want to typically add another 20 percent or so to this.

In Fig. 6.9, we can read off the peak current at 90 Vac for a load of 100 W (assuming full efficiency). We see that we get a peak of 2.15 A for a $V_{OR}$ of 100 V, but the peak *decreases* to 1.8 A for a $V_{OR}$ of 140 V. Note that the other curves in the figure will be discussed later.

*This was probably the intuitive "reason" why both the recommended operating duty cycle (at 90 Vac) and the $D_{max}$ were set higher for the new* *generation IP-Switch*. The physical reasoning behind the lower peak current is that by going to a high $V_{OR}$ we get a higher $D$. We also know that input power is virtually constant,

$$P_{IN} = I_{IN} \times V_{IN} = (I_C \times D) \times V_{IN}$$

The term in brackets above must remain constant at a given input; thus if $D$ increases, the center of the primary-side (reflected) current ramp $I_C$ must come down, along with its peak value. Reducing the peak value of the current was apparently important for marketing reasons, since the published "power capability" of the IP-Switch was based purely on current limit (which in turn was predicated by the die size). In general, achieving the published power capability figure of such off-line integrated switchers is possible in principle, but there is no holdup time available in that condition, and it may not even be practical in terms of the massive heatsinking required.

Now we illustrate how to use the curves presented so far for finding the zener dissipation for a more general case.

**Example 6.2**   Consider a worldwide input flyback that has 60 turns on the primary and three turns on the secondary. The output is 5 V at 2 A, at an efficiency of 69 percent. The switching frequency is 130 kHz. What is the zener dissipation if the effective (lumped) leakage inductance as seen by the switch is 40 $\mu$H?

The worst-case dissipation occurs at 90 Vac (maximum peak current), which we take as the minimum line input. Let us set $V_{OR}$ to 100 V (e.g., a turns ratio $n_P/n_S$ of 20 for a 5 V output). In the earlier discussions we had assumed 100 percent efficiency. Now we account for reality in the following manner. The output power here is 10 W. We increase it as per $P_O/\eta$ to $10/0.7 = 14.5$ W and this becomes the equivalent (effective) load power at 100 percent efficiency (from the viewpoint of the switch). Note that we need to first calculate the peak current from Fig. 6.9, before we can use Fig. 6.7 since the latter is in terms of $I_{PK}$. We know that for 100 W, the peak current is 2.15 A, and so in our case

$$I_{PK} = 2.15 \times \frac{14.5}{100} = 0.31 \text{ A}$$

Assuming we have a zener of 200 V, we see from Fig. 6.7 that we get a zener dissipation of 0.1 W under the following conditions: $V_{OR}$ of 100 V, 1 $\mu$H leakage, 1 A peak current, and frequency of 100 kHz. So for this example, the zener dissipation must scale as

$$P_Z = 0.1 \text{ W} \times \frac{130 \text{ kHz}}{100 \text{ kHz}} \times \left(\frac{0.31 \text{ A}}{1 \text{ A}}\right)^2 \times \frac{40 \text{ }\mu\text{H}}{1 \text{ }\mu\text{H}} = 0.5 \text{ W}$$

We see that this accounts for $0.5/4.5 = 11$ percent of the total 4.5 W dissipation. Setting a higher $V_{OR}$ (and lower zener clamp) could have increased the zener dissipation to over 1 W.

**Note:** In actual bench measurements we will usually see that the peak current flowing *into the zener* is only about 70 to 80 percent of $I_{PK}$, which is the calculated peak current through the primary winding. The fact is that the model for a transformer is not so trivial. Some of the inductor current flowing just prior to turn-off freewheels into the parasitic capacitances of the transformer and the switch. This could typically end up reducing the dissipation in the zener to almost half of the theoretical prediction above. But there is no easy equation for this. Therefore the engineer is advised to actually place a current probe in the zener path and measure the peak current flowing into the zener. Otherwise we may end up overestimating the size (and cost) of the zener.

**Note:** For higher output currents, the zener dissipation will become almost intolerable. Then an RCD clamp should be used to improve efficiency. The advantage of an RCD clamp is that it automatically tends to increase the clamp voltage level at low-line, but at high-line the clamp-level subsides on account of the smaller energy residing in the leakage inductance (lower primary-side peak current). Thus we get a high clamping voltage level where we need it the most (dissipation at low-line) and also a lower clamping level where we need that the most (switch voltage rating at high-line).

**Note:** An RCD clamp is *virtually independent of the value of capacitance of C*, provided it is not too small and we are operating in steady state. But during overloads or other transient conditions, the C of the RCD can get overcharged momentarily. Therefore, to protect the switch, either we have to increase the C substantially, or we need to retain a zener clamp in parallel to the RCD clamp. The zener clamp level must be set higher than the steady state clamping afforded by the RCD clamp over the *entire input range*, and we should thus ensure that the RCD is the only effective working clamp under normal operating conditions. The best way to fix the $R$ is on the bench. Here we operate the power supply at maximum load at highest input voltage (e.g., 270 Vac), and ensure that the $R$ is set high enough to just maintain the drain to source voltage to a little less than its rating (including any small derating margin). This gives the most optimum and efficient RCD clamp. The $C$ is typically increased to about 33 nF (for a 70 W universal input power supply), simply because under overloads we don't want it to get charged up faster than our current limiting protection circuitry can act.

**Step 5: Holdup time.** We have seen that at 90 Vac the duty cycle is 0.44 for a $V_{OR}$ of 100 V, and it is 0.525 for a $V_{OR}$ of 140 V. So, why do we need to set a maximum duty cycle much higher than this in the first place? The main reason is that we need some *holdup time* as explained in Chap. 5. Now we see why the maximum duty cycle $D_{max}$ is so important. As the input voltage on the bulk capacitor terminals slews down, the duty cycle increases. But the converter must continue to function for the specified holdup time without losing output regulation. Clearly the amount of holdup time available depends on the size of the input capacitance. If $C_{IN}$ is very large, it will hold the voltage up much longer and we won't require very high duty cycles (or wide duty cycle limit). It also depends inversely on the output power since a high output demand will drain the input capacitor faster. *A common "economical" choice for the input capacitance of a worldwide input power supply is 3 $\mu F / W$.*

This $\mu F/W$ figure is, or at least should be, the capacitance per *input* power. It is often wrongly referred to as being the output power. This misinterpretation seems somewhat deliberate in some cases, because in effect it leads to small input capacitors and "better-looking" evaluation boards for some integrated power switcher vendors to tout. But we have to realize that the input capacitor "knows" nothing about how much power is being delivered at the output. It is only conscious of the input power, for that is what it is handling. Sure, the two are related, but that link is through the efficiency of the power supply. And efficiency can change greatly from supply to supply and application to application, throwing any such input-output correlation overboard.

For example, if the input power drawn by the power supply is 5 W, we need a 15 $\mu F$ capacitor. It will then have the same holdup time (yet to be calculated) as a supply drawing 10 W with a 30 $\mu F$ input capacitor, all else being the same. The required equation is

$$\mu F \big/ W = \frac{2 \times t_{holdup} \times (D_{max} \times D)^2 \times 10^6}{V_{OR}^2 (D_{max} - D) \times (D_{max} + D - 2D_{max}D)}$$

where $D$ is the duty cycle at the normal minimum input voltage (e.g., at 90 Vac) and $t_{holdup}$ is in seconds.

In Fig. 6.10 we have plotted this equation out for 90 Vac and 20 ms (measured from the maxima of the input ripple). Some related points are as follows:

- For a $D_{max}$ of 0.67 (older IP-Switch), with $V_{OR}$ set to 80 V or 100 V or 140 V, we would need 2.75 $\mu F/W$, 2.9 $\mu F/W$, and 3.5 $\mu F/W$ respectively. So, a low $V_{OR}$ *would actually seem to help in reducing the size of the input capacitor*. We can understand this because we have seen that at 90 Vac, the steady state duty cycle is 0.44 for a $V_{OR}$ of 100 V, and is 0.525 for a $V_{OR}$ of 140 V. So a lower $V_{OR}$ allows for a larger

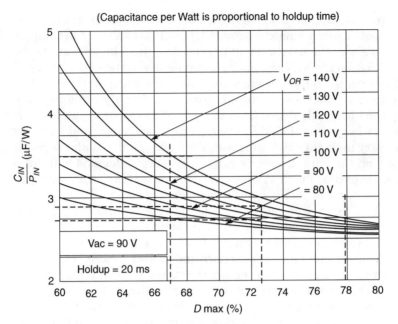

**Figure 6.10**  Holdup time design curves for flyback.

*change* in duty cycle. And it is this "headroom" that is even more important than the absolute values of $D$ and $D_{max}$.

- We can also reverse the question to: what if we don't want to incur any added cost by increasing the input capacitance beyond the value dictated by $V_{OR} = 100$ V and $D_{max} = 0.67$, nor do we want to compromise on the holdup time, but we still want to increase the $V_{OR}$ (for some reason), say to 130 V (this being the recommendation for the new IP-Switch). From Fig. 6.10 we see that we can do this, but *we now need a* $D_{max}$ *of about 72.5 percent*. In effect, we are saying that granted we somehow have chosen a high operating $D$, at least let us allow more headroom for it to *change*, so that we can maintain the holdup time without having to increase the capacitance. We note however that for some hitherto mysterious reason the IP-Switch $D_{max}$ is set not to around the seemingly more optimal 0.725, but to 0.78. Could that help in reducing the input capacitance further? It actually does because we have more headroom. But looking at Fig. 6.10 we can also see that it won't help a great deal because *at high $D_{max}$ all the $V_{OR}$ curves start bunching up together*.

- We can see that for a $V_{OR}$ of 140 V, with a $D_{max}$ of 0.78, we get a requirement of 2.75 $\mu$F/W, which is actually *the same as the $\mu$F/W for a $V_{OR}$ of 100 V using the older IP-Switch*. So what do we gain (or lose) even on a direct comparison? Basically, decreasing the operating $D$ at 90 Vac (i.e., choosing a smaller $V_{OR}$) for a given $D_{max}$ helps reduce the input capacitance, whereas increasing $D_{max}$ (for constant $D$ or fixed $V_{OR}$) also helps decrease the input capacitance. But *if we increase* $V_{OR}$, *(that is, increase* D*), and increase* $D_{max}$ *at the same time, we don't get any reduction in the input capacitance* because the headroom remained almost the same. This is one reason why, even after using the expert system software of the IP-Switch, we don't get any obvious improvement in the size of the input capacitor.

**Example 6.3**   How much input capacitance do we need for a worldwide input flyback with an output of 5 V@2 A, running at a measured 69 percent efficiency? The turns ratio is 20 and the maximum duty cycle is 0.67. We want to select the input capacitor for a 20 ms holdup time, but we also want to consider having a 40 ms holdup time capability for a certain critical application.

As in the previous example $P_{IN} = 14.5$ W and $V_{OR} = 100$ V. From Fig. 6.10, for a 20 ms holdup time we will need a $\mu$F/W of 2.9. Thus for our case

$$C_{IN} = P_{IN} \times {}^{\mu F}/_W = 14.5 \times 2.9 = 42 \ \mu F$$

**Note:**   We should account for typical capacitor tolerance of $\pm 20$ percent and also an end-of-life degradation of $\pm 20$ percent of the initial capacitance value. Therefore the selected nominal capacitance should be at least $1/(0.8 \times 0.8) = 1.56$ times the calculated value, i.e., 66 $\mu$F.

For a holdup time of 40 ms, we need to double the capacitance to at least $66 \times 2 = 132 \ \mu$F. The next (higher) standard value is 150 $\mu$F.

**Step 5: Inductor energy.**   We see that, for example, $D_{max} = 0.67$ with a $V_{OR}$ of 80 V gives us the same input capacitance (based on holdup time consideration) as for $D_{max} = 0.78$ with a $V_{OR}$ of 140 V. So which is better? What is the last remaining design consideration?

Returning our attention to Fig. 6.9, we look at the $D_{max}$ lines now. In the figure, we have shown the *peak* currents expected in the converter for a 100 W load (*here "peak" is at the end of holdup time period!*). Note that the curves are independent of the holdup time because we are assuming that the capacitance has been set accordingly. Let us actually compare the specific cases based on the following specific official recommendations for the IP-Switch i.e., a $V_{OR}$ of 100 V for the first generation ($D_{max} = 0.67$) and a $V_{OR}$ of 140 V for the second generation ($D_{max} = 0.78$). As we go from the normal operating duty cycle at 90 Vac to $D_{max}$ during a line disturbance, we see the following:

- For the older generation we will get a peak current varying from 2.15 A to 3.6 A.

- For the new generation device we will get a variation from 1.8 A (sounds good!) to 3.9 A (what?).

We know that the transformer's physical size is virtually proportional to the peak energy stored in it, which is $1/2 \times L \times I_{PK}^2$. So, for every doubling of peak current we would need to increase the core size by about four times. We are implicitly assuming that the transformer design is saturation-limited not core-loss limited, as is usually the case with ferrites. Therefore comparing a $V_{OR}$ of 100 V to a $V_{OR}$ of 140 V, the size of the inductor, calculated *only on the basis of steady operation* at 90 Vac (and ignoring holdup for now), does go *down* by $(2.15^2 - 1.8^2)/2.15^2 = 30$ percent. A higher $V_{OR}$ seems to help. But we also note that

$$L = \frac{V_{OR}}{I_{OR} \times r \times f}(1 - D)^2 \times 10^6 \ \mu H$$

Using $P_O = V_{OR}I_{OR}$ and $D = V_{OR}/(V_{IN} + V_{OR})$, we get Fig. 6.11. We see that *for a high* $V_{OR}$, *though the size of the inductor has come down, the required inductance has increased.* This would usually require more turns, so the copper losses will increase. But let us see what happens as we slew down to higher duty cycles as the input falls (as during a holdup event)? For a low $V_{OR}$ we go up to only 3.6 A (while meeting the holdup time requirement) whereas for the higher $V_{OR}$ we go up to 3.9 A (albeit with a slightly smaller capacitance of 2.75 $\mu F/W$ as compared to 2.9 $\mu F/W$). So, in fact, for the higher $V_{OR}$ we need a core

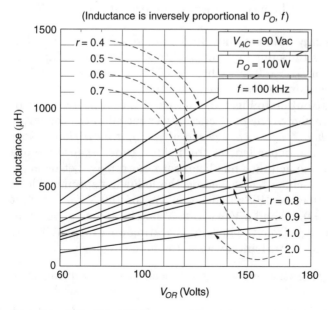

**Figure 6.11** Inductance selector.

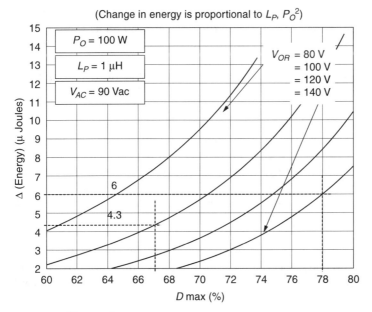

(Change in energy is proportional to $L_P$, $P_O^2$)

**Figure 6.12**   Energy of core.

size greater by $(3.9^2 - 3.6^2)/3.6^2 = 17$ percent. Yes, we may have got a 5 percent reduction in input capacitance from 2.9 $\mu$F/W to 2.75 $\mu$F/W in the process (and that is *assuming efficiency didn't change!*), but we now need to increase the core size by 17 percent. Though we can also see *that had we kept to a* $D_{max}$ *of 72.5 percent then with a* $V_{OR}$ *of 130 V we could have actually reduced the peak current compared to the case of* $V_{OR} = 100$ *V and* $D_{max} = 67$ *percent.*

We have plotted this out in Fig. 6.12. We can see that the *change* in the energy stored in the core (*in going from* D *to* $D_{max}$) is 4.3 $\mu$J for a $V_{OR}$ of 100 V ($D_{max} = 0.67$) but is 6 $\mu$J for a $V_{OR}$ of 140 V ($D_{max} = 0.78$), for every $1\mu$H of primary inductance and 100 W output power. So, though the normal operating energy in the core may have been less with a higher $V_{OR}$ (higher $D$) on account of the lower peak current, at $D_{max}$ the energy requirement may be large enough to completely swamp out any "advantage."

**Note:** We may have, theoretically speaking, sized our input capacitance to meet the required holdup time at a calculated $D_{max}$ of say 72.5 percent. But if we don't fix that value as a *hard limit* in the controller, the duty cycle will continue to increase to the maximum possible $D_{max}$ as the line voltage slews down. This could lead to higher and higher peak currents than expected. We could attempt to rely on current limiting to "rescue" us. Here we note that most integrated off-line switcher ICs incorporate fixed internal current limiting. To partially overcome this rather noticeable lack of flexibility, they are offered as a "family" with a set of discrete set of choices of current limits (and inversely proportional die sizes). But since the maximum allowable current is directly related to the size of the magnetics required, experienced designers always prefer to have full flexibility in setting both the duty cycle and the current limit. This is usually available only when using controller ICs with external FETs. With that, the highest degree of optimization and reliability can be achieved as follows: we must first correctly estimate the peak duty cycle and the peak instantaneous current needed to meet the required holdup time (with a cost-effective choice of input capacitance and transformer). Thereafter, we should set both the $D_{max}$ and the current limit *slightly higher* than the respective calculated peak values. This inequality should be guaranteed across all tolerances. Remember that the size of the transformer is always calculated based on the *upper* tolerance limit (maximum) of the current limit, whereas the maximum output power capability (and holdup time) must be guaranteed with respect to the *lower* tolerance limit (minimum) of the current limit.

We are now clear how our reasoning must proceed as we try to optimize the flyback. The ultimate choice will also depend on the *break points* in the standard component values available. So, for example, if we choose the input capacitance to be 2.75 $\mu$F/W instead of 2.9 $\mu$F/W, it may mean nothing on a practical level if in either case we will eventually pick the same closest standard value. In that case we might as well try to optimize the other parameters and component values.

### 6.2.9   Dissipation estimates and efficiency

We show a first-order iterative calculation process because otherwise we would actually require a complex math simulation file.

**Example 6.4**   A flyback has 60 turns on the primary and three turns on the secondary. The output voltage is set to 5 V. The load is 20 A ($P_O = 100$ W, $V_{OR} = 100$ V). What are the losses at 90 Vac ($V_{IN}$ 90 $\times \sqrt{2} = 127$ V)?

We start with an assumption of 100 percent efficiency. Then we know from Fig. 6.9 that for this condition the peak current is 2.15 A (with an assumed $r = 0.4$). So the center of the switch ramp must be $2.15/1.2 = 1.79$ A. Now, since $D = 0.44$ from Fig. 6.8, the average input current is $1.79 \times 0.44 = 0.79$ A. The 1.79 A center point reflects to the secondary side according to the turns ratio of 20 to become 1.79 A $\times 20 = 35.8$ A. So, the average current through the diode is $35.8 \times (1 - 0.44) = 20$ A. *Assuming that the switch is an FET and has a specified forward drop of 10 V at 1 A*, we get its drain to source resistance ($rds$) to be

$$rds = \frac{10\ \text{V}}{1\ \text{A}} = 10\ \Omega$$

The drop across the switch during the *on*-time is

$$V_{SW} = 10\ \Omega \times 1.79\ \text{A} = 17.9\ \text{V}$$

The dissipation in the FET is

$$P_{SW} = 1.79\ \text{A}^2 \times 10\ \Omega \times 0.44 = 14.1\ \text{W}$$

The diode dissipation is

$$P_D = V_D \times I_{D\_AVG} = V_D \times I_O = 0.6 \times 20 = 12\ \text{W}$$

Therefore, the total dissipation due to the switch and diode drops is about 26 W. The input power is therefore not 100 W but is 126 W (a 26 percent increase). We have ignored zener dissipation and other parasitics. But we can now quickly *recalculate some of the key parameters based on our new assumptions*. For example, the switch current under steady operation is actually

$$I_{PK} = 126\% \times 1.79 = 2.25\ \text{A}$$

Using simple math applied to the equivalent primary buck-boost model when the switch is *on* and the equivalent secondary buck-boost model when the switch is *off*, we can include the resistance of the primary and secondary windings to get a more accurate estimate of the losses in a flyback. We thus get Table 6.1.

Note that at high duty cycle (as at low $V_{IN}$ or/and high $V_{OR}$), the dissipation in the output capacitor will go up steeply. In fact, from a $V_{OR}$ of 100 V to a $V_{OR}$ of 130 V, the heating in the output capacitors (and its resulting temperature) will increase by over 25 percent.

Also note that the dissipation in the input capacitor too has a $D/(1-D)$ dependency; thus the remarks about the output capacitor generally apply here also. However, if $V_{OR}$ is increased, this helps lower the input rms current more than the $D/(1-D)$ term tries to increase it. Overall, a high $V_{OR}$ will help here.

The equations provided in the table assume a small current ripple ratio $r$. See Chap. 1 for more on the flat-top approximation, and where this creates a larger error, and where it doesn't.

**Note:** The input capacitor rms current is

$$I_{CIN\_RMS} = \left( \frac{P_O}{V_{OR}} \right) \times \sqrt{\frac{D}{1-D}}$$

**TABLE 6.1  Estimating Dissipation Inside the Flyback.**

$$D = \frac{V_{OR} + \left(\frac{V_{OR}}{V_O}\right)V_D + \left(\frac{V_{OR}}{V_O}\right)\left(1 + \frac{V_{OR}}{V_{IN}}\right)\frac{P_O}{V_O}R_S}{V_{IN} + V_{OR} + \left(\frac{V_{OR}}{V_O}\right)V_D - rds\left(1 + \frac{V_{OR}}{V_{IN}}\right)\left(\frac{P_O}{V_{OR}}\right) + \left(\frac{V_{OR}}{V_O}\right)\left(1 + \frac{V_{OR}}{V_{IN}}\right)\frac{P_O}{V_O}R_S - \left(\frac{V_O}{V_{OR}}\right)\left(1 + \frac{V_{OR}}{V_{IN}}\right)\frac{P_O}{V_O}R_P}$$

| Flyback loss terms | |
|---|---|
| Primary/switch peak current | $I_{PK} = \dfrac{1.2 \times \left(\frac{P_O}{V_{OR}}\right)}{1 - D}$ |
| Dissipation in primary winding | $P_P = \left(\dfrac{\frac{P_O}{V_{OR}}}{1 - D}\right)^2 \bullet R_P \bullet D$ |
| Dissipation in secondary winding | $P_S = \dfrac{\left(\frac{P_O}{V_O}\right)^2 R_S}{1 - D}$ |
| Dissipation in zener clamp | $P_Z = \dfrac{1}{2} \bullet I_{PK}^2 \bullet \dfrac{V_Z}{V_Z - V_{OR}}$ $\bullet f \bullet L_{LK} \bullet 10^{-6*}$ |
| Dissipation in switch (crossover loss) | $P_{CROSS} = \left(\dfrac{P_O}{V_{OR}}\right) \bullet V_{IN} \bullet t_{CROSS} \bullet f$ |
| Dissipation in switch (conduction loss) | $P_{SW\_COND} = \left[\dfrac{P_O}{V_{OR}\left(1 - D\right)}\right]^2 \bullet rds \bullet D$ |
| Dissipation in diode | $P_D = \left(\dfrac{P_O}{V_O}\right) \bullet V_D$ |
| Dissipation in the output capacitor | $P_{COUT} = \left(\dfrac{P_O}{V_O}\right)^2 \bullet \dfrac{D}{1 - D} \bullet ESR_{OUT}$ |
| Dissipation in the input capacitor | $P_{CIN} = \left(\dfrac{P_O}{V_{OR}}\right)^2 \bullet \dfrac{D}{1 - D} \bullet ESR_{IN}$ |

*$L_{LK}$ is the lumped effective leakage in $\mu$H, as seen by the switch.

and the output capacitor rms current is

$$I_{COUT\_RMS} = \left(\frac{P_O}{V_O}\right) \times \sqrt{\frac{D}{1 - D}}$$

Clearly, if the turns ratio is 1, or this is a dc-dc buck-boost converter (single winding inductor), the input and output rms currents are the same. This is just an interesting coincidence reported here.

In Fig. 6.13 we have provided the results of a sample calculation based on the IP-Switch, with converter parameters chosen as indicated. The inductance is assumed large (flat-top approximation). We see that the zener dissipation is the main loss component (besides the switch and

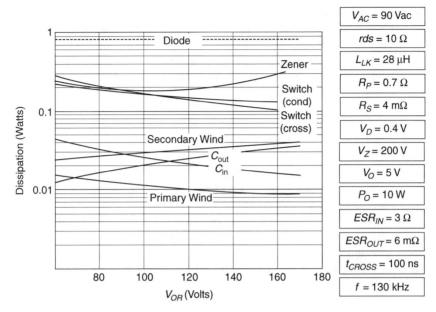

| $V_{AC}$ = 90 Vac |
| $rds$ = 10 Ω |
| $L_{LK}$ = 28 µH |
| $R_P$ = 0.7 Ω |
| $R_S$ = 4 mΩ |
| $V_D$ = 0.4 V |
| $V_Z$ = 200 V |
| $V_O$ = 5 V |
| $P_O$ = 10 W |
| $ESR_{IN}$ = 3 Ω |
| $ESR_{OUT}$ = 6 mΩ |
| $t_{CROSS}$ = 100 ns |
| $f$ = 130 kHz |

**Figure 6.13**  A sample calculation of parameters for various $V_{OR}$.

diode dissipations). Further, as we expected, it increases steeply as $V_{OR}$ increases. In Fig. 6.14 we have plotted the efficiency for various load conditions. We see that because of the zener dissipation term, *an "ideal"* $V_{OR}$ *would be around 110 V to 115 V purely on the grounds of efficiency.*

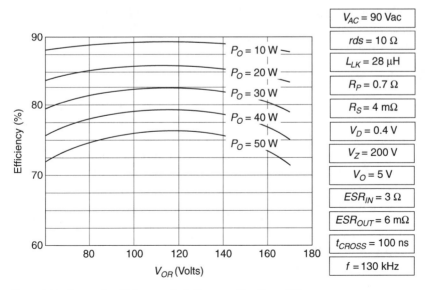

| $V_{AC}$ = 90 Vac |
| $rds$ = 10 Ω |
| $L_{LK}$ = 28 µH |
| $R_P$ = 0.7 Ω |
| $R_S$ = 4 mΩ |
| $V_D$ = 0.4 V |
| $V_Z$ = 200 V |
| $V_O$ = 5 V |
| $ESR_{IN}$ = 3 Ω |
| $ESR_{OUT}$ = 6 mΩ |
| $t_{CROSS}$ = 100 ns |
| $f$ = 130 kHz |

**Figure 6.14**  A sample efficiency calculation as a function of $V_{OR}$.

The *concave bell in the zener dissipation curve* of Fig. 6.13 is clearly responsible for the *convex bell in the efficiency curves* of Fig. 6.14. The reason for this is that *as we increase* $V_{OR}$, *up to some point we actually benefit because the peak current falls in the process. But after that point, the zener dissipation climbs again due to the* $V_Z/(V_Z - V_{OR})$ *term* in the zener dissipation equation given earlier in this chapter.

### 6.2.10   Practical flyback designs using 600-V switches

Just for our information, let us now create a typical profile for a cost-effective, universal input, commercial flyback designed for an IT (information technology) application.

Everything so far points to lower performance and no significant reduction in cost if we increase $V_{OR}$ without due thought. But as mentioned previously, some of the best designed commercial flybacks use only 600 V FETs. This is achieved by setting even a lower $V_{OR}$ (around 70 V), besides incorporating current limiting for the FET (on the primary side) and for each individual output (on the secondary side). Properly implemented input feedforward is also necessary, and the efficiency is optimized with an RCD clamp in place of a zener clamp. The typical operating drain to source voltage is about 450 V at 270 Vac input with a leakage spike of 50 V. This represents a good 85 percent derating margin. Under overloads and startup into short-circuited outputs, the drain to source voltage doesn't exceed about 585 V. This is considered acceptable.

*Such power supplies typically fix the turns ratio at around 14 for a 5-V output. This translates to a* $V_{OR}$ *of about 70 V.* The actual turns used are typically 85T:6T, or 60T:5T (T stands for turns here). The choice of whether to have 5T for the secondary versus say 6T depends largely on the "other" secondary output we are trying to set up. For example, if that is a 12 V output coming from a series pass stage (linear post-regulator), we will use 6T for the 5 V winding. Then assuming about 1 V drop coming from the 5 V output Schottky diode and its winding resistance, we are in effect getting 5 V per turn for the entire transformer. For the 12 V winding we then use $15\frac{1}{2}$ turns to get 15.5 V at this winding. If we use an ultrafast low-drop diode for this output (e.g., the FEP6AT through FEP6DT series from *www.vishay.com*), we can assume a, roughly, 1 V diode drop for this too. Thus we get 14.5 V at the input of the 12 V post-regulator. The post-regulator can be a series pass stage built around a popular and cost-effective FET like the MTP3055. This drops the voltage down to 12 V. This series FET works well up to about 2 to 3 A load on the 12 V output. A typical headroom of about 2.5 V is thus required across the series pass regulator because of cross-regulation limitations.

Sometimes we may need to go up to 16T for the 12 V winding just to deal with "starve" conditions, i.e., cases where one output is fully loaded whereas the other is almost unloaded, and specifications requiring that both outputs be fairly well regulated nevertheless. Generally, if the minimum specified load on the regulated 5 V output is zero, we will also need to put in a certain carefully calculated preload on the 5 V output (inside the power supply), so as to maintain a certain minimum voltseconds across the windings to help keep the other output within regulation.

**Note:** Sometimes the series pass FET can be eliminated by a technique called *double-point sensing* in which *both* the outputs are sensed through resistors to the LM431 (or TL431). One output is however "weighted" much more than the other. This means that the resistor coming from the main (regulated) output injects a much higher current into the node of the voltage divider than the resistor from the other output rail. In effect, this technique attempts to carefully compromise the regulation of the main output slightly, in favor of the other one.

### 6.2.11 Diode rating for higher $V_{OR}$

The minimum voltage rating of the output diode is

$$V_{D\_rating} = V_O \times \left(1 + \frac{V_{IN}}{V_{OR}}\right)$$

We will see that if the $V_{OR}$ is 100 V, we can safely use a 60 V Schottky diode even for a 12 V output. But to be able to use a 60 V diode for outputs as high as 15 V output we need to set the $V_{OR}$ to about 140 V! This is much too high for all the reasons discussed previously.

A corollary is that with 600 V switches, which require a lower $V_{OR}$ setting, we will not be able to use any commonly available Schottky diode for 12 V or –12 V outputs.

### 6.2.12 Pulse-skipping and required preload

At some lighter load we may reach the minimum duty cycle capability of the controller. Thereafter the controller will typically respond by randomly trying to omit cycles (unless designed specifically to handle this situation). What happens is that the energy being pushed into the converter even in one (minimum) pulse width is in excess of the output power requirement. The error amplifier then basically behaves in an almost random hysteretic mode thereafter, causing increased output voltage ripple and poor transient response. We may want to avoid this. But especially with current mode control, we always need a certain *blanking time* of around 50 to 150 ns to avoid jitter in the switching waveform (arising from the leading edge current spike). That translates to a minimum pulse width of the same duration. Clearly, at high switching frequencies, 150 ns may lead to an unacceptably large minimum duty cycle. A sure way to avoid the resulting random-skipping/pseudo-hysteretic mode is to *maintain a certain minimum external load on the converter output, i.e., a preload if necessary*. But how much is enough? We don't want to waste more power than necessary here.

The first question that arises is: as we increase input voltage, at what point does the converter enter discontinuous mode? The duty cycle in discontinuous conduction mode is (see Chap. 4):

$$D = \frac{\sqrt{2P_O L f}}{V_{IN} \times 10^3}$$

where $L$ is in $\mu$H, $f$ in Hz, and $P_O$ in watts. By equating the duty cycle equations for continuous and discontinuous modes we can show that the transition from CCM to DCM occurs at the following input voltage

$$V_{IN\_CRIT} = V_{OR} \times \left[ \sqrt{\frac{1}{\gamma^2} + \frac{1}{\gamma}} + \frac{1}{\gamma} \right]$$

where

$$\gamma = \frac{V_{OR}^2}{2P_O f L} \times 10^6 - 1$$

$L$ is in $\mu$H above. Note that this equation gives us "gamma" (an arbitrary parameter). It is not to be mistaken for the current ripple ratio $r$. Also, quick design curves are provided in Fig. 6.15.

**Example 6.5** We have a flyback switching at 100 kHz designed for a 10 W maximum load over the range 90 Vac to 275 Vac. If the $V_{OR}$ is 140 V, and we set $r = 0.9$ at minimum input, at what input voltage will it turn discontinuous? What is the minimum duty cycle we need to handle if the load is reduced to 1 W? Consider another design choice of $V_{OR} = 100$ V, implemented with a suitably designed transformer, everything else remaining the same. What is the minimum duty cycle now?

Since $V_{OR} = 140$ V, and $V_{OR} \times I_{OR} = 10$ W, we get $I_{OR} = 10/140 = 0.071$ A. At 90 Vac (ignoring diode and switch drops)

$$D = \frac{V_{OR}}{V_{OR} + V_{IN}} = \frac{140}{140 + 127} = 0.524$$

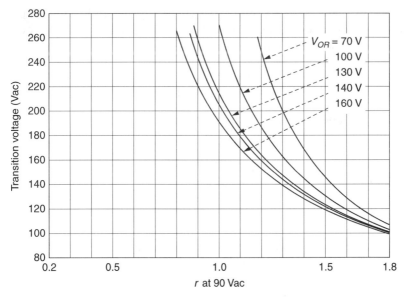

**Figure 6.15** Voltage at which a universal input flyback will enter DCM.

$L$ for a buck-boost (see Table 3.1 in Chap. 3) is

$$L = \frac{V_O + V_D}{I_O \bullet r \bullet f} \bullet (1 - D)^2 \bullet 10^6 \ \mu\text{H}$$

So since we can set $V_O$ to $V_{OR}$ and $I_O$ to $I_{OR}$

$$L = \frac{140}{0.071 \times 0.9 \times 10^5}(1 - 0.524)^2 \times 10^6 = 4964 \ \mu\text{H}$$

We calculate $\gamma$

$$\gamma = \frac{V_{OR}^2}{2P_O f L} \times 10^6 - 1 = \frac{140^2}{2 \times 10 \times 10^5 \times 4964} \times 10^6 - 1 = 0.974$$

Note that if $\gamma$ was negative, no solution would exist (i.e., the system could never go discontinuous by simply increasing the input voltage, we would need to also decrease the load current). So, in our case

$$V_{IN\_CRIT} = V_{OR} \times \left[ \sqrt{\frac{1}{\gamma^2} + \frac{1}{\gamma}} + \frac{1}{\gamma} \right]$$

$$= 140 \times \left[ \sqrt{\frac{1}{0.974^2} + \frac{1}{0.974}} + \frac{1}{0.974} \right] = 346 \text{ V}$$

This is equivalent to $346/(2)^{1/2} = 245$ Vac.

So, we will first go at maximum load from $V_{IN\_MIN}$ to $V_{IN\_CRIT}$. Then we will reduce the load and increase the voltage further. The duty cycle at the critical boundary (at maximum load) is

$$D_{CRIT} = \frac{V_{OR}}{V_{OR} + V_{IN\_CRIT}} = \frac{140}{140 + 346} = 0.29$$

Now, if we decrease the load and increase the input voltage we can see from

$$D = \frac{\sqrt{2P_O L f}}{V_{IN} \times 10^3}$$

that $D$ will vary as

$$\frac{D_{MIN}}{D_{CRIT}} = \frac{\sqrt{P_{O\_MIN}}}{\sqrt{P_{O\_CRIT}}} \times \frac{V_{IN\_CRIT}}{V_{IN\_MAX}}$$

So, finally,

$$D_{MIN} = \frac{\sqrt{1}}{\sqrt{10}} \times \frac{346}{389} \times 0.29 = 0.082$$

$D_{min}$ must therefore be less than 8.2 percent.

Now let us change $V_{OR}$ to 100 V. This time $I_{OR} = 10/100 = 0.1$ A. We also know that $D = 0.44$ at 90 Vac with $V_{OR} = 100$ V. $L$ is thus recalculated for the required $r$ and we get $L = 3484$ $\mu$H. Then

$$\gamma = \frac{V_{OR}^2}{2P_O f L} \times 10^6 - 1 = \frac{100^2}{2 \times 10 \times 10^5 \times 3484} \times 10^6 - 1 = 0.435$$

So,

$$V_{IN\_CRIT} = V_{OR} \times \left[ \sqrt{\frac{1}{\gamma^2} + \frac{1}{\gamma}} + \frac{1}{\gamma} \right]$$

$$= 100 \times \left[ \sqrt{\frac{1}{0.435^2} + \frac{1}{0.435}} + \frac{1}{0.435} \right] = 505 \text{ V}$$

This is equivalent to 357 Vac, well outside our operating range. *Therefore at 275 Vac, with maximum load, we are still running continuously.* The duty cycle at this point is

$$D_{VIN\_MAX} = \frac{100}{100 + 389} = 0.2$$

The current ripple ratio at this point is therefore

$$r_{VIN\_MAX} = \frac{V_{OR}}{I_{OR}Lf} \times (1 - D_{VIN\_MAX})^2 \times 10^6$$

$$= \frac{100}{0.1 \times 3484 \times 10^5} \times (1 - 0.2)^2 \times 10^6 = 1.84$$

Clearly, if now we decrease the load, we will go critical at

$$I_{CRIT} = \frac{r}{2} \times I_{OR} = \frac{1.84}{2} \times 0.1 = 0.092 \text{ A}$$

This is equivalent to an output power of $V_{OR} \times I_{OR} = 100 \times 0.092 = 9.2$ W. As we decrease the load further, we are in discontinuous mode. Now, since the input voltage does not change, the duty cycle will shrink going from 9.2 W to 1 W as per

$$\frac{D_{MIN}}{D_{CRIT}} = \frac{\sqrt{P_{O\_MIN}}}{\sqrt{P_{O\_CRIT}}} = \frac{\sqrt{1}}{\sqrt{9.2}}$$

So, finally,

$$D_{MIN} = 0.2 \times \frac{1}{3.03} = 0.066$$

The $D_{min}$ requirement is thus 6.6 percent in this case. At 100 kHz switching frequency this is equivalent to a minimum pulse width requirement of 660 ns. With a minimum load of 0.2 W we would have required less than 300 ns.

We can see that a high $V_{OR}$ does help somewhat in avoiding pulse-skipping at very light loads. Finally, we can also conversely calculate that given a minimum pulse width of the controller, what is the minimum load we need to ensure so as to avoid the random pulse-skipping mode.

### 6.2.13 Overload protection

One of the "tricks" we do with the popular 3842/3844 series is to place a resistor from $V_{IN}$ to the $I_{SENSE}$ pin. The purpose of this is to reduce the current limit and thus indirectly limit the maximum available duty cycle at high-line. At high-line the steady operating duty cycle is naturally lower, and so is the peak current. But under overloads, the converter will try to hit any available brickwall, looking for whichever "stop" comes up first. So, it will try to reach maximum duty cycle or hit the current limit, all this happening till the slower secondary-side current limit can start to work and limit the output power. But even during this interval, damage can occur. So we should limit the energy drawn from the mains by lowering the current limit at high-line and/or clamping the duty cycle.

Clamping the maximum duty cycle progressively to smaller and smaller values as the input increases is called "input feedforward." However we note that this term has a different connotation in dc-dc converters where the ramp of the pulse width modulator stage is made progressively steeper with increasing input voltage. The difference is that the latter type of input feedforward does not really change $D_{max}$. It just helps quickly reduce the operating $D$ in response to a sudden increase in input voltage. It thus affords no overload protection, only better line transient response for voltage-mode control ICs (bringing them virtually on par with current-mode control ICs). An alternative to reducing $D_{max}$ is to lower the current limit progressively from low-line to high-line, as we do for the 3842/3844 series.

Basically, by introducing a *feedforward* resistor $R_{FF}$, the current sense signal is dc shifted a little higher by an amount $R_{FF} \times I_{FF}$; thus it will hit the current limit a little quicker. Note that we do *not* want to affect the current limit down to very low input voltages because we still need holdup time. If $R_{BL}$ is the resistor normally connected between the sense resistor and the sense pin of the IC—typically 1 kΩ or so—the current limit at high-line $V_{IN\_MAX}$ as a ratio of the current limit at $V_{IN\_LO}$ (e.g., 60 Vac, i.e., 85 Vdc) is

$$\frac{CLIM_{VIN\_MAX}}{CLIM_{VIN\_LO}} = \frac{V_{CLIM} - \left(V_{IN\_MAX}/R_{FF} \times R_{BL}\right)}{V_{CLIM} - \left(V_{IN\_LO}/R_{FF} \times R_{BL}\right)}$$

where $V_{CLIM}$ is the voltage on the current sense pin at the current limit condition. So if, for example, $V_{IN\_MAX} = 389$ Vdc, $V_{IN\_LO} = 85$ Vdc, $R_{BL} = 1$ kΩ, and $V_{CLIM} = 1$ V, we get the required value of the feedforward resistor for reducing the current limit threshold from 1 V to 0.75 V from

$$0.75 = \frac{1 - \left(389/R_{FF} \times 1000\right)}{1 - \left(85/R_{FF} \times 1000\right)}$$

Solving, we get

$$R_{FF} = 1.3 \text{ M}\Omega$$

At 60 Vac this will also lower the current limit threshold slightly below 1 V. The current $I_{FF}$ is 85/1.3 M = 65 $\mu$A. Passing through the blanking resistor 1 kΩ it causes a drop of 0.065 V. So the current limit threshold is now $1 - 0.065 = 0.935$ V. Note that the current limit threshold actually has a typical tolerance of ±10 percent. So, we must assume that the worst-case threshold is actually $0.9 - 0.065 = 0.835$ V. Knowing this we can correctly set $R_{SENSE}$.

There are some additional subtleties that we should recognize here. For one, why should we be very careful when operating current-mode control ICs at very high switching frequencies? This situation cannot be understood in terms of any steady state scenario. Suppose the output voltage is forced to zero (perfect short). Then the down-slope of the primary-side inductor current $V_{OR}/L_P$ (or equivalently the down-slope of the secondary-side current $V_O/L_S$) is almost zero. Actually the diode drop helps us here since the latter term is actually $(V_O + V_D)/L_S$. Therefore the current can never quite reach the level it started the cycle with, and therefore a "steady state," by definition, cannot exist. So we will ultimately get a "flux-walking" (or "current staircasing") condition till we hit the current limit. But now, though the peak primary-side peak current is supposedly fast and limited precisely by the sense resistor, and should therefore protect the switch and the transformer from saturation, this does not always happen in practice. We know that the blanking time requirement of all current-mode control ICs translates to a minimum pulse width. It can also be shown that in this minimum duty cycle condition, *any current limit gets effectively bypassed.* Current limit, whenever reached, can only respond by commanding the controller to limit the duty cycle further. But the duty cycle being already as small as possible, what else can the current limit or control circuitry do? Now even during this minimum pulse width (of 150 ns or so), the slope of the up-ramp $V_{IN}/L$ is very high, and there is also virtually no down-ramp anymore because the output has dropped. So the current will actually continue to increase a little bit every cycle, possibly "staircasing" well beyond the supposed current limit. That means we actually have no current limit anymore!

Many modern current-mode dc-dc controllers/switchers respond by initiating a *frequency foldback* whenever the voltage on the feedback pin falls below a certain threshold. In doing so they effectively reduce the duty cycle under current limit condition, and this extends the off-time by a typical factor of 4 to 6. This gives enough time for the current to ramp down to a value less than what it started the cycle with, thereby quashing "staircasing." But note that the effectiveness of this technique depends largely on the diode drop! So, "good" diodes (i.e., with lower forward drops) actually make the fault currents even more severe, as they do not provide enough down-slope during the off-time.

We see that at high switching frequencies, a certain blanking time corresponds to a higher minimum duty cycle. This can also severely affect the range of output voltages possible. For example, in a buck converter where we want to step-down from a very high voltage to a very low output voltage, we need to be able to reduce the normal operating duty cycle sufficiently. And for current-mode control this is more likely to exceed the capabilities of the device since blanking time cannot just be wished away!

We may like to keep all these factors in mind the next time we have to decide whether to go in for voltage-mode control or for current-mode control. We need to look at their impact on each topology very carefully.

# Concepts in Magnetics

Magnetics is the holy grail of power conversion. It seems very hard to understand though at the same time we realize it can be equally useful to know really well. Actually, there are abundant misconceptions and myths prevailing, some quite basic and fundamental. In this chapter we present the basic concepts in a rather unconventional but straightforward way. Hopefully, this takes away some of the mystique behind magnetics. Then we will get to the practical design aspects in the following chapters.

Unless otherwise stated, the reader can safely assume we are using MKS (i.e., SI) units. For simplicity, we will also ignore signs unless really unavoidable. So generally, we are talking only in terms of *magnitudes*.

## 7.1  Basic Magnetic Concepts and Definitions (MKS Units)

1. *H-field*. Also called *field strength*, *field intensity*, *magnetizing force*, *applied field*, and the like. Units are ampere/meter.

2. *B-field*. Also called *flux density* or *magnetic induction*. Units of $B$ are tesla (T), or webers per square meter ($Wb/m^2$).

3. *Flux* is the integral over surface area $\phi = \int_S BdS$ (webers). If $B$ is constant over the surface we get the more common form $\phi = BA$. The integral of $B$ over a closed surface is zero since flux lines do not start or end at any given point, but are continuous.

4. $B$ is related to $H$ *at any given point* by the equation $B = \mu H$, where $\mu$ is the permeability of the material. Note that $\mu$ is the absolute (not relative) permeability in this equation.

5. In air $\mu$ is denoted by $\mu_0$, which is the permeability of free space. $\mu_0 = 4\pi \times 10^{-7}$ *henry/meter (in MKS units)*. In CGS units it equals 1.

6. Faraday's law of induction (also called Lenz's law) relates the induced voltage V that is developed across the ends of a coil (of say N turns), due to a (time varying) B-field passing through it. So

$$V = N\frac{\mathrm{d}\phi}{\mathrm{d}t} = NA\frac{\mathrm{d}B}{\mathrm{d}t}$$

7. Therefore, a uniform change of one weber per second in the flux linking a single-turn winding will generate one volt across it. Note that we are ignoring signs here. We can just keep in mind that the induced voltage is always in such a direction that it opposes its cause.

8. The inertia of a coil to a change in flux through it due to its own time varying current is its inductance L, defined as

$$L = \frac{N\phi}{I} \quad \text{(henries)}$$

Thus L is the number of *flux linkages* $N\phi$ divided by the current causing the (time varying) flux $\phi$. This expression is deceptive. L is actually $\propto N^2$ since $\phi \propto N$.

9. The proportionality constant is called the *inductance index* and is denoted by $A_L$. This is usually expressed as nH/turns² (though sometimes it is considered to be mH/1000 turns², both being numerically the same). So

$$L = A_L \times N^2 \times 10^{-9} \quad \text{(henries)}$$

10. When H is integrated over a *closed loop*, we get the current enclosed by the loop

$$\int_{CL} H\mathrm{d}l = I \text{ (amperes)}$$

where *CL* stands for *closed loop*. This is also called *Ampere's circuital law*.

## 7.2   The Inductor Equation

Combining Faraday's law with the definition of L, we get the most commonly used equation of power conversion (the *inductor equation*)

$$V = L\frac{\mathrm{d}I}{\mathrm{d}t}$$

Comparing this with the *law of induction*, as discussed earlier, and combining all the equations we get the following basic reference set of design equations

$$V = N\frac{d\phi}{dt} = NA\frac{dB}{dt} = L\frac{dI}{dt}$$

## 7.3 The Voltage-Independent Equation

Eliminating $V$ from Faraday's law and the inductor equation we get a key equation used in transformer/inductor design

$$NA\frac{dB}{dt} = L\frac{dI}{dt}$$

or

$$\Delta B = \frac{L\Delta I}{NA}$$

It so happens that for soft ferrites and most other magnetic materials used in power conversion there is almost no field remaining in the core if the current in the windings goes to zero (the *remanence* is almost zero). Therefore *since B is zero when I is zero* we can also write it as

$$B_{PK} = \frac{LI_{PK}}{NA}$$

From Fig. 7.1 we can see that in fact this holds for any *instantaneous* values of $B$ and $I$. So generically,

$$B = \frac{LI}{NA}$$

This is referred to as the *voltage-independent equation* by us. Note that the inductance can be considered to be virtually a constant provided we are not close to saturation. Then very simply

$$B \propto I$$

We see that the current ripple ratio $r$ therefore applies equally to current as to the field. If for example $r$ is 0.4, then $B_{AC} \cong 0.2 \times B_{DC}$. So, the current ripple ratio can be considered to be a *field ripple ratio* too.

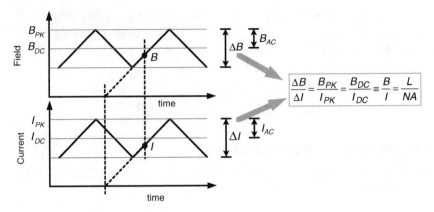

**Figure 7.1**  How $B$ and $I$ are related if remanence is zero.

Their equations are similarly related. For example, we can write

$$B_{DC} = \frac{2 \times B_{PK}}{r+2} \quad \text{or} \quad I_{DC} = \frac{2 \times I_{PK}}{r+2}$$

$$B_{AC} = \frac{r \times B_{PK}}{r+2} \quad \text{or} \quad I_{AC} = \frac{r \times I_{PK}}{r+2}$$

**Example 7.1**  The voltage-independent equation is useful if, for example, we want to do a quick check on saturation of a core. If we have wound 40 turns on a core with a datasheet specified value of $A = 2$ cm$^2$ (effective area, also called $A_e$), its measured inductance being 200 $\mu$H, and the peak current (measured or estimated) being 10 A, then the peak flux density is

$$B_{PK} = \frac{200 \times 10^{-6} \times 10}{40 \times 2 \times 10^{-4}} = 0.25$$

For a ferrite, this is quite close to its saturation flux density, which is typically around 0.3 T. Therefore, peak currents greater than 10 A should not be passed through this inductor. *Nor should the turns be increased any further.* This latter observation actually seems contradictory to the observation that in $B = LI/NA$, $N$ is in the denominator, which seems to indicate that increasing $N$ should help in reducing the peak field. However, that is not so. The reason is that $L$ in the numerator increases according to $N^2$, and so the field is effectively directly *proportional* (not inversely so) to the number of turns (and also to the current passing through it). We must keep in mind that this implies that for a given core there is a *maximum ampere-turns it can handle before it saturates* and becomes ineffective.

## 7.4   The Voltage-Dependent Equation

There are so many forms of the *voltage-dependent equation* (as we are calling it), that this is the point where the average designer probably loses all interest in magnetics. Let us therefore follow this through.

The basic design equation set is

$$V = N\frac{d\phi}{dt} = NA\frac{dB}{dt} = L\frac{dI}{dt}$$

or

$$\boxed{\Delta B = \frac{V_{AVG}\Delta t}{NA}}$$

where in the latter form we have, generically speaking, taken the *average value* of the applied voltage $V_{AVG}$ just in case it is varying over the evaluation interval. In square wave (nonresonant) switching converters it is a constant and is either $V_{ON}$ or $V_{OFF}$, as applicable. $V_{ON}$ is the voltage applied across the inductor during the switch on-time and $V_{OFF}$ the voltage during the switch off-time. Since $B_{AC} = 2\Delta B$ we get the first form of the voltage-dependent equation

$$\boxed{B_{AC} = \frac{V_{AVG}\Delta t}{2 \times NA}}$$

For either continuous conduction mode (CCM) or discontinuous conduction mode (DCM)

$$\boxed{B_{AC} = \frac{V_{ON}D}{2 \times NAf}}$$

And for CCM (only), $t_{OFF} = (1 - D)/f$, so

$$\boxed{B_{AC} = \frac{V_{OFF}(1 - D)}{2 \times NAf}}$$

where $f$ is the switching frequency in Hz.

If, and only if, a *square wave voltage* is applied to the core (defined as implying $t_{ON} = t_{OFF} = T/2$), then

$$\Delta t = \frac{1}{2f}$$

and so

$$B_{AC} = \frac{B_{PP}}{2} = \frac{V_{AVG}}{4 \times NAf} \quad \text{(square voltage waveform)}$$

Manufacturers of magnetic components often express the voltage-dependent equation in terms of *the applied rms voltage*. By the volt-seconds law, for any topology in continuous conduction mode condition, the following will hold true

$$V_{OFF} = V_{ON}\frac{D}{1-D} \quad \text{(any topology)}$$

By the basic definition of rms of a waveform, we can show that the rms of this applied voltage waveform (calculated over the entire cycle) is

$$V_{RMS} = V_{ON}\sqrt{\frac{D}{1-D}} \quad \text{(CCM)}$$

or, equivalently,

$$V_{RMS} = V_{OFF}\sqrt{\frac{1-D}{D}} \quad \text{(CCM)}$$

*Notice that if, and only if,* D = 0.5 *(square wave), we get* $V_{RMS} = V_{ON} = V_{OFF}$. That's how we get one of the several *common forms* of the voltage-dependent equation

$$B_{AC} = \frac{V_{RMS}}{4 \times NAf} \quad \text{(square wave voltage, CCM)}$$

But let us also be clear about what the shape of the corresponding current waveform is under the square wave voltage case. In reality, *the voltage-dependent equation says nothing about the absolute value of the current (or field)*. It only defines the *change* when a certain voltage is applied for a certain time. The actual value of the current and field in the inductor depends on the external application conditions and the actual details of the schematic. The dc level of the current waveform could have been almost anything for the same $\Delta I$. A zero dc level is just one of the many possibilities, and to achieve it requires that we deliberately create the appropriate external conditions, schematically and electrically (e.g., capacitive coupling), for this to happen. Vendors

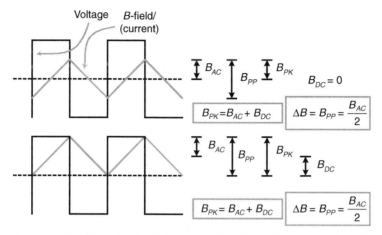

**Figure 7.2**  How changing the dc level can affect the peak current and field.

of magnetic materials often test the core under a symmetric excitation (square or sine wave) but *also* implicitly (without stating this clearly) *create external conditions to keep the current symmetric* around zero (i.e., bidirectional and with $B_{DC} = 0$). This is not necessarily what happens in most switching topologies. Note that most topologies are *unipolar* or *unidirectional*. So current can flow through the coil in only one direction. This forms a special case of a nonsymmetric excitation, one in which the current waveform is not just offset ($I_{DC} \neq 0$), but restricted entirely to the upper (or lower) half of the graph. All this can change the relationships between peak current (or field) and the peak-to-peak values, and the like. See Fig. 7.2 for an example of what can change, and what doesn't.

So only for true symmetric excitation can we assume that

$$B_{AC} = B_{PK} \text{ (symmetric excitation)}$$

Then we get another common form of the voltage-dependent equation

$$\boxed{B_{PK} = \frac{V_{RMS}}{4 \times NAf}} \text{ (square wave voltage, symmetric excitation)}$$

Note that in all cases the equation $B_{PP} = 2 \times B_{AC}$ holds true. Thus we can write another form of our voltage-dependent equation

$$\boxed{B_{PP} = \frac{V_{RMS}}{2 \times NAf}} \text{ (square wave voltage, symmetric or nonsymmetric)}$$

Magnetics vendors may provide data for a core under sine wave excitation. Here the current will also be a sine wave, and so will be the corresponding $B$-field. We assume that the on-time here is the first half-cycle (0 to $\pi$ radians), followed by an off-time (from $\pi$ to $2\pi$ radians). The voltage level is not "flat" in this case, and we now need to calculate the average voltage during the first half-cycle. The basic form of the voltage-dependent equation is

$$\Delta B = \frac{V_{AVG}\, \Delta t}{NA}$$

We can write the average voltage as

$$V_{AVG} = \frac{V_{AVG}}{V_{PK}} \times \frac{V_{PK}}{V_{RMS}} \times V_{RMS}$$

Realizing that $V_{AVG}$ calculated over each half-cycle will be the same as the well-known forms applicable to a standard rectified ac sine wave, we get

$$\frac{V_{AVG}}{V_{PK}} = \frac{2}{\pi}$$

and

$$\frac{V_{RMS}}{V_{PK}} = \frac{1}{\sqrt{2}}$$

So

$$V_{AVG} = \frac{2}{\pi} \times \frac{\sqrt{2}}{1} \times V_{RMS} = 0.9 \times V_{RMS}$$

The voltage-dependent equation for a symmetrical sine wave excitation becomes

$$\Delta B = \frac{V_{AVG}\, \Delta t}{NA} = \frac{0.9 \times V_{RMS}}{2 \times NAf}$$

So the additional forms of the voltage-dependent equation are

$$\boxed{\begin{aligned} B_{AC} = B_{PK} &= \frac{V_{RMS}}{\pi\sqrt{2} \times NAf} \\ &= \frac{0.9 \times V_{RMS}}{4 \times NAf} \\ &= \frac{V_{RMS}}{4.4428 \times NAf} \end{aligned}}$$   (sine wave voltage, symmetric excitation)

Also, using $V_{PK} = V_{RMS} \times (2)^{1/2}$

$$B_{AC} = B_{PK} = \frac{V_{PK}}{2\pi \times NAf}$$

or equivalently, since $B_{PP} = 2 \times B_{AC}$,

$$B_{PP} = \frac{V_{RMS}}{2.222 \times NAf}$$

For a sine wave excitation, sometimes the rms of the $B$-field may be used in the above equations. Since the $B$-field is a sine wave (like the current), we can use

$$B_{PK} = \frac{B_{PK}}{B_{RMS}} \times B_{RMS} = \sqrt{2} \times B_{RMS}$$

*The bottom-line is that in switching converters we must be very cautious in using the commonly available forms of the rather loosely called* Faraday's Law (*i.e., the voltage-dependent equation*) *as seen in magnetic material datasheets.* We must go back to the basic initial equation if necessary, to avoid confusion. See Table 7.1 for a summary of the equations discussed so far.

## 7.5 Units in Magnetics

Here lies another source of massive confusion. There are several systems of units in circulation. The one we have used so far is the modern, international (or *rationalized*) system based on meters, kilograms, seconds, and amperes (called the MKS or MKSA system, or the SI system, or the Georgi system). The older, but still commonly used one is the CGS system (for centimeters, grams, seconds). But there are many other engineers (mainly in the United States) who are still using inches instead of meters or centimeters. And there are others who even mix different systems into one equation. For example, the voltage-dependent equation we have derived is actually

$$B_{AC\_Teslas} = \frac{V_{AVG}}{4 \times N \times A_{sq.meter} \times f_{Hz}}$$

where $V$ is in Volts, $A$ is in m², $f$ is in Hz, and $B$ is in teslas. But if $B$ is in teslas (MKS) and we want to use cm² (CGS) instead of m², we get

$$B_{AC\_Teslas} = \frac{V_{AVG} \times 10^4}{4 \times N \times A_{sq.centimeter} \times f_{Hz}}$$

**TABLE 7.1  Basic Design Table for Magnetics (MKS units).**

**General design equations**

$$V = N\frac{d\phi}{dt} = NA\frac{dB}{dt} = L\frac{dI}{dt}$$

**Voltage-independent equation**

$$B = \frac{LI}{NA}$$

(connects instantaneous values, provided remanence is zero)

**Voltage-dependent equation**

| | |
|---|---|
| General forms | $\Delta B = 2B_{AC} = \dfrac{V_{ON}t_{ON}}{NA} = \dfrac{V_{OFF}t_{OFF}}{NA}$ |
| | $B_{AC} = \dfrac{V_{ON}D}{2 \times NAf}$ |
| | $B_{AC} = \dfrac{V_{OFF}(1-D)}{2 \times NAf}$ <br> (CCM only) |
| | $B_{AC} = \dfrac{\Delta B}{2} = \dfrac{B_{PP}}{2} = \dfrac{V}{4 \times NAf}$ <br> where $V \equiv V_{ON} = V_{OFF}$ <br> (square wave only) |
| Square wave | $B_{AC} = \dfrac{B_{PP}}{2} = \dfrac{V_{RMS}}{4 \times NAf}$ |
| Sine wave, excitation symmetric | $B_{PK} = B_{AC} = \dfrac{B_{PP}}{2} = \dfrac{V_{RMS}}{4.4428 \times NAf}$ |

We can also convert the $B$-field to the CGS system by using the conversions in Table 7.2.

So, finally, we get one of the commonly seen forms of the voltage-dependent equation

$$B_{AC\_Gauss} = \frac{V_{AVG} \times 10^8}{4 \times NA_{\text{sq.centimeter}}\, f_{\text{Hz}}}$$

For symmetric excitation we saw that $B_{AC} = B_{PK}$. So we could write

$$B_{PK\_Gauss} = \frac{V_{AVG} \times 10^8}{4 \times NA_{\text{sq.centimeter}}\, f_{\text{Hz}}}$$

TABLE 7.2  **Magnetic Units: Conversions.**

| Property | CGS units | MKS units | Conversion |
|---|---|---|---|
| Magnetic flux | Line (or Maxwell) | Weber | 1 Wb $= 10^6$ lines |
| Flux density (B) | Gauss | Tesla (or Wb/m$^2$) | 1 T $= 10^4$ G |
| Magnetomotive force | Gilbert | Ampere-turn | 1 Gb $= 0.796$ ampere-turn |
| Magnetizing force field (H) | Oersted | Ampere-turn/ meter | 1 Oe $= 1000/4\pi = 79.577$ A/m |
| Permeability | Gauss/Oersted | Weber/m-ampere-turn | $\mu_{MKS} = \mu_{CGS} \times (4\pi \times 10^{-7})$ |

*In general, when in doubt, we should stick to a system of units* (prefer-ably MKS), and *only at the very end should we use conversions* as those provided in Table 7.2 to convert the final results if desired (for example from tesla to gauss).

## 7.6  The Magnetomotive Force (mmf) Equation

In Fig. 7.3 we have an exploded view of a toroidal core with an air gap. The flux lines are also shown. The coil that creates this flux has $N$ turns and carries a current $I$. Integrating over the path indicated, and applying Ampere's law, we get

$$H_c l_c + H_g l_g = NI$$

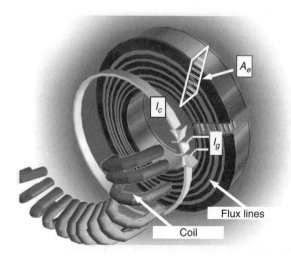

**Figure 7.3**  Exploded view of toroid with gap.

where the subscript $c$ stands for the core and $g$ refers to the (air) gap. Note that from Ampere's law we get the total current *enclosed* (by the path we integrated over), which is $NI$ here.

Writing $H = B/\mu$ and $B = \phi/A$ we get $H = \phi/A\mu$, so

$$\frac{\phi_c l_m}{A_c \mu_c} + \frac{\phi_g l_g}{A_g \mu_g} = NI$$

Note that usually (for small gaps) $A_c$ is almost the same as $A_g$. But sometimes it may be quite different. We will take that up in the section on *fringing flux* later in this chapter. However, flux lines are continuous and assuming we have adjusted the area, if necessary, so that we don't exclude any flux lines as we move from core to gap to core, then $\phi_c = \phi_g$, so

$$\boxed{NI = \phi \times \left( \frac{l_c}{\mu_c A_c} + \frac{l_g}{\mu_g A_g} \right)}$$

In general, this is expressed as the *magnetomotive force equation*

$$\boxed{\phi = \frac{mmf}{\Re}}$$

where mmf is the magnetomotive force, equal to NI, and $\Re$ is the *reluctance* given by

$$\boxed{\Re = \frac{l_c}{\mu_c A_c} + \frac{l_g}{\mu_g A_g} = \Re_c + \Re_g}$$

We observe the analogy with $V = IR$ (Ohm's law) where $V$, the voltage is also called the *electromotive force* (emf). Here we have $(NI) = \phi\Re$. So, in a *magnetic circuit* flux plays the same role as current does in an electrical circuit.

In our case, we have two materials in series and so the total reluctance is the *series sum* of two separate reluctances, one for each material. Reluctance behaves in a completely analogous manner to electrical resistance which is

$$\text{Resistance} = \frac{1}{\text{Conductance}} = \frac{l}{\sigma A}$$

where $\sigma$ is the electrical conductivity (inverse of resistivity) and is a property of the material itself. The complete analogy is thus

$$V = I \times R \Longleftrightarrow mmf = \phi \times \Re$$

or

$$(NI) = \phi \times \Re$$

We interpret this as saying that the *magnetic voltage* (the magneto-motive force, *NI*) is responsible for producing a *magnetic current* (the flux $\phi$) whose magnitude depends on the *magnetic resistance* (the reluctance $\Re$). Note that all the current flowing into an electrical node must leave, just as flux also behaves.

## 7.7 Effective Area and Effective Length

To a very close approximation, effective area (the average area available for the flux lines) is simply the geometric cross-sectional area of the core. It is called the effective area $A_e$.

Similarly the effective length ($l_e$) of a toroidal core (without a gap) is almost exactly equal to its geometrical circumference. Since the core has an OD (outer diameter) and an ID (inner diameter) we need to take the arithmetic average of the two. So, using the equation for the circumference of a circle

$$l_e = \pi \times \text{Diameter} = \pi \times \left( \frac{OD + ID}{2} \right)$$

*Better agreement with the more accurate equation can be obtained if the 2 in the denominator is 2.1 instead.* We can compare the simple estimates with the more accurate forms, and we can see that unless the OD is more than twice the ID (a rather unusual core if such a one exists), the agreement is very good (see Fig. 7.4). So we can usually just stick to the quick (geometrical) estimate.

When we come to E-cores, we must realize that the flux in the center limb splits equally into the outer limbs. In that case, $A_e$ is (very close) to the geometrical cross-section of the center limb. *Most E-cores are designed so that the sum of the cross-sectional areas of the two outer limbs equals the area of the center limb.* In rare cases (nonstandard cores) we will find that for some strange reason the sum of the outer limbs is less than the center limb. To account for these additional possibilities, in general, we may use for $A_e$ either the area of the center limb, or the sum of the areas of the outer limbs, *whichever is less*. However, this is not really an exact rule, just an approximation.

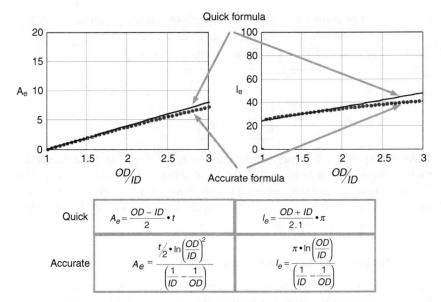

**Figure 7.4**  Effective area and effective length equations for a toroid of thickness $t$.

Note that in all cases the effective volume of the core $V_e$ is

$$V_e = A_e \times l_e$$

**Note:**  When in doubt we can just turn to the published values for $A_e$ and $l_e$ but we should be cautious about the units. Some manufacturers give these values in $mm^2$ and mm (almost as a default sometimes, without even caring to write the units explicitly), and some may give these in $m^2$ and m. Either way, if we are sticking to MKS units, we must convert these to $m^2$ and m as and when required.

## 7.8   The Effect of the Air Gap

Returning to

$$NI = \phi \times \left( \frac{l_c}{\mu_c A_c} + \frac{l_g}{\mu_g A_g} \right)$$

we see that we can write $l_c = l_e$ (the effective length of the core before gapping). Also, $\mu_g$ which is basically the permeability of air, is called $\mu_o$. In MKS units its value is $\mu_o = 4\pi \times 10^{-7}$ (weber/ampere-meter *or* henry/meter *or* joule/Amp²-meter)

We also define the *relative permeability* of the material as

$$\mu = \frac{\mu_c}{\mu_g} \equiv \frac{\mu_c}{\mu_o}$$

**Note:** Do not confuse the $\mu$ here with the $\mu$ originally written in the equation $B = \mu H$. In that equation, $\mu$ was in a sense symbolic. It actually referred to the material's *absolute* permeability (not its relative permeability). Now we know that the absolute permeability of the material can also be written as $\mu \times \mu_O$. In CGS units we should note that since $\mu_O = 1$, there is no difference between the numerical values of the absolute permeability of a material and its relative permeability. So, $B = \mu H$ is unambiguous in that case.

We also assume we have a uniform area $A_e = A_c$ for the entire structure (no fringing). So, we get

$$\phi = \frac{NIA_e\mu_c}{l_e + \mu l_g}$$

that is,

$$B = \frac{\mu_c NI}{l_e + \mu l_g}$$

or

$$\phi = \frac{NIA_e\mu_e}{l_e} \text{ where } \mu_e = \frac{\mu_c}{\left(1 + \mu l_g / l_e\right)}, \text{ that is, } B = \frac{\mu_e NI}{l_e} \text{ with gap}$$

Note that *had there been no air gap*, we would have got

$$\phi = \frac{NIA_e\mu_c}{l_e}, \text{ that is, } B = \frac{\mu_c NI}{l_e} \text{ without gap}$$

We, therefore, realize that we can look at the air gap in two ways:

1. We can think of the air gap as having decreased the field because of an increase in the magnetic path length—from $l_e$ to $l_e + \mu l_g$. So,

$$\phi = \frac{NIA_e\mu_c}{l_e} \Rightarrow \frac{NIA_e\mu_c}{l_e + \mu l_g}$$

This means that though the gap introduced was only $l_g$, it contributed $\mu$ times the geometric length into the effective magnetic path length. Note that the factor $\mu$ is the *relative* permeability of the material *surrounding the gap*!

2. We can alternatively think of the effect of the gap as having caused the field to decrease by changing the effective permeability from $\mu_m$ to $\mu_e$. So,

$$\phi = \frac{NIA_e\mu_m}{l_e} \Rightarrow \frac{NIA_e\mu_e}{l_e}$$

Note that the two perspectives described above are equivalent but *alternative* ways of looking at the effects of the gap. *Therefore only one or the other should be utilized, not both.* Further, this mathematical trick should be used only for finding $B$, because $B$ is the same in the core as in the gap. Thereafter, $H$ should be derived, using $H = B/$permeability.

## 7.9   The Gap Factor z

This factor is being introduced in this book for the first time as the author observed that it can serve as a useful mathematical and intuitive tool for simplifying the rather complex looking magnetic equations that throw most designers over the edge. For example, we defined

$$\mu_e = \frac{\mu_c}{\left(1 + \mu l_g / l_e\right)}$$

This is simply

$$\mu_e = \frac{\mu_c}{\left(\frac{l_e + \mu l_g}{l_e}\right)}$$

So,

$$\boxed{\mu_e = \frac{\mu_c}{z}}$$

Thus, permeability fell by $1/z$, where the $z$ factor is the ratio

$$\boxed{z = \frac{l_e + \mu l_g}{l_e}}$$

It is the ratio of the new path length $l_g + l_e$ to the old (ungapped) length $l_e$. It is always equal to or greater than 1. *A value of z = 1 means an ungapped core.*

Note that as mentioned, we can use the $z$-factor only once—either to adjust the permeability or the length, but not both. Applying this rule to the equation for $B$ with no gap, we can easily get the equation with a gap as follows

$$B = \frac{\mu_c NI}{l_e} \Rightarrow \frac{\mu_c NI}{(z \times l_e)} \equiv \frac{\mu_c/z NI}{(l_e)} = \frac{\mu_e NI}{l_e}$$

**Note:**  For E-cores, if the center limb is ground to create an air gap, and so no gap exists on the outer limbs, the gap length is equal to the physical dimensions of the gap in the center limb. Alternatively, if spacers are used on the outer limbs (center limb not being ground), then the effective gap length is *twice* the gap set on each limb.

## 7.10   The Origin and Significance of z

Flux lines are continuous. Therefore the $B$-field everywhere in the core and gap is the same. But $B = \mu H$, and since $\mu$ changes from core to gap, the $H$-field is not the same. That is surprising to some who are trained to think of $H$ as a driving field and $B$ the response. *When a gapped core is present, it is more helpful to think in terms of B instead of H.* Let us calculate the $B$ and $H$ fields in the core and in the gap.

**B in the core and gap**

$$B_c = B_g = \frac{\mu_c NI}{(zl_e)} \equiv \frac{\mu_e NI}{l_e}$$

**H in the core**

$$H_c = \frac{B}{\mu_c} = \frac{NI}{(zl_e)}$$

**H in the gap**

$$H_g = \frac{B}{\mu_g} = \left(\frac{\mu_c}{\mu_g}\right) \times \frac{NI}{(zl_e)} = \frac{\mu NI}{(zl_e)}$$

We can see that

$$\boxed{H_g = \mu \times H_c}$$

Since ferrite, for example, can have a relative permeability between 1000 and 5000, *the H field in the air gap is that many times larger than the H field in the magnetic core!* We remember that the $B$ field is continuous on the other hand.

**Question:**  How does the $H$ field compare to the field that *would* have been present if the magnetic core was not present (i.e., core replaced by free air)? We assume that the coil and current remain the same.

Applying Ampere's law to an air-cored coil we would get

$$H = \frac{NI}{l_e}$$

So, we can conclude the following:

- The field in the core of the gapped magnetic core is significantly less than the air-core case by the factor $1/z$.
- The field in the gap of the gapped magnetic core is much greater than the air-core case by the factor $\mu/z$, where $\mu$ is the relative permeability of the core material.

As for the $B$-field, for the air-core case we get

$$B = \mu_o H = \frac{\mu_o NI}{l_e}$$

compared with the gapped core case

$$B_c = B_g = \frac{\mu_e NI}{l_e}$$

We can see that B has increased substantially in the gap and core as compared to the air-core coil.

See a comparison in Table 7.3 for a summary of all the cases discussed above.

**Caution:** $\mu_c$ is the *permeability* of the core. $\mu_o$ is the permeability of air. $\mu_o$ is numerically a small number ($4\pi \times 10^{-7}$) in MKS units (though

**TABLE 7.3  *B* and *H* Fields Inside the Core (and Gap).**

| MKS units | | $B$ | $H$ |
|---|---|---|---|
| Ungapped core<br>$\mu_e = \mu_c, z = 1$ | | $\dfrac{\mu_c NI}{l_e}$ | $\dfrac{NI}{l_e}$ |
| Gapped core<br><br>$\mu_e, z > 1$ | Core | $\dfrac{\mu_c NI}{zl_e} \equiv \dfrac{\mu\mu_0 NI}{zl_e} \equiv \dfrac{\mu_e NI}{l_e} \equiv \dfrac{\mu_e NI}{l_e}$ | $\dfrac{NI}{zl_e}$ |
| | Gap | | $\dfrac{\mu NI}{zl_e}$ |
| Air-cored coil<br>$\mu_e = \mu_0, z = 1$ | | $\dfrac{\mu_0 NI}{l_e}$ | $\dfrac{NI}{l_e}$ |

$z$ is $(l_e + \mu l_g)/l_e$
$\mu$ is relative permeability $= \mu_{core}/\mu_o$,
$\mu_e$ is effective permeability $= \mu_{core}/z$

it equals 1 in CGS units). The effective permeability $\mu_e$ is equal to $\mu_c/z$, and in MKS units it is typically a very small number too. We have also defined *relative permeability* $\mu$ as $\mu_c/\mu_o$. In either system of units the relative permeability is the (large) dimensionless number that we typically read from ferrite datasheets for example (2000 or 5000). We could also, in principle, define the following large number in either system of units: $\mu/z$, that is, $(\mu_c/\mu_o)/z$ which we can call the *effective relative permeability* (this would be dimensionless too). It would express the effective permeability (with gap) as compared to the permeability of air. We remind ourselves that numerically speaking, in CGS units there is no difference between relative permeability and permeability, but in MKS units there is a big difference.

## 7.11   Relating *B* to *H*

The equation $B =$ permeability $\times\, H$ applies to a *point* in space, not to the *entire* structure. Therefore no equation of the symbolic form $B = \mu H$ can actually be written out as somehow being an averaged relationship valid for the entire gapped core. That is because though $B$ is a constant as we go through the core and gap, $H$ changes in discrete steps, and so there is no *average H*. Yes, we can symbolically write

$$B = \mu_e H$$

but we must remember that this means either of the following

$$B_{gapped\_core} = \mu_e \times H_{air\_core} \quad \text{(OK)}$$

or

$$B_{gapped\_core} = \mu_e \times H_{ungapped\_core} \quad \text{(OK)}$$

but *not*

$$B_{gapped\_core} = \mu_e \times H_{gapped\_core} \quad \text{(Not OK)}$$

## 7.12   E-cores

E-cores are topologically equivalent to the toroid. The area of the center limb of the E-core becomes equivalent to the effective area of the toroid. The effective length measured through the center limb of the E-core passing through either side-limb becomes the effective length of the toroid. The total gap in the *center* limb of the E-core becomes the gap length of the toroid.

However, we must take note of the following

- If both halves of an $E$-$E$-core are equally ground to achieve a certain air gap length $l_g$, then each half must be ground to only $l_g/2$.
- If only one half of an $E$-$E$-core is ground to achieve a certain air gap length $l_g$, then it must be ground to $l_g$.
- If the center limbs of the $E$-$E$-cores are not going to be ground, and the method of achieving a certain gap length $l_g$ is by means of identical spacers on the outer limbs, then the spacer thickness must be set to $l_g/2$.

## 7.13   Energy Storage Considerations

The energy stored in the inductor is the same as the energy associated with its field. That is, as per the laws of electromagnetism

$$E = \frac{1}{2}BH \times Volume$$

Since, in general, we write $B = \mu H$, (where $\mu$ here is *not* the relative permeability, but the *absolute* permeability of the material), we get the following energy relationships for the gapped toroid.

For the core

$$E_c = \frac{1}{2}\frac{B^2}{\mu_c} \times A_e l_e \quad \text{(Joules)}$$

and for the gap

$$E_g = \frac{1}{2}\frac{B^2}{\mu_0} \times A_e l_g \quad \text{(Joules)}$$

So, the total stored energy is

$$E = \frac{1}{2}B^2 A_e \left[ \frac{l_e}{\mu_c} + \frac{l_g}{\mu_0} \right]$$

Therefore,

$$E = \frac{1}{2}B^2 A_e l_e \frac{z}{\mu_c}$$

or since by definition $V_e = A_e \times l_e$,

$$E = \frac{1}{2} B^2 V_e \frac{z}{\mu_c}$$

We also get several interesting ratio relationships

$$\frac{E_g}{E_c} = z - 1 = \frac{\mu l_g}{l_e}$$

We can see that the amount of energy in the gap can rival the energy in the core simply because though its dimensions are small, from the magnetics viewpoint, it gets multiplied by the *relative* permeability of the *surrounding* material. We can thus write

$$\frac{E_g}{E_c + E_g} \equiv \frac{E_g}{E} = \frac{\mu l_g}{l_e + \mu l_g}$$

where $E$ is the total energy stored in the structure (i.e., $E_c + E_g$). Then using $z = (l_e + \mu l_g)/l_e$ we get

$$\frac{E_g}{E} = \frac{\mu l_g}{l_e + \mu l_g} = 1 - \frac{1}{z}$$

This is the ratio of the energy stored in the gap to the total energy.

We can also eliminate $B$ from the earlier equation for $E$ by using the relevant equation from Table 7.3, that is,

$$B = \frac{\mu \mu_0 N I}{z l_e}$$

We get

$$E = \frac{\mu \mu_0 N^2 I^2 A_e}{2 l_e} \times \frac{1}{z}$$

Similarly, the energy stored in the core alone (excluding gap) is

$$E_c = \frac{\mu \mu_0 N^2 I^2 A_e}{2 l_e} \times \frac{1}{z^2}$$

and the energy in the gap is

$$E_g = \frac{\mu \mu_0 N^2 I^2 A_e}{2 l_e} \times \frac{1}{z} \left( 1 - \frac{1}{z} \right)$$

There are several ways of looking at the equations above and presenting the analysis that follows below. A lot depends on *what we are keeping constant as we vary the gap.* Since grave confusion can arise, this needs to be looked at quite closely.

### First study: ampere-turns constant, ($N = $ constant)

The current $I$ depends essentially on the application, i.e., the load and the input/output voltages. Since these conditions are not being changed here, $I$ too can be considered virtually constant.

Now if we simply kept the number of turns unchanged as we increase the gap, then $NI$ is also virtually fixed. So, from the equation of $B$ for a gapped core we can see that now $B$ will *fall in inverse proportion to the increase in z.*

From the energy equations above, we see that now the total stored energy $E$ will *decrease rather than increase!*

If we plot this out, we will see that *for a constant ampere-turns* (NI), the energy in the total structure *falls as we increase z.* The energy in the core actually falls off at even a faster rate, but that is partly compensated for by the *initial increase* in the energy of the gap. If we look at the *ratio* of the energy in the gap to the total energy, it reaches 50 percent by the time $z = 2$, continuing to rise thereafter. But from this point onward the absolute value of the energy in the gap starts decreasing, and the total energy drops off rapidly. Note that $z = 2$ corresponds to $l_g = l_e / \mu$.

### Second study: $B$ constant, ($N \propto z$)

We can easily see from the previous equations that the energy in the total gapped structure will increase if *a constant B is somehow maintained as we increase the gap.*

One way to keep $B$ constant is to increase $N$ in direct proportion to $z$ (assuming $I$ is held virtually constant, as under normal steady operation). So,

$$\boxed{N \propto z} \quad (B \text{ constant, } E \propto z)$$

## Third study: *L* constant, ($N \propto z^{1/2}$)

At this point we can read the next section first if we want to know how we derived the following equation for *L*.

$$L = \frac{1}{z}\left(\frac{\mu\mu_0 A_e}{l_e}\right)N^2$$

Assuming this equation is correct for now, we see that for a given coil (*N* fixed), *L* varies with air gap as $1/z$. But in an actual power supply we usually have a design target for *L*, based on the optimal suggestion of $r = 0.4$ (a current ripple ratio of ±20 percent). So, what we really want to do is to keep L *fixed for whatever value of z we decide* to set.

Since *L* varies as $N^2/z$, the only way we can keep *L* constant is to increase the number of turns according to

$$\boxed{N \propto \sqrt{z}} \quad (L = constant, E = constant)$$

With this change the energy terms vary as

$$E_c = \frac{\mu\mu_0 N^2 I^2 A_e}{2l_e} \times \frac{1}{z^2} \propto \frac{1}{z}$$

$$E_g = \frac{\mu\mu_0 N^2 I^2 A_e}{2l_e} \times \frac{1}{z}\left(1 - \frac{1}{z}\right) \propto \left(1 - \frac{1}{z}\right)$$

$$E = \frac{\mu\mu_0 N^2 I^2 A_e}{2l_e} \times \frac{1}{z} \Rightarrow constant$$

Therefore, the energy in the core drops but is exactly compensated for by an increase in the energy in the gap. Eventually, for very large gaps, almost the entire energy stored in the structure will reside in the gap. However, as we increase *z*, the total energy of the gapped structure remains constant.

If we look at the equation for *B*, we will see that it is proportional to $N/z$. So, in this study, *B* will vary as $1/z^{1/2}$. The operating *B* field has thus *decreased* and we have more "headroom" before we hit $B_{SAT}$ (which does *not* change with gap).

We see that we didn't decrease the *energy storage requirement* by our procedure here. But what we do change by introducing an air gap is the *energy storage capability* of the core. This distinction will be become clearer in the next section.

## 7.14  How an Air Gap Helps

We know that cores are limited to a certain saturation flux density $B_{SAT}$ beyond which the inductance starts to fall off quite rapidly. So, the maximum energy $E_{SAT}$ (in Joules) that can be stored in a gapped core is

$$E_{SAT} = \frac{1}{2} B_{SAT}^2 A_e l_e \frac{z}{\mu_c}$$

We can see that for a given $B_{SAT}$ *the higher the z, the higher is the total energy that can be stored in the structure.* We also see that decreasing the core material permeability, not increasing it, will increase $E_{SAT}$ (for a given $B_{SAT}$). So, why not just use an air-cored coil (lowest permeability)? That's simply because creating such a large field would require an unrealistically large number of turns (and lead to enormous copper losses too).

One intuitive reason why we add an air gap is that air never saturates. So, by "mixing" some air into the structure, the structure starts taking on some of the useful properties of air. The *BH* curve "flattens" out. Though *this does nothing to change the saturation flux density* $B_{SAT}$ *of the structure,* it does "soften" the tendency to enter saturation *abruptly*—in the sense that now a larger change in current (ampere-turns) is required to cause a certain change in *B*. This feature helps a lot because it gives enough time for most controllers and switches to react (by means of their current sense for example) and to be able to turn off fast enough to prevent switch destruction.

For gapped cores, the final $A_L$ value also starts to become virtually independent of the permeability of the material and will thereafter start depending almost completely on the air gap. Thus the design becomes relatively insensitive to variations in the material characteristics. And so long as we can control the air gap tolerance well, we can automatically control the inductance well too.

But we said that the air gap also affects the energy storage capability. This will be clearer now. Let us recap some of the equations that went into the earlier analyses. We have the general equations

1. $E = \dfrac{\mu \mu_0 N^2 I^2 A_e}{2 l_e} \times \dfrac{1}{z}$

2. $B = \dfrac{\mu \mu_0 N I}{z l_e}$

3. $E = \dfrac{1}{2} B^2 V_e \dfrac{z}{\mu \mu_0}$

4. $L = \dfrac{1}{z} \left( \dfrac{\mu \mu_0 A_e}{l_e} \right) N^2$

5. $\mu_e = \dfrac{\mu_c}{z}$

Let us study some specific cases to exemplify the possible variations:

If, for example, we have set an air gap such that the (relative) permeability falls from 2000 to 200, that is, by 10 times, then from the fifth equation above, $z$ must have gone from 1 (no air gap) to 10. From the second equation above we see that to keep the operating $B$-field unaltered (say as when we are operating close to $B_{SAT}$), we can safely increase the ampere-turns $NI$ by 10 times (i.e., the number of turns, since current is virtually predetermined). From the first equation above, we see that the energy stored in the core increases by a factor $10^2/10 = 10$ times. From the fourth equation, inductance has also been increased in the process by $10^2/10 = 10$ times. So, in this case $N \propto z$, Energy $\propto z$, $B = constant$, $L \propto z$.

However, we know that for an optimum converter design we want to fix a certain $r$ (usually at 0.4). At a given frequency and for a fixed application condition this means we have a *specific L* that we want to achieve. No more, no less. So, in power supply design what we really want to do is described next.

From the fourth equation above we see that to be able to keep the inductance fixed as $z$ goes from 1 to 10, we only need to increase $N$ by $10^{1/2} = 3.2$ times. Therefore, from the second equation above we can see that if $z$ went from 1 to 10, but $NI$ was increased only by a factor of 3.2 (so as to keep $L$ fixed), then the operating $B$-field would be *reduced* to 1/3rd of its original value. From the first equation, the energy stored in the core has remained unaltered in the process, though from the third equation we can see that its *overload capability* (i.e., measured up to a certain $B_{SAT}$) has certainly increased 10 times. So, any headroom as measured from the operating $B$-field value to the saturation level ($B_{SAT}$), or from the operating energy storage level to the peak energy handling capability must have increased considerably—even though inductance has been kept a constant in this case. All this could translate to a much higher field reliability where the converter will likely encounter severe abnormal or transient line/load conditions. However, *if all the bells and whistles are present* in the design of the control circuitry (e.g., voltage feedforward, primary/secondary current limit, duty cycle clamp and the like), and they serve to protect the converter adequately against any such abnormal conditions, this gives us a great *opportunity to select a smaller core* for the same power level. In doing so we would be essentially returning to the point of *optimum core size* where the operating $B_{PK}$ is close to $B_{SAT}$. That is the essential advantage accruing from the presence of an air gap.

Note that the two cases discussed in the above paragraph happen to be the last two "studies" we conducted previously. These studies are summarized below in the order in which they were discussed

- If $N =$ constant, then energy $\propto 1/z$ ($B \propto 1/z$, $L \propto 1/z$)
- If $N \propto z$, then energy $\propto z$ ($B = constant$, $L \propto z$)
- If $N \propto z^{1/2}$, then energy $=$ constant ($B \propto 1/z^{1/2}$, $L$ *constant*)

## 7.15   Understanding *L*

We saw that in the core and gap,

$$B = \frac{\mu \mu_0 NI}{z l_e}$$

That means that $B$ falls as $1/z$, *just as E does*. So, though we write $E$ as being apparently proportional to $B^2$, the energy in the entire magnetic structure is actually proportional to $B$. We can see this by combining the two equations discussed above

$$E = \frac{\mu \mu_0 N^2 I^2 A_e}{2 l_e} \times \frac{1}{z}$$

and

$$B = \frac{\mu \mu_0 NI}{z l_e}$$

We get

$$\boxed{E = \frac{1}{2} \times B A_e I N}$$

So, for a given coil (fixed $N$, $I$, and $A_e$), energy is proportional to $B$. *It does not even depend directly on the permeability*. The permeability is important only indirectly, because it will affect our ability to develop a sufficiently large $B$-field. However, note that even if we cannot develop a large enough $B$-field, we can still develop enough energy if we simply increase $N$. Our ability to store more energy in the magnetic structure is therefore based on the following:

- The permeability affects our ability to increase $B$ to the desired level.
- If the material saturates we cannot achieve the theoretically calculated $B$-field from Table 7.3.
- We can try to compensate the permeability by putting in more turns, but that may be limited by the available window. In addition, we could also end up making the structure lossy due to an increase in ac and dc losses in the copper windings.
- Note that most engineers already know that $E = \frac{1}{2}(LI^2)$, and since $I$ is essentially determined by the application conditions, not by design, we expect that *all the previous three statements must be merely equivalent to our ability to achieve a high enough inductance.*

And they are. That is the importance of defining a quantity called $L$. It virtually summarizes all we know about the core and gap. Equating the equation for $E$ as expressed in terms of $A_e$, $l_e$, and the like to the standard $E = \frac{1}{2}(LI^2)$ we can actually derive $L$

$$E = \frac{\mu\mu_0 N^2 I^2 A_e}{2l_e} \times \frac{1}{z} = \frac{1}{2}LI^2$$

$$\boxed{L = \frac{1}{z}\left(\frac{\mu\mu_0 A_e}{l_e}\right) \bullet N^2} \text{ (henry)}$$

We see that since $A_e$ is being considered equal to $A_g$, the reluctance

$$\Re = \frac{l_c}{\mu_c A_c} + \frac{l_g}{\mu_g A_g}$$

becomes

$$\Re = \frac{l_e}{\mu\mu_0 A_e} + \frac{l_g}{\mu_0 A_e} = \left(\frac{l_e}{\mu\mu_0 A_e}\right)\left[1 + \frac{\mu l_g}{l_e}\right]$$

But the term in square brackets is $z$. So,

$$\boxed{L = \left(\frac{1}{\Re}\right) \bullet N^2 \equiv P \bullet N^2}$$

where $P$ is defined as the *permeance* (the magnetic analog of the electrical parameter known as *conductance*).

The inductance is thus the product of several terms. The first is a property of the core geometry and material. This is multiplied by a term in $z$, which includes the effect of the air gap. Lastly, we have the

$N^2$ term, which comes from the self-inductance effect caused by the number of flux lines produced being proportional to $N$, and the linkages produced by any given line of flux, also being proportional to $N$, giving us an overall $N^2$ dependency for the total number of flux linkages produced.

Since we had previously defined

$$L = A_L \times N^2 \times 10^{-9} \text{ (henry)}$$

we get

$$A_L = \frac{1}{z} \left( \frac{\mu\mu_0 A_e}{l_e} \right) \times 10^9 \text{ (nH/turns}^2\text{)}$$

We also get

$$\boxed{A_L = P \times 10^9}$$

So, $A_L$ is, physically speaking, just the permeance. If $A_L$ is nanohenries per turns squared, permeance can be thought of as being henries per turns squared.

## 7.16    Difference between an Inductor and (Flyback) Transformer

In Fig. 7.5 typical voltage and current waveforms are presented to indicate more clearly how they will appear for an inductor as compared to transformers with different turns ratios. The horizontal axis is time. We must remember that as far as the core is concerned, it is totally unaware whether it is has several windings or just one. If the current in one of its windings stops, the other takes over, and so continuity in the ampere-turns is maintained. *And that is all the core cares about.* So, what is the difference between an inductor and a flyback transformer? The difference is in the *current through each winding.* If there was only one winding, the current through it would be smooth and undulating, and thus considered to be essentially dc. But if there are several windings (being switched around), the current in each winding is necessarily going to be a "choppy" trapezoidal type of waveform—one that will have very high frequency harmonic content. So, if we do not consider the effect of *skin depth*, we will probably get very high ac resistance losses. Thus, in general, the core loss (which depends only on the excitation of the core) will not be affected, but the copper loss will always be much higher for a transformer as opposed to an inductor.

To understand the concept of scaling better, see also Chap. 8.

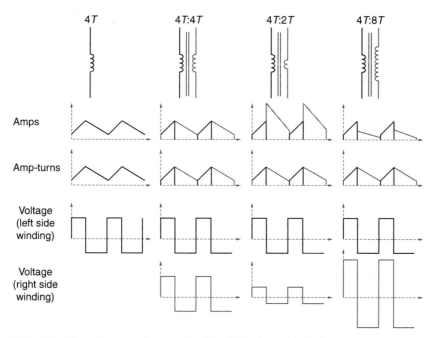

**Figure 7.5**  How voltages and currents will look like in multiwinding structures.

## 7.17  Transformers

*Flyback transformers* are actually just multiwinding inductors. When the primary winding conducts, the secondaries do not conduct, and vice versa. Therefore, the basic purpose of a flyback transformer is identical to that of any inductor—storage of energy during the switch on-time for subsequent delivery to the output when the switch turns *off*. The turns ratio provides step-up or step-down as an added bonus. And we can also get primary to secondary isolation. However, when we come to the forward converter and similar topologies, we use a genuine transformer. Its main purpose then is not energy storage, but purely transformer action (i.e., for stepping up or down the input voltage) as required. For example, if we didn't step the rectified ac input voltage down by using a step-down turns ratio, we would need an extremely small and impractical duty cycle to achieve the usual low output voltage levels. We need to get help from both the duty cycle and the turns ratio. So, for example, the dc transfer function for a forward converter (in CCM) combines the effects of *both n and D*

$$V_o = V_{IN} \times \frac{D}{n}$$

The transformer first carries out the *n conversion*, and this applies an effective input voltage of $V_{INR} = V_{IN}/n$ at one end of the output inductor. The inductor-diode essentially form a buck cell to carry out the final $D$ *conversion*, that is, $V_o = V_{INR} \times D$.

We can always identify an actual transformer-based topology (as opposed to one that uses a multiwinding inductor) by the fact that *when the primary winding conducts, so does the secondary.*

Though the energy storage ability of an inductor is, and will always be, its main selection criterion, the energy stored in the transformer is purely activation or *excitation energy*. It is not related to the load current at all, and it varies only when the input voltage changes for any reason. It therefore suits us to try to keep this energy term small, since all we are going to do with it anyway is to dump it back into the input bulk capacitor every cycle. For that reason too, *the transformer of a forward converter, for example, is invariably operated in discontinuous conduction mode.* Why put in more energy if we are going to recover and recycle it anyway? There is also a problem with ensuring reset of transformer, which is why it is kept to duty cycles less than 50 percent. Also, in principle, a forward converter transformer has no air gap since energy storage is not required. But *a small gap may be introduced to make the design more stable* and more tolerant of variations in the permeability of the material.

As we can begin to see, all these factors make the selection criteria and design of a forward converter transformer very different from that of a flyback transformer.

A basic underlying difference can be understood by looking at Fig. 7.6. Here at time $t = 0$, we suddenly apply a step voltage across the primary. We first consider the case where $R$ is not connected (for this case the parameters are primed below). Since $I'_S$ is zero (no conduction), for all practical purposes the secondary winding does not exist. It produces no flux and cannot in any way affect what happens elsewhere. In essence, we just have a simple inductor constituted around the primary winding, and so all the basic laws of magnetics presented earlier apply here too. In particular, the flux at any moment in the core is related to the instantaneous current by

$$\phi' = \frac{n_P \times I'_P}{\mathfrak{R}}$$

where $\mathfrak{R}$ is the reluctance of the magnetic core. But the rate of change of flux is related both to the causative voltage $V_P$, and the induced voltage $V_S$. We now stop disregarding the signs as we have done so far. So,

$$V_P = -n_P \frac{d\phi}{dt}$$

$$V_S = -n_S \frac{d\phi}{dt}$$

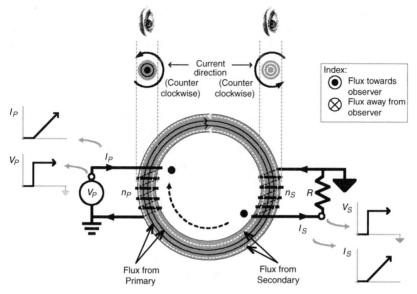

**Figure 7.6** Understanding transformer action.

The primes have been omitted above, since the above two equations are always going to hold regardless of current (the value of $R$). We thus get the familiar transformer scaling rule for voltages

$$\frac{n_P}{n_S} \equiv n = \frac{V_P}{V_S}$$

For instructional purposes we will now start thinking in terms of the flux *components* arising from each winding. The net flux $\phi$ is then the algebraic sum of the components.

**Note:** Yes, there is, in fact, a voltage developed across the floating secondary winding, though it does nothing if the load resistor $R$ is not connected. It is interesting to note, however, that this induced voltage sets *itself* up with such a polarity that *had there been an available path* for current to flow, it *would* have produced an induced current. And this current *would* have been in such a direction that it *would* have produced a flux, which *would* have been in such a direction as to oppose the *change* that caused everything!

**Note:** The direction of the induced flux component is such that it opposes the causative *change*. It will do whatever is necessary (*and possible*) to try to keep the (previously) existing flux level in the core *unchanged*. It does not necessarily need to be in an opposite *direction* to the primary-side flux. So, if the original flux tended to increase (as in our example), the induced flux correspondingly increases in magnitude in an opposite direction in an attempt to cancel the increase. But if the secondary winding had a previous dc bias, it could very well have *decreased* in magnitude while being in the *same* direction as the flux from the primary. Either way, it would have opposed the *change*. This concept must be very clear to the designer, especially when some relatively exotic topologies are encountered.

Now let's consider the case where a certain finite, nonzero resistive load $R$ is present. The secondary winding will now start to conduct at $t = 0$. But the voltage across the secondary winding is already fixed by the voltage scaling rule. Therefore the secondary current too is known from $V = IR$. From the magnetic circuit we thus get

$$\phi_P = \frac{n_P \times I_P}{\mathfrak{R}}$$

and

$$\phi_S = \frac{n_S \times I_S}{\mathfrak{R}}$$

But the flux from the secondary is in an opposite direction to the primary flux. So the net flux is

$$\phi = \frac{n_P \times I_P}{\mathfrak{R}} - \frac{n_S \times I_S}{\mathfrak{R}}$$

However, we reach an apparent paradox here. How can we have a different flux in the core while continuing to always and unconditionally satisfy the previous $V = -n \times d\phi/dt$ equations? Since the applied voltage on the primary has not changed, neither can the $d\phi/dt$ or the voltage induced on the secondary. The only possible way out of this quagmire is if the *net flux itself remained the same* (regardless of whether we have connected $R$ or left it open). Therefore

$$\frac{n_P I_P}{\mathfrak{R}} - \frac{n_S I_S}{\mathfrak{R}} = \frac{n_P I'_P}{\mathfrak{R}}$$

or

$$n_P I_P = n_P I'_P + n_S I_S$$

The *currents* in the windings have however changed from what they were when $R$ was not connected. We thus get the following current scaling rule

$$\left| \frac{I_P - I'_P}{I_S} \right| = \frac{n_S}{n_P}$$

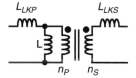

**Figure 7.7** Transformer model.

So, what happens to the flux? We have

$$\phi = \frac{n_P \times I_P}{\Re} - \frac{n_S \times I_S}{\Re} = \frac{n_P \times I_P}{\Re} - \frac{n_P(I_P - I_P')}{\Re}$$

So,

$$\boxed{\phi = \frac{n_P \times I_P'}{\Re}}$$

Therefore the *flux in the core does not change*, whether R is connected or not (or in fact whatever value R takes, provided it is nonzero).

Ignoring the signs between $I_S$ and $I_P$ and denoting the current with R not connected as the *magnetization current, $I_M$*, we get

$$I_P = I_M + {I_S}/{n}$$

where $n = n_P/n_S$ is the turns ratio. $I_S/n$ is the reflected secondary current $I_{SR}$.

The equivalent transformer diagram is thus represented as in Fig. 7.7. The primary and secondary-side leakage inductances, $L_{LKP}$ and $L_{LKS}$, are also shown.

## 7.18  Fringing Flux Correction

We have assumed so far that the area available to the flux lines in the magnetic material is the same as the area in the gap. Clearly that is not so. Particularly in the gap, the flux balloons out. How does this change the results? We introduce a *fringing flux correction* that effectively averages this over the entire gapped structure. It turns out that this fringing flux term is actually beneficial in many ways, as it is almost equivalent to increasing the effective area by a dimensionless factor *FF*. The effective area gets modified to

$$A_{ef} = A_e \times FF$$

There are harmful effects of fringing flux, especially in the form of increased eddy current losses in the windings adjoining the gap and the like. But by effectively increasing the effective area, literally for free, a significant advantage accrues in that the energy storage capability of

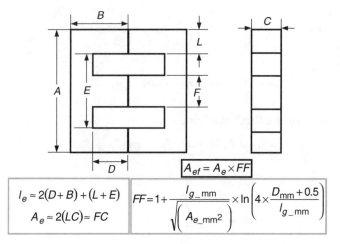

$$A_{ef} = A_e \times FF$$

$$l_e \approx 2(D+B) + (L+E)$$

$$A_e \approx 2(LC) \approx FC$$

$$FF = 1 + \frac{l_{g\_mm}}{\sqrt{\left( A_{e\_mm2} \right)}} \times \ln\left( 4 \times \frac{D_{mm} + 0.5}{l_{g\_mm}} \right)$$

**Figure 7.8** Effective area, effective length, and fringing flux estimates.

the core goes up. If we don't realize this, we will almost certainly end up unnecessarily increasing the size of the core (whenever energy storage is the purpose).

FF is described as

$$FF = 1 + \frac{l_g}{\sqrt{(100 \times A_e)}} \times \ln\left( 4 \times \frac{D + 0.5}{l_g} \right)$$

where $l_g$ is the air gap in mm, $A_e$ is expressed in cm$^2$, and $D$ is the dimension shown in Fig. 7.8 (in mm).

By introducing the fringing flux correction, reluctance *decreases* by the factor FF

$$\Re = \frac{l_c}{\mu_c A_c} + \frac{l_g}{\mu_g A_g} \Rightarrow \frac{l_c}{\mu_c A_{ef}} + \frac{l_g}{\mu_g A_{ef}}$$

Note that reluctance would also decrease had we interpreted the fringing flux correction as having decreased the lengths involved rather than increasing the area. But lengths are geometrically fixed, it is the area available for the flux lines that can and does change. We can go back to all the earlier equations and simply replace $A_e$ with $A_{ef}$. We will see that by introducing the fringing flux correction

- $B$ and $H$ fields are not going to change (see Table 7.3)
- $z$ does not change
- $\mu_e$ does not change

- Inductance is $1/\Re$ so it increases by factor FF
- Energy $\propto$ Area $\times$ length, so this increases by factor FF

If $L$ increases, it means that the value of $L$ we calculated ignoring fringing flux was actually an underestimation, because in reality when we wind the given number of turns on the given gapped core, we will measure a higher inductance, called $L_f$ here. This leads to an interesting error if we check the core for saturation. Let us consider the basic voltage-independent equation

$$B = \frac{LI}{NA}$$

So, without fringing flux considered

$$B = \frac{L_{calculated} \times I}{N \times A_e}$$

With fringing flux considered

$$B_f = \frac{L_{measured} \times I}{N \times A_{ef}} = \frac{(FF \bullet L_{calculated}) \times I}{N \times (FF \bullet A_e)} = B$$

So, the $B$-field does not change, as we expected from Table 7.3 too. However, what we do in practice is we typically first calculate the number of turns, and then when it is built we measure the inductance and use it to set production limits. But if we now use the *measured* value of inductance and try to see if our core is saturating by using the datasheet value of effective area of the core, we get an error:

$$B_{check} = \frac{L_{measured} \times I}{N \times A_e} = FF \times B$$

Since *FF* can be as large as 1.4 (for some large gapped cores), we will end up thinking that our $B$-field is not say 3000 G as we had initially calculated, but 4200 G. This will certainly throw everyone into a state of needless panic: "the core is saturating. . . ." Actually it is not! We can certainly use the voltage-independent equation to find $B$, but provided we plug in $A_e$ with the theoretically calculated $L$, *or* we use $A_{ef}$ with the measured $L$ (or $L_f$). In effect, this means that either we must decide to include the effects of fringing flux completely, or we must consciously ignore it altogether. We cannot "mix-and-match," especially because measured values would certainly include the effects of fringing (if applicable).

Failure to understand this will likely lead us to take the wrong design decision of using a larger core than really necessary.

## 7.19  Worked Example

We have a 42/42/15 core set. The material is 3C85. We wind 40 turns on it. We introduce an air gap by means of two spacers of 0.5 mm thickness each on both sides. What is the value of $L$? Also validate the energy equations given in this chapter. Assume we are passing an instantaneous current of 1.2 A through it.

Note that the core is two halves joined together; thus it is sometimes referred to in terms of the size of only one half, and is thus often called 42/21/15. However, note that all vendors state its relevant parameters (like $V_e$ and $l_e$) with the implicit assumption that two halves are joined together in a set. In Chap. 11 we have provided a table for some popular core sizes. We have for this material $\mu = 2000$, and for this core set

| Core | $V_e$ (cm$^3$) | $A_e$ (cm$^2$) | $l_e$ (mm) | $2 \times D$ |
|------|------|------|------|------|
| E 42/21/15 | 17.6 | 1.82 | 97 | 29.6 |

Note that by geometry, if the area is 1.82 cm$^2$, and the depth is 15 mm as is obvious from the part number of the core, the center limb is $1.82/1.5 = 1.21$ cm wide. The way E-cores are usually made, this center width splits up into the two equal side limb widths. So, each limb is $1.21/2 = 0.61$ cm wide. If we subtract this from the length along each side (i.e., 21 mm) we should get $D$. So, our quick-estimate of $D$ is $21 - 6.1 = 14.9$ mm or $2 \times D = 29.8$ mm, which is almost in complete agreement with the data provided by the vendor. *So, we don't really need the vendor to tell us* $2 \times D$ *usually.*

Therefore, the calculations proceed as follows (gap being 1 mm):

$$\boxed{FF = 1 + \frac{l_g}{\sqrt{(100 \times A_e)}} \times \ln\left(4 \times \frac{D + 0.5}{l_g}\right)}$$

$$FF = 1 + \frac{1}{\sqrt{(100 \times 1.82)}} \times \ln\left(4 \times \frac{14.8 + 0.5}{1}\right) = 1.305$$

So,

$$\boxed{z = \frac{l_e + \mu l_g}{l_e}}$$

$$z = \frac{97 + 2000 \times 1}{97} = 21.62$$

$$\boxed{A_L = \frac{1}{z}\left(\frac{\mu\mu_0 A_e}{l_e}\right) \times 10^9 \times FF} \quad \text{(nH/turns}^2\text{)}$$

$$A_L = \frac{1}{21.62}\left(\frac{2000 \times 4\pi \times 10^{-7} \times \frac{1.82}{10^4}}{\frac{97}{10^3}}\right) \times 10^9 \times 1.305 = 284.6 \text{ nH/turns}^2$$

$$\boxed{L = A_L \times N^2 \times 10^{-9}} \quad \text{(henry)}$$

$$L = 284.6 \times 40^2 \times 10^{-9} \times 10^6 = 455.4 \ \mu\text{H}$$

$$\boxed{E = \frac{\mu\mu_0 N^2 I^2 A_e}{2l_e} \times \frac{1}{z} \times FF} \quad \text{(joules)}$$

$$E = \frac{2000 \times 4\pi \times 10^{-7} \times 40^2 \times 1.2^2 \times \frac{1.82}{10^4}}{2 \times \frac{97}{10^3}} \times \frac{1}{21.62} \times 1.305 \times 10^6 = 327.9 \ \mu\text{J}$$

Comparing with the alternate form of stored energy

$$E = \frac{1}{2} \times L \times I^2 = \frac{1}{2} \times 455.4 \times 1.2^2 = 327.9 \ \mu\text{J}$$

We have complete agreement, thus validating our design equations.

# Tapped-Inductor
# Topologies

## 8.1 The Tapped-Inductor Buck

We have learned the voltseconds rule in Chap.1, and know that we can
use it to derive the duty cycle equation for any topology (that can exist).
But so far we have been largely dealing with a single winding on a core.
To transition to more complex structures like transformers, a better
way is to take the intermediate path of tapped-inductor topologies.

As shown in Fig. 8.1, we have a tapped-inductor buck. Let us say it
has an input of 34 V, and its output is set to 10 V. We can understand
this topology better by going through the following numerical steps.

- The current flows through all eight turns when the switch is *on*. We
  call this the *primary* winding here.

- The current flows through five turns when the switch turns *off*. These
  turns are said to constitute the *secondary* winding.

- *The (algebraic) sum of the ampere-turns (summed over all windings
  present on the same core) must be preserved at any given moment.
  The total ampere-turns cannot change in a discontinuous fashion at
  any moment.* The reason becomes clearer if we write out the voltage-
  independent equation for a multiwinding inductor as

$$B = \frac{LI}{NA} = \frac{A_L NI}{A} 10^{-9} \Rightarrow \left( \frac{A_L}{A} 10^{-9} \right) \bullet \sum N_i I_i$$

The term in brackets is a characteristic, not of the windings but of the
core geometry and its material. The summation is over the index $i$
where each winding is arbitrarily numbered $i = 1, 2, 3$ and the like. So

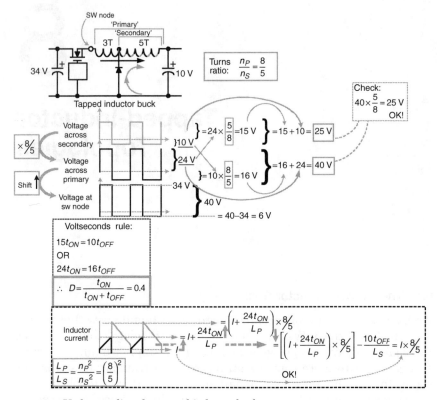

**Figure 8.1** Understanding the tapped-inductor buck.

a sudden jump in the net ampere-turns would imply a sudden change in the field. The field however corresponds to the energy stored in the core, and energy cannot suddenly disappear and appear. Therefore we get

$$I_S = I_P \times \frac{n_P}{n_S}$$

- According to this rule, we can conclude that *at* the moment of either crossover transition (turn-on or turn-off), the ampere-turns of the inductor must be preserved as the current changes its route from primary to secondary (and back). So if, for example, 1 A was flowing through the primary, we have eight ampere-turns just before the turn-off transition. Therefore, the current that develops in the *secondary* just after the transition must have the same ampere-turns. Hence, the secondary current is $8/5 = 1.6$ A.

- The voltage across the windings also scales according to the turns ratio, though in a manner opposite to the current. So,

$$V_S = V_P \times \frac{n_S}{n_P}$$

This follows from Faraday's law

$$V = N\frac{\mathrm{d}\phi}{\mathrm{d}t} = NA\frac{\mathrm{d}B}{\mathrm{d}t}$$

which when written for a multiwinding choke is

$$\frac{V_i}{N_i} = \frac{\mathrm{d}\phi}{\mathrm{d}t}$$

where the right-hand side is again a characteristic, not of the windings, but of the magnetic core on which the windings $i$ happen to be wound. Therefore the left-hand side too must apply to each and every winding (each considered separately). Therefore, we can now state a general rule: At any given moment *the volts per turn existing across any (and every) winding present on a core must be equal* (excluding any spikes due to uncoupled/leakage inductance elements present).

- *Transformer scaling* for voltages is equivalent to the Volt/turn statement. And the (inverse) transformer scaling for currents is equivalent to the ampere-turns statement.

- We must however remember that *scaling* applies to the voltages present *across windings*, not necessarily to voltages that may be present *at different points* in the primary and secondary sections. Similarly (inverse) scaling applies only to currents actually passing *through windings*.

- Looking at the figure, we can ask: what do we already know about the voltages present across the windings? For one, we know that the voltage across the primary winding (eight turns) is $34 - 10 = 24$ V *when the switch is on*. We also know that the voltage across the secondary winding (five turns) is 10 V when the switch is *off*. Note that these are underlined in the figure.

- Applying scaling to each of the above, we can get the voltage across the secondary winding when the switch is *on*, and across the primary winding when the switch is *off*. Therefore we get the total peak-to-peak voltage swing across each winding. These are respectively 40 V and 25 V, as shown in the figure.

- The voltage across the primary winding is basically the voltage at the switching node measured with respect to the output voltage rail. Therefore, if we want to know the voltage at the switching node (which is by definition referenced to ground) we just need to "y-shift" the primary voltage waveform by 10 V. Note that this is in conformity with the observation that when the switch is *on*, the switching (SW) node must be equal to the input voltage (34 V).

- We find that if each winding is considered separately, *the voltseconds law applies to any chosen winding*. That is what can be seen from the dot-outlined block in the figure. *So, a calculation based on either winding will give us* D = 0.4.

- For analyzing the current waveforms further, let us assume that the primary inductor current ramp starts at a reference value designated $I$ on its waveform. From there, it rises according to $V = LdI/dt$. $V$ is the voltage across the chosen winding. $L$ is the inductance of that winding (assuming that the other windings are open when it is measured).

- The primary-side current therefore ramps up to the value

$$= I + \frac{24t_{ON}}{L_P}$$

- At that point we have a crossover, and so the current jumps at the transition as per the scaling law. Therefore, the ramp-down of the secondary current starts at

$$= \left( I + \frac{24t_{ON}}{L_P} \right) \times {}^8\!/_5$$

- It ramps down according to $V = LdI/dt$, where $V$ is the voltage during the off-time across the secondary winding, and $L$ is its inductance. We will see that the inductance scales from primary-side to secondary-side according to the square of the turns ratio. Finally the secondary current ramps down to

$$= \left[ \left( I + \frac{24t_{ON}}{L_P} \right) \times {}^8\!/_5 \right] - \frac{10t_{OFF}}{L_S}$$

- Then a crossover transition occurs again, and the current in the primary must be (as per the scaling rule)

$$= \left\{ \left[ \left( I + \frac{24t_{ON}}{L_P} \right) \times {}^8\!/_5 \right] - \frac{10t_{OFF}}{L_S} \right\} \times {}^5\!/_8$$

But this must return the current to the starting value it had (i.e., $I$), because that is the definition of a *steady state*. So, by using

$$\frac{L_P}{L_S} = \frac{n_P^2}{n_S^2} = \left( \frac{8}{5} \right)^2$$

and simplifying, we get

$$t_{ON} = t_{OFF} \times {}^2\!/_3$$

This is equivalent to saying that $D = 0.4$, which is what we derived in the figure on the basis of the voltseconds law.

- A more generic derivation for the tapped-inductor buck gives the following equation for $D$

$$D = \frac{(V_O + V_D) \bullet n_P}{(V_{IN} - V_{SW}) \bullet n_S + V_O \bullet (n_P - n_S) + V_D \bullet n_P}$$

where $V_D$ and $V_{SW}$ are their forward drops across the diode and switch respectively.

Ignoring the diode and switch drops and writing

$$n = \frac{n_P}{n_S}$$

we get an easier to remember form

$$D = \frac{nV_O}{V_{IN} + V_O(n - 1)}$$

- The average inductor current (primary and secondary currents combined) must be equal to load current (as is required of any buck type of converter). Actually, the expression is simpler for the center of the ramp. For a conventional buck converter we know that the center is the load current. For the tapped-inductor buck, the center of the primary current ramp is

$$I_C = \frac{n_S}{D(n_S - n_P) + n_P} I_O$$

The center of the secondary current ramp is by the current scaling rule

$$I_{CS} = I_C \frac{n_P}{n_S}$$

The swing $\Delta I$ for the primary side is of magnitude

$$\Delta I = \frac{V_{IN} - V_{SW} - V_O}{L} \times \frac{D}{f}$$

where $L$ is the inductance of the entire coil (eight turns) and f is the frequency in Hz. The secondary-side swing is

$$\Delta I_S = \Delta I \frac{n_P}{n_S}$$

The same rationale and reasoning that applies to the tapped-inductor applies equally to transformers. Note that only the leakage inductance has been ignored so far. In reality, *to avoid damaging the switch from spikes arising from this leakage, we would usually need an RC or zener snubber/clamp for all tapped-inductor configurations.*

**Note:** The tapped-inductor buck topology was presented mainly to illustrate transformer principles more clearly. However, in reality, very few controllers or integrated switchers exist that can even handle a tapped-inductor topology. For example, most switcher ICs are by their very design restricted in that the SW node is not allowed to go more than about 1 V negative. But for the tapped-inductor buck topology, the negative voltage is of magnitude $V_O \times (n_P/n_S - 1)$. Therefore, this can exceed the ratings of the device.

## 8.2   Other Tapped-Inductor Stages and Duty Cycle

We can also have a tapped-inductor buck stage wired up differently from the one discussed so far. This is shown in Fig. 8.2. And, we can also have tapped-inductor boost stages, as are also shown. The input to output transfer function for each of these is also provided in the figure.

In Fig. 8.3 we see how the duty cycle is altered from that of the *standard dc-dc converter* as we try to garner any advantage that may accrue from varying the turns ratio. But, as mentioned, because of maximum voltage ratings of the pins, integrated switchers usually cannot accept tapped-inductor stages unless they were explicitly designed to do so. But controllers should usually work just fine.

Tapped-inductor stages throw up several interesting possibilities that we should be on the lookout for. For example, if we have a "5 A buck switcher IC," it can usually handle only a maximum of 5 A load current. But suppose the IC was designed to handle a negative voltage swing in excess of 1 V on the switching pin. We could then use it in a tapped-inductor buck stage (set up with primary turns greater than secondary turns), and thus we could even get 10 A of load current under certain input-output conditions. The designer may also be able to optimize efficiency by changing the duty cycle so as to reduce the time the

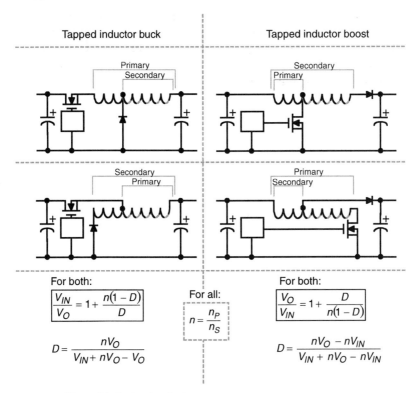

**Figure 8.2**  Tapped-inductor topologies.

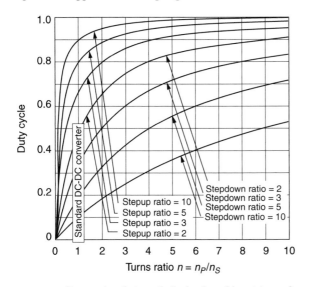

**Figure 8.3**  Comparing duty cycle for buck and boost tapped-inductor topologies.

current spends in the more dissipative element out of the switch and diode. Tapped-inductor stages can also help overcome any minimum or maximum duty cycle limits, especially under extreme input-output conversion ratios. There could be several other uses in particular applications, besides those mentioned here.

# Selecting Inductors
# for DC–DC Converters

## 9.1 Introduction

Power supply engineers don't usually *design* an inductor for dc-dc converter applications. The preferred approach here is to use an easily available off-the-shelf inductor. In the first pass, the current rating of the inductor is chosen to be roughly equal to the maximum continuous current passing through it in the particular application. Its inductance is chosen based on a preferred design value of about 0.4 for the current ripple ratio $r$, which by definition is (for any topology)

$$r = \frac{\Delta I}{I_C}$$

$I_C$ is the center of the inductor-current waveform at maximum load and $\Delta I$ is the swing. This definition of $r$ applies only if the converter is in continuous conduction mode, as is our basic assumption throughout this chapter. Therefore, $r$ can vary between 0 (for a very large inductance) and 2 (for a "critical" inductance, i.e., for which the inductor operates at the boundary between continuous conduction and discontinuous conduction modes). A calculated value of $r$ greater than 2 simply implies discontinuous mode.

The worst-case inductor current depends on the topology. So, for a buck we know that the average value (geometric center) of the inductor current waveform equals the load current, irrespective of the input voltage. And this center remains constant as we change the input voltage. However, as we increase the input, the peak current increases. Therefore, for magnetics design, a high input voltage forms the worst-case condition for a buck converter. But for a boost or a buck-boost the center shifts upward as we increase the duty cycle (decrease $V_{IN}$). The equations are

$$I_C = I_O \text{ (Buck)}$$

$$I_C = \frac{I_O}{1 - D} \text{ (Boost/buck-boost)}$$

These follow from the fact that in a buck energy is being transferred to the load during the entire cycle; thus the average inductor current must equal the load current. For a boost or a buck-boost, energy is pushed into the output only when the diode conducts. So the average diode current $I_C \times (1 - D)$ must equal the load current.

*For all topologies, a high duty cycle implies a low input voltage.* So for a boost or a buck-boost we get the maximum (average) inductor current at low input condition $V_{INMIN}$. Therefore, for magnetics design, a low input voltage forms the worst-case condition for a boost or buck-boost converter.

The reader can brush up on his or her knowledge of basic waveform analysis from Chap. 1. We see that the peak current (for the switch, diode, and inductor) is

$$I_{PK} = I_C \times \left(1 + \frac{r}{2}\right)$$

The reader can also prove from the method indicated for an arbitrary waveform that the rms of the inductor current waveform (for all topologies) is

$$I_{L\_RMS} = I_C \sqrt{1 + \frac{r^2}{12}}$$

Note that the rms is very close to the dc (average) value of $I_C$. Even if $r$ is at its maximum value of 2, the rms value is only 15 percent higher than the average value. For our typical design goal of $r = 0.4$, the difference is less than 1 percent. Therefore we should not be too concerned if when looking at a vendor's datasheet we find that he or she has chosen to specify the current rating of the inductor in terms of $I_{DC}$ instead of $I_{RMS}$. In either case this refers to the continuous *heating current* that we can pass through the inductor. Clearly, it will depend on the expected temperature rise when mounted on a printed circuit board (PCB). So manufacturers will also specify the dc/rms current rating at a *specified temperature rise over ambient* $\Delta T$ in °C. This is usually in the range of 30 to 50°C. When comparing inductors from different vendors, this rated temperature rise should be noted.

So far, we are talking only about the copper loss. The core loss component is usually only 5 to 10 percent of the total loss when dealing with off-the-shelf inductors made of ferrites, but we should remember that if the material used happens to be powdered iron for example, this number could be as high as 20 to 30 percent. And that could increase the temperature rise in a typical switching environment over what we expect purely on the basis of the dc/rms current rating number. Core loss depends on the *flux swing* in our specific application (see Chap. 7 and Chap. 11), and we need to calculate this too. That is a key reason why we need to go from our first pass design selection, based on only the dc current rating, to a more accurate estimate. Another reason is that our frequency may be different. We will come to the detailed procedure soon.

Talking about flux swings and $B$-fields, we also do not want to *saturate* the inductor. After all, we can simply pass a dc current in the coil and measure its heating current capability, but we still don't know what the magnetic state of the core in the process was. In a pure dc current test, the core will not contribute anything to the total loss. But in a real converter we do care deeply, not only about the total loss due to ac and dc effects, but also the need to avoid saturation. Therefore, clearly, the $I_{DC}$ rating of an inductor just does not suffice.

For ruling out saturation we need to know the *peak* flux density (i.e., the peak $B$-field) that the inductor can tolerate. Vendors sometimes provide this figure, but rather indirectly so, in terms of the "safe" current or $I_{SAT}$ that we can pass without saturating the core. This may or may not be the same as the $I_{DC}$ rating. Some of the more clever designers/ vendors of standard magnetic components have probably realized that there is no point putting thicker copper if the core is saturating at that point, *nor putting less copper*. The latter will simply be tantamount to failing to fully exploit the available energy storage capability of the core. So they usually try to set $I_{DC} = I_{SAT}$ and thus state only one current rating figure. This means that at the current at which the heating test is performed, the core happens to be almost at the point of saturation too. In general, if two disparate current ratings are provided by the vendor, one for heating and one for saturation, we should simply ignore the higher of the two numbers. The effective "current rating" of the inductor, for all practical purposes, is the *lesser* of the two (whichever one it is).

There is, historically speaking, a sense of justification in having some headroom available before saturation occurs. This means that some designers may consciously want to have an $I_{SAT}$ rating that is higher than the continuous dc rating $I_{DC}$. But if we have set a certain current limit for the switch that is considered "correct," then by definition the switch cannot be destroyed (immediately) at this peak current, *irrespective of whether the inductor was saturating or not* at the moment when the limit was reached. For long-term protection we would usually have *OTP* (over temperature protection). In the author's experience, at least up to an input voltage of 40 V, we need not worry about damage occurring due to saturation, especially when using FET-based integrated switchers, since these have inherently very fast-acting current limiting. So, it is commonplace to find that the inductor rating is always based on the maximum load current of the application, *not the current limit*, which may be much higher. With this deliberately underdesigned inductor we realize that under start-up or sudden step line/load changes, the inductor will almost certainly saturate because the current will hit the current limit (momentarily) as it tries to correct the output. But there is no cause for alarm since we are assuming that the current limit is set correctly according to the switch capabilities. But above 40 V, especially when using BJT-based switchers, some failures were seen under severe testing on the bench. We will see that the current now rises much more steeply as the inductor starts to saturate, and the IC may not act *fast enough to enforce the set current limit* before switch destruction occurs. The situation can clearly get worse if we are using a controller IC (i.e., an external switch with trace inductances which tend to slow the responses) and/or if we have current mode control (minimum pulse-width and other blanking time issues). In such cases, it may make much sense to maintain the headroom we talked about above. So now, two current ratings may make economical sense if we set the dc rating as per the load and the saturation rating as per the current limit. Of course, if the inductor provided has only a single current rating, then below 40 V of input we should ensure this is equal to or higher than the load current, and above 40 V we should ensure it is equal to or higher than the current limit. The 40 V threshold is actually a thumb rule of sorts and therefore judiciousness must be applied. We may even need to verify an acceptable current rating by bench testing on a case-to-case basis. However, in a high-voltage application, e.g., an off-line power supply, the transformer must *always* be sized according to the current limit (or the current limit adjusted), since we just cannot afford to saturate the transformer even for a moment here. We do acknowledge the actual maximum load current, but only for the purpose of selecting the thickness of the copper and estimating operating core loss (i.e., all heating effects). Finally, note that in all cases the set current limit has a certain tolerance. When talking about magnetics, we should always concern ourselves with the upper limit (maximum) of this tolerance range.

## 9.2   Specifying the Current Ripple Ratio *r*

*r* is the starting point of any converter design as mentioned in Chap. 1. By definition, *r* is a constant for a given converter/application (even though the actual load on the output varies). That is because *r* is simply the current ripple ratio *at maximum load*. The input voltage at which we set $r$ depends on the topology. So, for a buck we set it at $V_{INMAX}$ simply because though the average inductor remains constant as we vary the input voltage, the peak current does get higher at higher input voltages. For a boost or a buck-boost we know that the worst case for the inductor current is at the lowest input voltage, because the average current goes as $1/(1-D)$; thus we set $r$ at $V_{INMIN}$ for these.

A high inductance reduces $\Delta I$ and results in a lower $r$ (and lower rms current in the input and/or output capacitors) but may result in a very large and impractical inductor. So, typically, for most buck regulators, $r$ is chosen to be in the range of 0.3 to 0.5 (at the maximum rated load). Once the inductance is selected, as we decrease the load on the converter (keeping input voltage constant), $\Delta I$ remains fixed but the dc level decreases, and so the current ripple ratio increases. Ultimately, at the point of transition to discontinuous mode of operation the dc level is $\Delta I/2$. We conclude the following:

- The current ripple ratio at the point of transition to discontinuous mode is 2. Therefore, for continuous conduction mode, $r$ ranges from 0 (an extremely large inductance) to 2 (*critical inductance*).

- The load at which transition happens can be shown by simple geometry to be $r/2$ times the maximum-rated load. For example, if the inductance is chosen to be such that $r$ is 0.3 at a maximum load of 2 A, the transition to discontinuous mode of operation will occur at 0.15 times 2 A, which is 300 mA. This may be an additional consideration when choosing the inductance.

## 9.3  Mapping the Inductor

The vendor of the off-the-shelf inductor would necessarily have used certain specific test conditions to characterize and evaluate the component. The practical problem in adjudging this component to be suitable for our application is that the test conditions may be very different from our particular application conditions. We cannot have a countless number of custom parts for countless application conditions. We must therefore learn to evaluate how *any given component is likely to perform in any given application*, despite the fact that the application conditions may seem very different from its test conditions, at least at first sight. So, in this chapter our main purpose is to take the vendor's test conditions and *map* them to our application. In doing so we will know more precisely how that inductor will behave in our specific application. We might even find that we can change the frequency (or some other condition the inductor was originally designed for) and not lose much in the process. This will help us pick the most cost-effective inductor from a much wider array of candidates. The only obstacle may be the provision of insufficient data for this exercise by the vendor. But he or she can usually be prodded to produce much more information than available in the datasheet.

## 9.4 Voltseconds

We know from Chap. 1 that the duty cycle of any converter follows from the voltseconds law. This law is merely an expression of a *steady state* in which the change in current during the switch on-time is exactly equal (in magnitude) to the change during the switch off-time, thus returning us exactly to the same point we started the cycle with. In particular, from the basic inductor equation $V = LdI/dt$ we get

$$V_{sec\_ON} = L\Delta I_{ON} = V_{sec\_OFF} = L\Delta I_{OFF}$$

So, generically speaking,

$$\boxed{V_{sec} = L\Delta I}$$

Sometimes engineers prefer to talk in terms of voltμsecs instead of voltseconds as this gives a more *manageable number*. This is likewise simply the voltage across the winding of the inductor times the duration *in* μsecs for which it is applied. This calculation too can be expected to give the same result if performed for the on-time as for the off-time. If it doesn't, the engineer needs to recheck the equation being used for duty cycle.

Voltseconds ($V_{sec}$) completely defines the *current swing*. Together with $I_{DC}$, (that is, $I_C$), it actually completely defines the application from a magnetics viewpoint. $V_{sec}$ determines the ac component, and $I_{DC}$ is the dc component. As a corollary, all applications with the same $V_{sec}$ and $I_{DC}$ can be considered to be the "same" application from the viewpoint of inductor design/selection. *An inductor designed for a given* $V_{sec}$ *and current can be cross-utilized across all applications with the same* $V_{sec}$ *and* $I_C$ *without question.* Even frequency doesn't matter, except to a second order since it indirectly enters into our core loss calculation. But we can be assured that the core certainly will not saturate.

For example, it doesn't matter if we are applying 5 V for 2 μs or 10 V for 1 μs. For a given inductance, we see from the equation above that $\Delta I$ will remain the same. So, the core loss will be the same too (at least that part of it that depends on flux swing). And if the dc current is the same, the $I^2R$ losses will not change and neither will the peak current change since it is simply $I_{DC} + \Delta I/2$. The question arises only when we attempt to use the inductor for a different $V_{sec}$, or $I_C$, than its original design. For that we need the formal mapping procedure presented in this chapter.

Note that the voltseconds depends only on the input and output voltages for a given topology. As long as it is in continuous conduction mode, changing load current, $L$, or even $r$ does not affect $V_{sec}$.

## 9.5  Choosing *r* and *L*

In Table 9.1 we have the relevant parameters for the three topologies. Look at the first column for the basic definition of the terms. We have indicated the parasitic terms by horizontal curly brackets. These are $V_{SW}$—the forward drop across the switch, and $V_D$—the forward diode drop. The idea is that we can ignore these curly bracket terms if the parasitic drops are small with respect to the input and output rails. We can also ignore them for a quick estimate or for exploring optimization possibilities rather than getting bogged down by terms that don't affect the larger picture.

In the table we may observe that we have expressed $r$ in terms of $L$ and also $L$ in terms of $r$. So, how do we know how to select either of them? We turn to the row which tells us the required *energy handling capability* E

$$ E = \frac{1}{2} L I_{PK}^2 \ \text{J} $$

Using the basic definitions in the first column of the table we can carry out a simple derivation as follows

$$ E = \frac{1}{2} L I_C^2 \left( 1 + \frac{r}{2} \right)^2 = \frac{1}{2} \frac{V_{sec}}{\Delta I} I_C^2 \left( 1 + \frac{r}{2} \right)^2 $$

$$ E = \frac{1}{2} \frac{V_{sec}}{\Delta I} I_C^2 \left( 1 + \frac{r}{2} \right)^2 $$

$$ E = \frac{1}{2} \frac{I_C V_{sec}}{r} \left( 1 + \frac{r}{2} \right)^2 $$

$$ E = \frac{1}{2} I_C V_{sec} \frac{1}{r} \left( 1 + \frac{r^2}{4} + r \right) $$

$$ E = \frac{1}{8} I_C V_{sec} \frac{r^2}{r} \left( \frac{4}{r^2} + 1 + \frac{4}{r} \right) $$

$$ \boxed{ E = \frac{I_C V_{sec}}{8} \left[ r \left( \frac{2}{r} + 1 \right) \right]^2 } \ \text{J} $$

TABLE 9.1   Basic Design Table of Parameters Relevant to Magnetics.

| Parameters | Buck | Boost | Buck-boost |
|---|---|---|---|
| Average current in inductor $I_C$ (A) | $I_O$ | $\dfrac{I_O}{1-D}$ | $\dfrac{I_O}{1-D}$ |
| RMS current in inductor $I_{L\_rms}$ (A) | $I_O\sqrt{1+\dfrac{r^2}{12}}$ | $\dfrac{I_O}{1-D}\sqrt{1+\dfrac{r^2}{12}}$ | $\dfrac{I_O}{1-D}\sqrt{1+\dfrac{r^2}{12}}$ |
| $V_{ON}$ (Volts) | $V_{IN}-V_O-\overbrace{V_{SW}}$ | $V_{IN}-\overbrace{V_{SW}}$ | $V_{IN}-\overbrace{V_{SW}}$ |
| $V_{OFF}$ (Volts) | $V_O+\overbrace{V_D}$ | $V_O-V_{IN}+\overbrace{V_D}$ | $V_O+\overbrace{V_D}$ |
| Duty cycle $D$ $=\dfrac{V_{OFF}}{V_{OFF}+V_{ON}}$ | $\dfrac{V_O+\overbrace{V_D}}{\underbrace{V_{IN}-V_{SW}+V_D}}$ | $\dfrac{V_O-V_{IN}+\overbrace{V_D}}{\underbrace{V_O-V_{SW}+V_D}}$ | $\dfrac{V_O+\overbrace{V_D}}{\underbrace{V_{IN}+V_O-V_{SW}+V_D}}$ |
| $V_{sec}$ (Volt-sec) $=V_{ON}t_{ON}$ $=V_{OFF}t_{OFF}$ | $\dfrac{V_O+\overbrace{V_D}}{f}(1-D)$ | $\dfrac{V_O-\overbrace{V_{SW}+V_D}}{f}D(1-D)$ | $\dfrac{V_O+\overbrace{V_D}}{f}(1-D)$ |
| $\Delta I$ (A) $=\dfrac{V_{sec}}{L}$ | $\dfrac{V_O+\overbrace{V_D}}{Lf}(1-D)$ | $\dfrac{V_O-\overbrace{V_{SW}+V_D}}{Lf}D(1-D)$ | $\dfrac{V_O+\overbrace{V_D}}{Lf}(1-D)$ |
| $r=\dfrac{\Delta I}{I_C}$ | $\dfrac{V_O+\overbrace{V_D}}{I_OLf}(1-D)$ | $\dfrac{V_O-\overbrace{V_{SW}+V_D}}{I_OLf}D(1-D)^2$ | $\dfrac{V_O+\overbrace{V_D}}{I_OLf}(1-D)^2$ |
| $I_{PK}$ (A) $=I_C\left(1+\dfrac{r}{2}\right)$ | $I_O\left(1+\dfrac{r}{2}\right)$ | $\dfrac{I_O}{1-D}\left(1+\dfrac{r}{2}\right)$ | $\dfrac{I_O}{1-D}\left(1+\dfrac{r}{2}\right)$ |
| $L$ (H) $=\dfrac{V_{sec}}{\Delta I}$ | $\dfrac{V_O+\overbrace{V_D}}{I_Orf}(1-D)$ | $\dfrac{V_O-\overbrace{V_{SW}+V_D}}{I_Orf}D(1-D)^2$ | $\dfrac{V_O+\overbrace{V_D}}{I_Orf}(1-D)^2$ |
| Energy handling capability $E$ (joules) | $\dfrac{I_CV_{sec}}{8}\left[r\left(\dfrac{2}{r}+1\right)^2\right]$ | $\dfrac{I_CV_{sec}}{8}\left[r\left(\dfrac{2}{r}+1\right)^2\right]$ | $\dfrac{I_CV_{sec}}{8}\left[r\left(\dfrac{2}{r}+1\right)^2\right]$ |

TABLE 9.2    Extended Design Table of Parameters for a Buck Converter.

| Buck converter parameters | |
| --- | --- |
| RMS current in output cap (*Amps*) | $I_O \dfrac{r}{\sqrt{12}}$ |
| RMS current in input cap (*Amps*) | $I_O \sqrt{D\left[1 - D + \dfrac{r^2}{12}\right]}$ |
| RMS current in inductor (*Amps*) | $I_O \sqrt{1 + \dfrac{r^2}{12}}$ |
| RMS current in switch (*Amps*) | $I_O \sqrt{D\left[1 + \dfrac{r^2}{12}\right]}$ |
| Average current in switch (*Amps*) | $I_O D$ |
| Average current in diode (*Amps*) | $I_O(1 - D)$ |

This expression is valid for all topologies, *though the relationship of $I_C$ to load current may be different.* But the term in the square brackets has a particular shape with respect to $r$ and this provides the clue for describing an optimum setting for $r$. To gauge the significance of the energy term vis-à-vis the other parameters in a power supply, we plot the results out using the extended table for a buck in Table 9.2 as an example. This is shown in Fig. 9.1. Here we are varying $r$ to see what happens to the current components. This is tantamount to changing the inductance for a given frequency. We observe the following:

- The energy curve has a "knee" at around 0.4.
- Too low of an $r$ will significantly increase the physical size of the core, because the size is proportional to its energy handling capability.
- Too high an $r$ will not significantly decrease the size of the core since we reach a point of diminishing returns.
- As we increase $r$, the related currents in the capacitors and switch increase. This in turn may demand larger capacitors and the like.
- However, to put it in balance, what we are showing in Fig. 9.1 are the *normalized* values. So, a large percentage *variation* may not be very significant if we consider that the absolute value of the concerned parameter may be very small. So, the rms current through the output capacitor is actually very small for a buck converter, even though its apparent variation with respect to $r$ seems the most prominent. Further, the basic criterion of its selection may not even be the rms

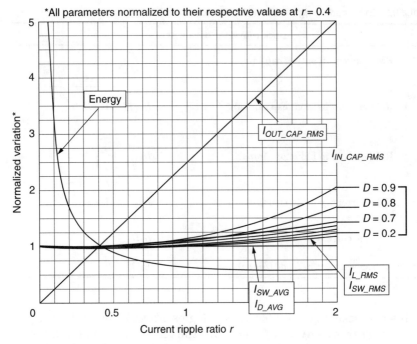

**Figure 9.1** Variation of parameters for a buck converter as $r$ varies.

current through it. In this case, what is probably more important is its effect on the output voltage ripple and/or the feedback control loop characteristics. For example, with voltage mode control, the ESR (equivalent series resistance) of the output capacitor provides the required zero in its loop response, and most engineers know all too well that selecting a very low ESR capacitor may cause instability.

- Of great importance for a buck converter is its input capacitor selection procedure. The physical size of this component may be determined purely by the ripple current through it. So, a doubling of the rms current will typically increase its size by four times.

- Generally speaking, the optimum setting is $r \approx 0.4$ *for all topologies* since it coincides with the "knee" of the energy curve.

### 9.6  *B* in Terms of Current

It is suggested that Chap. 7 be revisited before reading the following section. We start with the familiar voltage-independent law, now expressed in terms of voltseconds

$$\Delta B = \frac{L \times \Delta I}{N \times A} = \frac{V_{\text{sec}}}{N \times A} \text{ (tesla)}$$

$A$ is in m$^3$. Writing this in terms of cm$^3$, gauss, and voltμseconds, we get

$$\Delta B = \frac{100 \times V_{\mu\,\text{sec}}}{N \times A_{\text{cm}^3}} \quad \text{(gauss)}$$

As clarified in Chap. 7, this equation also applies to instantaneous values. Thus at peak current we can write (in MKS units)

$$B_{PK} = \frac{L \times I_{PK}}{N \times A} \quad \text{(tesla)}$$

Note also that this seems to indicate that we cannot change the flux density swing $\Delta B$ by any other means except by changing the area of the core ($A$), the number of turns, or the voltseconds. But voltseconds cannot change if the input and output voltages are held fixed (frequency assumed fixed too). So for a given application condition, we conclude that though we may change the $L$ or the $\Delta I$ individually, since $L\Delta I$ equals $V_{\text{sec}}$ we still cannot change their product (or the flux swing) unless we change the number of turns and the area of the core (only). So for all practical purposes, if we try to optimize by changing cores and/or turns, the only applicable relationship is

$$\Delta B \propto \frac{1}{NA} \quad \text{(fixed) frequency}$$

In the preceding discussion we have assumed that the frequency is fixed. We will discuss the full effect of frequency later. For now we recognize that voltseconds is inversely proportional to frequency. So, we can extend the proportionality equation for the case of variable frequency by writing

$$\Delta B \propto \frac{1}{NAf}$$

Note that starting with

$$\Delta B = \frac{L \times \Delta I}{N \times A}$$

and using

$$L \propto \frac{N^2 A}{l}$$

we can simplify to get an *alternative, but equivalent proportionality statement* for $B$ that should also help in selecting optimum components

$$\Delta B \propto \frac{N\Delta I}{l}$$

In this discussion, $A$ is the same as the effective area of the core $A_e$ and $l$ is the same as effective length $l_e$.

We can see in the next section that the form of equation we use for $\Delta B$ is just a matter of convenience since both give the same result.

## 9.7 A Feel for Core Loss Optimization by Understanding Variations

Here we will test some popular myths associated with magnetic components and also carry out some thought experiments which will give us a better feel for the subject and thereby help us in selecting more optimum cores. The reader may like to refer to Chap. 11 for the units commonly used in expressing core loss.

**Counterintuition.**   The size of the inductor is related primarily to its energy handling capability. The energy handling capability $E$ is $1/2 \times L \times I_{PK}^2$. For a given application, if we reduce inductance, we may think that this would cause an increase in $I_{PK}$, and thus in $E$ too. So we may think that if we reduce inductance, the size of the core will need to be increased. However, an actual calculation shows the converse is true. $I_{PK}$ is equal to $I_C + \Delta I/2$, and so the effect of varying $\Delta I$ on $I_{PK}$ (and on $E$) is much less than the effect from the reduction in $L$. The energy handling requirement actually decreases as we reduce $L$. This is also clear from Fig. 9.1. Reality is counterintuitive here.

### Varying the volume of the core

*Increasing each dimension, N unchanged.*   Core loss is sometimes expressed as a power loss density, that is, W/cm$^3$. On the basis of this some feel that a larger volume will automatically increase the absolute value of this loss as expressed in watts. We need an example here to check this out. Suppose we increase *each* dimension of the core by a factor of 2. Its effective length $l_e$ doubles but its effective area $A_e$ increases four times. If we keep the number of turns unchanged, since $L$ is proportional to $A_e/l_e$, the inductance will double. Since $r$ is inversely proportional to $L$, $\Delta I$ halves. But core loss typically varies as $B^{2.5}$ for ferrites and powdered iron. Here $B$ is by industry practice meant to mean $B_{AC}$, which is *half* the total flux density swing $\Delta B$.

But, as discussed above, $\Delta B$ is proportional only to $1/NA_e$; thus the net effect on total core loss as expressed in watts is

$$P_{CORE} \propto \Delta B^{2.5} \times Volume = \left(\frac{1}{4}\right)^{2.5} \times 8 = 0.25$$

The total core loss has actually *decreased* by a factor of 4. So, larger cores may not be cost-effective, but they are not going to hurt either.

**Note:** If all dimensions are changed by a factor $x$, the inductance also changes by the same factor, because $A_e/l_e = x^2/x = x$.

**Note:** We could have used the alternative proportionality equation for $B$ in the previous section and we would get the same results as above.

*Increasing each dimension, L unchanged.* Now had we tried to keep the inductance constant ($r$ constant too) as we increased the volume by eight times (each dimension doubled), we would have to decrease the number of turns by a factor of $2^{0.5} = 1.414$ since $L$ is proportional to turns squared. This would have kept $\Delta I$ unchanged. So, now we get

$$P_{CORE} \propto \Delta B^{2.5} \times Volume = \left(\frac{1}{\frac{1}{\sqrt{2}} \times 4}\right)^{2.5} \times 8 = 0.595$$

So, core loss has decreased. Note that the exponent of $B$ in the core loss equation is 2.5 here. Had it been 2 (as for Kool Mu), we would have got a result of 1 above, implying *no change* in core loss. If the exponent were less than 2 (implying a better material) in fact we would have got more core loss with larger core volumes (for same inductance).

*Increasing area, N unchanged.* Suppose we had increased the volume by four times by increasing the effective area, not the effective length, with no change in the number of turns. Then the inductance would increase four times and $\Delta I$ would reduce by 4. Then using $B \propto 1/NA$

$$P_{CORE} \propto \Delta B^{2.5} \times Volume = \left(\frac{1}{4}\right)^{2.5} \times 4 = 0.125$$

So, core loss has decreased.

*Increasing area, L unchanged.* Now had we tried to keep the inductance constant ($r$ constant) as we increased the volume, we would need to decrease the number of turns by a factor of 2 since $L$ is proportional to turns squared and we are trying to keep it constant. But this would have kept $\Delta I$ unchanged. So, now we get

$$P_{CORE} \propto \Delta B^{2.5} \times Volume = \left( \frac{1}{1/2 \times 4} \right)^{2.5} \times 4 = 0.707$$

The core loss has decreased. Note that the exponent of $B$ in the core loss equation is 2.5 here. As in the earlier case, had the exponent of $B$ in the core loss equation been 2, we would again have got a result of 1 above, implying no change in core loss. If the exponent were less than 2 (implying a better material) in fact we would have got *more* core loss with larger core volumes (for same inductance). This is explained by the fact that a better material has lower losses to start with.

### Varying the frequency

*Increasing* f, L *unchanged.*   Since the core loss equation is the product of two terms—one related to flux swing and one to frequency—with the exponent of the frequency term being typically around 1.5, it is sometimes believed that increasing the frequency (for a given core and inductance) will automatically increase core losses. Here we use the full form of the $B$ proportionality equation, i.e., $\Delta B \propto 1/(NAf)$. We note that voltseconds is inversely proportional to frequency. So if we double the frequency, the variation of the core loss is

$$P_{CORE} \propto \Delta B^{2.5} \times f^{1.5} \times Volume = \left( \frac{1}{1 \times 1 \times 2} \right)^{2.5} \times 2^{1.5} \times 1 = 0.5$$

So, core loss will actually decrease. But note that we have kept inductance constant here.

*Increasing* f, r *unchanged,* N *decreased.*   If we keep inductance fixed as we increase frequency, $r$ will decrease. We know from a design optimization viewpoint that this is advisable to keep $r$ at around 0.4, irrespective of topology, frequency, or application conditions. So, let us decrease the number of turns accordingly, *while keeping to the same core size for now.* Therefore, if we increase the frequency four times, we need to reduce the inductance by the same factor, and that means that the number of turns has to be halved. So, now for the variation in core loss we get (using $\Delta B \propto 1/(NAf)$)

$$P_{CORE} \propto \Delta B^{2.5} \times f^{1.5} \times Volume = \left( \frac{1}{1/2 \times 1 \times 4} \right)^{2.5} \times 4^{1.5} \times 1 = 1.414$$

The core loss has increased.

*Increasing* f, r *unchanged,* N *unchanged, decreasing each dimension.* But the whole idea of using high switching frequencies is to be able to use *smaller* inductors. What if we increased frequency by four times, but achieved the reduction in *L* not by changing the number of turns but by making the core physically smaller. If we do so by altering all three dimensions of the core, then we know that to achieve a reduction of 4 in its inductance, we need to change each dimension by the same factor. So, $A_e$ will now change by a factor of $4 \times 4 = 16$. Applying this result to our core loss equation we get

$$P_{CORE} \propto \Delta B^{2.5} \times f^{1.5} \times Volume = \left( \frac{1}{1 \times \frac{1}{16} \times 4} \right)^{2.5}$$

$$\times 4^{1.5} \times \frac{1}{4 \times 4 \times 4} = 4$$

So, core loss has increased tremendously.

*Increasing* f, r *unchanged,* N *unchanged, area decreased.* What if we reduced the inductance by 4, not by changing the number of turns, neither changing $l_e$ but only by changing $A_e$, that is, by a factor of 4 (volume reduced by the same factor)? Then

$$P_{CORE} \propto \Delta B^{2.5} \times f^{1.5} \times Volume = \left( \frac{1}{1 \times \frac{1}{4} \times 4} \right)^{2.5}$$

$$\times 4^{1.5} \times \frac{1}{4 \times 4 \times 1} = 0.5$$

So, core loss has decreased.

*Our conclusion is that to reduce core loss it is not just enough to decrease inductance as we increase frequency but to actually reduce the physical size of the magnetic core.* However, decreasing the size of the core should be carried out carefully. There are optimum combinations of $A_e$, $l_e$, and $N$ that contribute to an optimum high-frequency magnetic design. Some don't!

But we must point out here that if we change inductance in inverse proportion to frequency, which is what we need to do to maintain $r$, then the energy requirement of the core also decreases roughly in the same proportion. This is fortuitous because at no time do we want to select a core size based on optimum core loss, which compromises the required energy handling capability. But we must also be careful that any $A_e$ and $l_e$ combination we pick must have enough core volume $V_e = A_e \times l_e$ to be able to do the job.

Magnetics vendors are constantly putting out new parts. Now the designers know that just as they often set disparate and sometimes meaningless current ratings, (e.g., one for heating and one for saturation) they don't necessarily always pick the most optimum combination of core characteristics and winding to achieve the most optimum results. All inductors are just not being created equal. So, especially in high-frequency designs, a careful evaluation of any off-the-shelf part is well deserved.

## 9.8   A Walk-Through Example

The input dc voltage is 24 V into a buck converter. The output is 12 V at a maximum load of 1 A. We wish to allow a current ripple ratio of 30 percent (at maximum load). We assume $V_{SW} = 1.5$ V, $V_D = 0.5$ V, and $f = 150,000$ Hz.

For a buck regulator, the duty cycle $D$ is

$$D = \frac{V_O + V_D}{V_{IN} - V_{SW} + V_D}$$

So, the switch on-time is

$$t_{ON} = \frac{D}{f} = \frac{(12 + 0.5) \times 10^6}{(24 - 1.5 + 0.5) \times 150000} \ \mu s$$

$$t_{ON} = 3.62 \ \mu s$$

So, the $V_{\mu sec}$ for the application is

$$V_{\mu sec} = (V_{IN} - V_{SW} - V_O) \times t_{ON} = (24 - 1.5 - 12) \times 3.62$$

$$V_{\mu sec} = 38.0$$

$$L = \frac{V_{\mu\, sec}}{r \times I_O} \ \mu H$$

$$L = \frac{38.0}{0.3 \times 1.0} \ \mu H$$

$$L = 127 \ \mu H$$

The required energy handling capability is calculated next. Every cycle, the peak current is

$$I_{PEAK} = I_O + \frac{Et}{2 \bullet L} = 1.0 + \frac{38}{2 \bullet 127} \ A$$

$$I_{PEAK} = 1.15 \ A$$

The required energy handling capability $E$ is

$$E = \frac{1}{2} \times L \times I_{PEAK}^2 \ (\mu J)$$

$$E = \frac{1}{2} \times 127 \times 1.15^2 = 84 \ \mu J$$

## 9.9  Choosing an Inductor

*Note that the vendor we are going to consider uses the symbol ET for voltµseconds. To avoid confusion we too will do the same from this point onward.*

Our first pass selection is based on the inductance calculated above and the dc current rating (maximum load). We tentatively select a part from Pulse Engineering (at *http://www.pulseeng.com*) simply because its $L$ and $I_{DC}$ are close to our requirements, even though the rest does not seem to fit our application very well, not at least at first sight (see Table 9.3). In particular, the frequency for which the inductor was designed is 250 kHz, but our application is 150 kHz. Note that we will dispel another intuitive myth—since we are decreasing the frequency of operation of the inductor, our flux swing will increase and our core loss will therefore go up, along with the peak flux density and energy. In fact, the reverse happens in our case, and that is one reason why it is important to follow the full mathematical procedure presented below.

**Note**

- For core loss equations it is conventional to use half the peak-to-peak flux swing. So, for most vendors $B$ in their datasheets actually refers to $\Delta B/2$. This must be kept in mind in the calculations that follow.

- The astute designer can recognize the exponents of $B$ and $f$ in the core loss equation in Table 9.3, as corresponding to a ferrite

**TABLE 9.3  Specifications of Choke to be Evaluated for Application.**

| Part number | Reference values | | | Control values | Calculation data |
|---|---|---|---|---|---|
| P0150 | $I_{DC}$ (Amps) | $L_{DC}$ (µH) | *ET* (Vµsecs) | *DCR* (nom) mΩ | $ET_{100}$ (Vµsecs) |
| | 0.99 | 137 | 59.4 | 387 | 10.12 |

The inductor is such that 380 mW dissipation corresponds to a 50°C rise in temperature.
The core loss equation for the core is $6.11 \times 10^{-18} \times B^{2.7} \times f^{2.04}$ mW where $f$ is in Hz and $B$ is in gauss.
The inductor was designed for a frequency of 250 kHz.
$Et_{100}$ is the Vµsecs at which $B$ is 100 G.

(which this is). Powdered iron material 52 (from Micrometals Inc. at *http://www. micrometals.com*) would for example have given us an equation of the form $\propto B^{2.11} \times f^{1.26}$).

- Most off-the-shelf inductors are designed for a temperature rise of 30 to 50°C.

## 9.10  Evaluating the Inductor in Our Application

We have the inductor operating under its test conditions. We will now map its performance to our specific application conditions. We follow the summary in Table 9.4. So set unprimed parameters to be the *design values* and the corresponding primed parameters as the *application values*. The overall mapping procedure is described hereafter.

The following are the *design conditions* of the inductor:

- $I_{DC}$
- $ET$
- $f$
- $T_{AMBIENT}$

The *application conditions* we will immerse the chosen inductor in are as follows:

- $I'_{DC}$
- $ET'$
- $f'$
- $T'_{AMBIENT}$

In going from the *design conditions* to the *application conditions* the following are considered virtually constant:

- $L$
- $DCR$
- $Rth$
- *The core loss equation*

Finally, to approve the inductor for our application we should certify that

1. $r$ is acceptable, that is, choice of $L$
2. $B_{PK}$ is OK (important in core saturation limited inductors)

**TABLE 9.4 Implementing Mapping Procedure for an Inductor (from left to right).**

| Design parameters | Design conditions $I_{DC}, ET, f,$ $T_{AMBIENT}$ | Application conditions $I'_{DC} \equiv I_C, ET',$ $f', T'_{AMBIENT}$ |
|---|---|---|
| Current swing (A) | $\Delta I = \dfrac{ET}{L}$ | $\Delta I' = \Delta I \bullet \left[\dfrac{ET'}{ET}\right]$ |
| Current ripple ratio | $r = \dfrac{ET}{L \bullet I_{DC}}$ | $r' = r \bullet \left[\dfrac{ET' \bullet I_{DC}}{ET \bullet I'_{DC}}\right]$ |
| Peak current (A) | $I_{PEAK} = I_{DC} + \dfrac{ET}{2 \bullet L}$ | $I'_{PEAK} = I_{PEAK} \bullet \left[\dfrac{(2 \bullet L \bullet I'_{DC}) + ET'}{(2 \bullet L \bullet I_{DC}) + ET}\right]$ |
| RMS current in inductor (A) | $I_{RMS} = \left[I_{DC}^2 + \dfrac{ET^2}{12 \bullet L^2}\right]^{1/2}$ | $I'_{RMS} = I_{RMS} \bullet \left[\dfrac{(12 \bullet I_{DC}^2 \bullet L^2) + ET'^2}{(12 \bullet I_{DC}^2 \bullet L^2) + ET^2}\right]^{1/2}$ |
| Flux density swing (gauss) | $\Delta B = \dfrac{ET}{ET_{100}} \bullet 200 = \dfrac{100 \bullet ET}{N \bullet A_e}$ | $\Delta B' = \Delta B \bullet \left[\dfrac{ET'}{ET}\right]$ |
| Peak flux density (gauss) | $B_{PEAK} = \dfrac{200}{ET_{100}} \bullet \left[(I_{DC} \bullet L) + \dfrac{ET}{2}\right]$ | $B'_{PEAK} = B_{PEAK} \bullet \left[\dfrac{2 \bullet L \bullet I'_{DC} + ET'}{2 \bullet L \bullet I_{DC} + ET}\right]$ |
| Copper losses (mW) | $P_{CU} = DCR \bullet \left[I_{DC}^2 + \dfrac{ET^2}{12 \bullet L^2}\right]$ | $P'_{CU} = P_{CU} \bullet \dfrac{(12 \bullet I_{DC}^2 \bullet L^2) + ET'^2}{(12 \bullet I_{DC}^2 \bullet L^2) + ET^2}$ |
| Core losses (mW) | $P_{CORE} = a \bullet \left[\dfrac{ET}{ET_{100}} \bullet 100\right]^b \bullet f^c$ | $P'_{CORE} = P_{CORE} \bullet \left[\left(\dfrac{ET'}{ET}\right)^b \bullet \left(\dfrac{f'}{f}\right)^c\right]$ |
| Energy in core (μJ) | $E = \dfrac{1}{2} \bullet L \bullet \left[I_{DC} + \dfrac{ET}{2 \bullet L}\right]^2$ | $E' = E \bullet \left[\dfrac{(2 \bullet L \bullet I'_{DC}) + ET'}{(2 \bullet L \bullet I_{DC}) + ET}\right]^2$ |
| Temperature rise (°C) | $\Delta T = Rth \bullet \dfrac{P_{CU} + P_{CORE}}{1000}$ | $\Delta T' = \Delta T \bullet \left[\dfrac{P'_{CU} + P'_{CORE}}{P_{CU} + P_{CORE}}\right]$ |

$ET$ in V$\mu$secs, $DCR$ in m$\Omega$, $L$ in $\mu$H, $f$ in Hz, $A_e$ in cm$^2$, and $N$ is number of turns.

3. $I_{PK} < I_{CLIM}$ (or the controller will be unable to provide the required power)
4. $\Delta T$ is OK (evaluate $P_{CU} + P_{CORE}$)

We assume that the vendor has provided all the following inputs:

- $ET$ (V$\mu$secs)
- $ET_{100}$ (V$\mu$secs per 100 G)
- $L$($\mu$H)
- $I_{DC}$ (Amps) (maximum rating)
- $DCR$ (m$\Omega$)

- $f$ (Hz)
- The form for core loss (in mW) as $a \times B^b \times f^c$, where $B$ is in gauss, $f$ in Hz. $B$ is half the peak-to-peak flux swing.
- Thermal resistance of inductor in free air (°C/W)

If any of these are unknown, the vendor should be contacted.
*We now proceed with the specific example (see Table 9.4)*
The design conditions of the inductor are:

- $ET = 59.4$ Vµs
- $f = 250{,}000$ Hz
- $I_{DC} = 0.99$ A

Our application conditions are

- $ET' = 38$ Vµs
- $f' = 150{,}000$ Hz
- $I'_{dc} = 1$ A

(We assume that $T_{AMBIENT}$ is unchanged, and so we have ignored.)

### a) Current ripple ratio
*By design*

$$r = \frac{ET}{L \bullet I_{DC}}$$

$$r = \frac{59.4}{137 \bullet 0.99}$$

$$r = 0.438$$

*In application*

$$r' = r \bullet \left[ \frac{ET' \bullet I_{DC}}{ET \bullet I'_{DC}} \right]$$

$$r' = 0.438 \bullet \left[ \frac{38 \bullet 0.99}{59.4 \bullet 1} \right]$$

$$r' = 0.277$$

We expected $r'$ to be slightly lower than 0.3 since the chosen inductor has a higher inductance than we required (137 µH instead of 127 µH).

**b) Peak flux density**

*By design*

$$B_{PK} = \frac{200}{ET_{100}} \bullet \left[ (I_{DC} \bullet L) + \frac{ET}{2} \right] \text{ G}$$

$$B_{PK} = \frac{200}{10.12} \bullet \left[ (0.99 \bullet 137) + \frac{59.4}{2} \right] \text{ G}$$

$$B_{PK} = 3267 \text{ G}$$

*In application*

$$B'_{PK} = B_{PK} \bullet \left[ \frac{2 \bullet L \bullet I'_{DC} + ET'}{2 \bullet L \bullet I_{DC} + ET} \right] \text{ G}$$

$$B'_{PK} = 3267 \bullet \left[ \frac{2 \bullet 137 \bullet 1 + 38}{2 \bullet 137 \bullet 0.99 + 59.4} \right] \text{ G}$$

$$B'_{PK} = 3084 \text{ G}$$

which is less than $B_{PK}$ and, therefore, acceptable.

**c) Peak current**   To ensure that the regulator will deliver rated load, we need to ensure that the peak current is less than the internal current limit of the switcher IC.

*By design*

$$I_{PK} = I_{DC} + \frac{ET}{2 \bullet L} \text{ A}$$

$$I_{PK} = 0.99 + \frac{59.4}{2 \bullet 137} \text{ A}$$

$$I_{PK} = 1.21 \text{ A}$$

This corresponds to a $B$-field of 3267 G as calculated earlier. This gives us an energy handling capability of

$$E = \frac{1}{2} \times L \times I_{PK}^2 \text{ (μJ)}$$

$$E = \frac{1}{2} \times 137 \times 1.21^2 \text{ (μJ)}$$

$$E = 100 \text{ (μJ)}$$

whereas we required at least 84 μJ. So, the inductor seems correctly sized and we can proceed with the analysis.

*By application*

$$I'_{PK} = I_{PK} \bullet \left[ \frac{(2 \bullet L \bullet I'_{DC}) + ET'}{(2 \bullet L \bullet I_{DC}) + ET} \right] \text{ (A)}$$

$$I'_{PK} = 1.21 \bullet \left[ \frac{(2 \bullet 137 \bullet 1.0) + 38}{(2 \bullet 137 \bullet 0.99) + 59.4} \right] \text{ A}$$

$$I'_{PK} = 1.14 \text{ A}$$

This corresponds to a *B*-field of 3084 G as just calculated. We check that the minimum value of the current limit over temperature and tolerance is 2.3 A. So, since the peak value $I'_{PK}$ is less than 2.3 A, the controller will be able to provide the desired output power (without hitting the current limit).

**d) Temperature rise**
*By design*

$$P_{CU} = DCR \bullet \left( I_{DC}^2 + \frac{ET^2}{12 \bullet L^2} \right) \text{ (mW)}$$

$$P_{CU} = 387 \bullet \left( 0.99^2 + \frac{59.4^2}{12 \bullet 137^2} \right) \text{ mW}$$

$$P_{CU} = 385 \text{ mW}$$

$$P_{CORE} = a \bullet \left( \frac{ET}{ET_{100}} \bullet 100 \right)^b \bullet f^c \text{ (mW)}$$

where the vendor has provided that $a = 6.11 \times 10^{-18}$, $b = 2.7$, and $c = 2.04$. So,

$$P_{CORE} = 6.11 \bullet 10^{-18} \bullet \left( \frac{59.4}{10.12} \bullet 100 \right)^{2.7} \bullet f^{2.04} \text{ mW}$$

$$P_{CORE} = 18.7 \text{ mW}$$

Hence,

$$\Delta T = Rth \bullet \frac{P_{CU} + P_{CORE}}{1000} \, ^\circ \text{C}$$

The vendor has stated that 380 mW dissipation corresponds to a 50°C rise in temperature. So, the thermal resistance of the inductor is

$$R_{TH} = \frac{50}{(380/1000)} = 131.6 ^\circ \text{C/W}$$

$$\Delta T = 131.6 \bullet \frac{385 + 18.7}{1000} = 53 ^\circ \text{C}$$

*In application*

$$P'_{CU} = P_{CU} \bullet \frac{(12 \bullet I'^2_{DC} \bullet L^2) + ET'^2}{(12 \bullet I^2_{DC} \bullet L^2) + ET^2} \text{ (mW)}$$

$$P'_{CU} = 385 \bullet \frac{(12 \bullet 1^2 \bullet 137^2) + 38^2}{(12 \bullet 0.99^2 \bullet 137^2) + 59.4^2} \text{ mW}$$

$$P'_{CU} = 389 \text{ mW}$$

$$P'_{CORE} = P_{CORE} \bullet \left[ \left( \frac{ET'}{ET} \right)^b \bullet \left( \frac{f'}{f} \right)^c \right] \text{ (mW)}$$

$$P'_{CORE} = 18.7 \bullet \left[ \left( \frac{38}{59.4} \right)^{2.7} \bullet \left( \frac{150000}{250000} \right)^{2.04} \right] \text{ mW}$$

$$P'_{CORE} = 2 \text{ mW}$$

So,

$$\Delta T' = \Delta T \bullet \left[ \frac{P'_{CU} + P'_{CORE}}{P_{CU} + P_{CORE}} \right] \text{ (}^\circ\text{C)}$$

$$\Delta T' = 53 \bullet \left[ \frac{389 + 2}{385 + 18.7} \right] = 51^\circ\text{C}$$

which was deemed to be acceptable in this application.

This completes the qualification analysis for the short-listed inductor. The summary is as follows:

- The inductor is designed for about 50°C rise in temperature over ambient at a load of 1 A. We will get 51°C.

- The copper loss (385 mW) predominates and the core loss is relatively small. We will get 389 mW.

- The peak flux density is about 3200 G, which occurs at a peak instantaneous current of 1.2 A. We will get in a total inductor loss of 3080 G at 1.15 A.

- The rated energy handling capability of the core is 100 μJ. We require 84 μJ.

# 10

# Flyback Transformer Design

We are using MKS units, unless otherwise stated.

## 10.1  Design Equations

**The voltage-dependent equation: a practical form.**  We discussed several forms of the voltage-dependent equation, also known as Faraday's law, in Chap. 7. Here we will provide a more practical form as applied to most switching converters.

The basic form of the equation is

$$B_{AC} = \frac{V_{AVG}\,\Delta t}{2 \times NA_e}$$

As we mentioned, this only talks about the *change* in the $B$-field. However, we can easily relate it to the peak field too for most magnetic materials. Let us first assume that the gapped inductor has finally been designed fairly optimally so that its peak field is very close to $B_{SAT}$. Then realizing that *current and B are proportional to each other* (though this is not entirely true for powdered iron), the relationship for the current ripple ratio $r$ must apply to the $B$-field too. Since $\Delta B = 2B_{AC}$ this gives us the relationship

$$\frac{2 \times B_{AC}}{B_{DC}} = r$$

But by definition, $B_{DC} = B_{PK} - B_{AC}$ is always true, so

$$\frac{2 \times B_{AC}}{B_{PK} - B_{AC}} = r$$

Simplifying, and combining with the voltage-dependent equation we get

$$N = \left(\frac{2}{r} + 1\right) \times \frac{V_{ON}D}{2 \times B_{PK}A_e f}$$

This will, for example, give us the primary number of turns for a flyback transformer, but it can also be used for any dc-dc converter inductor too. Note that this is an interesting relationship. It does not depend directly on the air gap, the effective length, or effective volume of the core, and not even on the permeability! It does not depend directly on load current either, and that is because, as implied by our use of $r$ (which is defined only for CCM), we are talking only of continuous conduction mode.

**Energy stored (as related to volume of core in ungapped core).** We have shown in Chap. 7 that

$$E_c = \frac{1}{2}\frac{B^2}{\mu_c} \times A_e l_e \text{ J}$$

i.e.,

$$E_c = \frac{1}{2}\frac{B^2}{\mu\mu_0} \times V_e \text{ J}$$

Here $\mu_c$ is the absolute permeability of the core, $\mu_0$ is the absolute permeability of air, and $\mu$ is the relative permeability of the core material.

Note that it is often more useful to talk in terms of energy per unit volume instead, i.e.,

$$\boxed{\frac{E_c}{V_e} = \frac{1}{2}\frac{B^2}{\mu\mu_0}} \text{ J/m}^3$$

For a typical ferrite, assuming the relative permeability is about $\mu = 2000$, and the saturation flux density $B_{SAT} = 0.3$ T (3000 G), we get (for most *ungapped* ferrite cores) a typical power density of

$$\frac{E_c}{V_e} = 17.91 \text{ J/m}$$

or

$$\boxed{\frac{E_c}{V_e} \approx 18} \text{ J/m}^3 (\mu = 2000, B_{SAT} = 0.3T)$$

This is the energy stored only inside the core, for either a gapped or an ungapped core. It is also the overall energy stored in an ungapped core.

**General energy relationships for a gapped core.** From Chap. 7, the summarized relationships of the energy components in terms of $z$ are

$$\boxed{\frac{E}{E_c} = z} \quad \boxed{\frac{E_g}{E_c} = z - 1} \quad \boxed{\frac{E_g}{E} = 1 - \frac{1}{z}} \text{ (general equations)}$$

$E$ is the total energy (gap plus core), $E_c$ is the energy in the core, $E_g$ is the energy in the gap, and $z$ is defined as

$$\boxed{z = \frac{l_e + \mu l_g}{l_e}} \text{ (definition)}$$

**General relationships for $A_L$ and $\mu$.** The inductance index varies as $1/z$

$$\boxed{A_L = \frac{1}{z}\left(\frac{\mu\mu_0 A_e}{l_e}\right) \times 10^9} \text{ (nH/Turns}^2\text{) (general equation)}$$

We can solve for the *relative* permeability of the material $\mu$

$$\boxed{\mu = \frac{A_{L\_nogap} \times l_{e\_mm} \times 10}{4\pi \times A_{e\_sqmm}}} \text{ (general equation)}$$

**Duty cycle of universal input flyback with $V_{OR} = 100$ V.** Now we will take the most typical design case for an off-line flyback and come up with equations that will help speed up the design process.

The inductor design must proceed at the *lowest input* for this topology (as for the boost too) since the peak currents are highest when $D$ is closer to 1. Assuming 100 percent efficiency, the average inductor current (in the *equivalent primary buck-boost* model, see Chap. 6) is

$$I_L = \frac{I_O/n}{1 - D}$$

where $n$ is the turns ratio $n_P/n_S$. The duty cycle is

$$D = \frac{nV_O}{nV_O + V_{IN}}$$

90 Vac is rectified to $90 \times 1.414 = 127$ V. This is $V_{IN}$. So

$$D = \frac{nV_O}{nV_O + V_{IN}} = \frac{20 \times 5}{20 \times 5 + 127} = 0.44$$

$$\boxed{D = 0.44} \quad (90 \text{ Vac}, \, V_{OR} = 100 \text{ V})$$

**The area × turns rule.** We know that a flyback transformer design is always done at minimum $V_{IN}$. Therefore, if such a power supply is designed for a worldwide input (90 to 270 Vac), and uses a reflected output voltage $V_{OR}$ (i.e., $nV_O$) of 100 V and is switching at a frequency of 100 kHz, with an optimum $r$ of 0.4, we get from the above equation

$$NA_e = \left(\frac{2}{r} + 1\right) \times \frac{V_{IN}D}{2 \times B_{PK} \times f} = \left(\frac{2}{0.4} + 1\right) \times \frac{127 \times 0.44}{2 \times 0.3 \times 10^5}$$

$$NA_e \cong 5.588 \times 10^{-3} \text{m}^2$$

So, irrespective of the core shape, core size, air gap, or even the output power, we need to set the following design target for any universal input flybacks, which are *optimally designed* $(r = 0.4)$, and use a ferrite transformer:

$$\boxed{(Area_{sqmm} \times Turns) \approx 5600} \quad (r = 0.4, V_{OR} = 100V, B_{SAT}$$

$$= 0.3 \text{ T}, \, f = 100 \text{ kHz}, 90 \text{ Vac})$$

Or equivalently, the rule for any general frequency $f$ in Hz

$$\boxed{(Area_{sqmeter} \times Turns \times f_{Hz}) \approx 560} \quad (r = 0.4, V_{OR} = 100 \text{ V}, B_{SAT}$$

$$= 0.3 \text{ T}, 90 \text{ Vac})$$

### 10.2   Worked Example (Part 1)

A 90 to 270 Vac (worldwide input) flyback with an output of 5 V@5 A and with a transformer turns ratio of 20 needs a core. Suggest a suitable candidate. The efficiency is 70 percent and the switching frequency is 100 kHz.

**Required L.** We know from Chap. 6 that assuming perfect efficiency, the center of the primary-side current ramp is

$$I_C = \frac{I_{OR}}{1 - D}$$

We now increase the current to account for the less than perfect efficiency. Note that the center of the switch current ramp is the same as the average inductor current of the equivalent primary buck-boost model. So

$$I_L \equiv I_C = \frac{5/20}{1 - 0.44} \times \frac{1}{70\%} = 0.64 \text{ A}$$

We set the current ripple to $\pm 20$ percent (i.e., $r = 0.4$). So the required ac ramp is

$$\Delta I = r \times I_L = 0.4 \times 0.64 = 0.26 \text{ A}$$

The required inductance is derived from $V = LdI/dt$ as applied to the *on* period

$$L = \frac{V_{IN} \times D/f}{\Delta I} = \frac{127 \times 0.44/100000}{0.26} \Rightarrow 2.15 \text{ mH}$$

$$\boxed{L = 2.15} \text{ mH}$$

This is the inductance calculated from the required $r$.

**Required energy.**   The peak operating current is given by

$$I_{PK} = I_L + \frac{\Delta I}{2} = 0.64 + 0.13 = 0.77 \text{ A}$$

So the required energy handling capability of the core (*based on this estimated peak current*) is

$$E_{PK} = \frac{1}{2} L I_{PK}^2 = \frac{1}{2} \times 2.15 \times 10^{-3} \times 0.77^2 \text{ J}$$

$$\boxed{E_{PK} = 6.37 \times 10^{-4}} \text{ J}$$

**Pick a core.**   Let us first just examine a popular "candidate" for this level of power, the E25/13/7 (EF25) core set. Later we will discuss more clearly what makes such a core a suitable choice. Its key parameters are given by its datasheet as

$$A_e = 52.5 \text{ mm}^2, l_e = 57.5 \text{ mm}, A_L = 2000 \text{ nH/Turns}^2 \Leftarrow \textbf{(EF25)}$$

**N from voltage-dependent equation.**   Our frequency is 100 kHz, and our $V_{OR} = V_O \times n = 5 \times 20 = 100$ V; thus for an optimal design we can simply use our typical equation

$$(Area_{sqmm} \times Turns) \approx 5600$$

$$N = \frac{5600}{52.5} = 107 \text{ Turns}$$

**Required $A_L$.** We have calculated that the required value of $L$ is 2.15 mH. Since $N$ is known, the calculated $A_L$ is

$$A_L = \frac{2.15 \times 10^6}{107^2} = 188\text{nH/Turns}^2$$

**Required $z$ from energy storage considerations.** For the core $V_e = A_e \times l_e = 3.0$ cm$^3$. So its energy handling capability (if ungapped) is

$$E_c = 18 \times \frac{3}{100^3} = 5.37 \times 10^{-5} \text{ J}$$

But we have calculated that we need to store at least $6.4 \times 10^{-4}$ J in our application (to avoid saturation at 90 Vac). So the balance of the energy, i.e., $6.37 - 0.537 \Rightarrow 5.833 \times 10^{-4}$ J must be able to reside in the air gap. So

$$z = \frac{E}{E_C} = \frac{6.37}{0.537} = 11.86$$

This is a reasonable and practical value since $z$ can *typically vary between the values of 1 and 10*, though in some cases it may even be set as high as 25.

In general, we may not be able to accept a certain core choice for several reasons. This could be because the amount of copper required just cannot be accommodated in the available window, or if the air gap is clearly an absurdly large (impractical) value. Then we will certainly need to look for other possible cores.

**Note:** To meet holdup-time requirements the peak energy handling capability (possibly the core size) *may have to be increased several times* for any flyback transformer (or boost choke), to prevent saturation under outages (or even under a normal power-down sequence). See Chap. 5 and Chap. 6 for a clearer understanding of holdup-time issues.

**Required $z$ from inductance considerations.** We know that $A_L$ *varies as* $1/z$, so, $z$ can be calculated from

$$z = \frac{A_{L\_nogap}}{A_{L\_gapped}} = \frac{2000}{188} = 10.64$$

We notice that we have slightly different values of $z$ emerging from the two calculations above. Which one is right? The energy relationship $E_C/V_e = 17.9$ J/m$^3$ was clearly a slight approximation and was provided only to speed up core selection (which we will take up next). In particular, it was based on an assumption that $\mu = 2000$. As a matter of fact, we can reverse-calculate the permeability from the more accurate value of $z$ and we will see that it is close to, but not quite equal to, 2000. That explains the slight discrepancy between the $z$ values.

Size of air gap:

$$l_g = \frac{l_e(z-1)}{\mu} = \frac{57.5 \times (10.64 - 1)}{1743} = 0.32 \text{ mm}$$

Note that the gap length is $0.32/57.5 \cong 0.6$ percent of the effective length here. This is a reasonable number for gapped cores in general (for any material). A gap length of up to 2 to 3 percent of effective length is considered practicable.

**Permeability of the core.**  From the equations derived above

$$\mu = \frac{A_{L\_nogap} \times l_{e\_mm} \times 10}{4\pi \times A_{e\_sqmm}} = \frac{2000 \times 57.5 \times 10}{4\pi \times 52.5}$$

$$\mu = 1743$$

If the designer is surprised that we seem to have first calculated the air gap, and now we are finding the permeability, he or she should remember that the permeability was in fact implicit in the $A_L$ value provided for the core and already used by us. So, the calculation performed here for relative permeability amounts, basically, to just a check.

We can now see that $\mu = 1743$ is a little different from the value of 2000 we had assumed for the quick energy estimate equation. But we can see that $E_C/V_e$ is inversely proportional to $\mu$. So had $\mu$ been taken as 1743 instead of 2000, we would get the energy estimate equation as

$$\boxed{\frac{E_c}{V_e} = 17.91 \times \frac{2000}{1743} = 20.55 \text{ J/m}^3 (\mu = 1743, \ B_{SAT} = 0.3T)}$$

For our example, the corresponding $z$ would be proportionally *less from the earlier calculated value*, i.e.,

$$z = 11.86 \times \frac{1743}{2000} = 10.34$$

The slight remaining discrepancy between $z = 10.34$ (calculated from energy considerations above) and $z = 10.64$ (calculated previously using inductance) is explained by the fact that the equation used to find $N$ was a little approximate too. Plus we had also rounded up to the nearest integral value of turns. But, either way, the calculated values are well within normal tolerances on air gap and $A_L$ anyway.

## 10.3  Some Finer Points of Optimization

We observe that the number of (primary) turns required varies as $A^{-1}$ i.e., it is inversely proportional to the effective area of the core.

We can also show that the total length of the primary winding then varies as $A^{-1/2}$ (copper losses will vary similarly).

We have been assuming that it is possible to set $r = 0.4$ and $B_{PK}$ close to 0.3 T (3000 G) *simultaneously*. That may however not be possible in practice.

In a flyback, with a turns ratio of say 20, we get a reflected output voltage $V_{OR}$ of 100 V for a 5 V output. We know from Chap. 6 that for this topology, the $V_{OR}$ is a major design goal, and is usually set in the range of 70 to 130 V for worldwide input off-line power supplies. But since the secondary number of turns must be an integer, the corresponding number of primary turns will vary in discrete steps of *20 turns for each additional secondary turn we decide to use,* thus denying us a smooth "continuum" of possible values. The inductance too will jump similarly *in steps* related to its $A_L$. Therefore, from the practical form of the voltage-dependent equation we can conclude the following:

For a given core, if we try to set $B_{PK}$ close to $B_{SAT}$, we can expect that $r$ will have to be allowed to jump in corresponding steps, and therefore we may or may not be able to set it close to the optimum value of 0.4.

If we try to set $r$ to say 0.4, $B_{PK}$ will necessarily jump in steps, and we may or may not be able to set it to the optimum value (close to $B_{SAT}$).

We may need to be a little flexible about the desired $V_{OR}$, as this will allow us similar flexibility in the desired turns ratio.

Notice that the voltage-dependent equation (practical form) has provided the required number of primary turns merely on the basis of the $A_e$ of the core. The $A_L$, $l_e$, and air gap have not even entered the picture!

As a corollary, since ultimately we will always be trying to operate close to $B_{SAT}$, the number of primary turns is essentially fixed for a given core in a given application, with a design target for $r$. *This asks us to put neither more nor less turns than so calculated.*

Note that though the number of turns required does not depend directly on the inductance, we will need to adjust the air gap to get the inductance to be such that $r$ is close to our design target of about 0.4.

If we allow for a different $r$, we can vary the number of primary turns. In fact, sometimes we may even be forced to do this if, for example, 107 turns cannot be physically accommodated in the available window area of the chosen core (and for some reason we don't want to go to a core with a larger window). We must, however, remember that if we do so, we have side effects from the higher $r$ on the rest of the converter as shown in Chap. 9.

Once $r$ is fixed, so is the required inductance for a given frequency (and given application). Therefore, from the basic definition of $L$ given previously, we can use our knowledge of $N$, $\mu$, $l_e$, and $A_e$ to calculate the required $z$ (and air gap).

We must finally check that the required number of turns can be physically accommodated in the available core window area. This will be discussed in more detail later in Chap. 11.

It is interesting to note that if $r$ is fixed by design, and $A_e$ is also fixed (i.e., we are sticking to a given core), then if we *also* try to fix $N$, we will see that the value of $B_{PK}$ is no longer directly in our hands! In such a case, we may end up not even operating close to $B_{SAT}$.

Therefore, the bottom line is that we just don't have so many available degrees of design freedom. *We have to compromise somewhere*—either on the design goal for $V_{OR}$, or on the $r$, or on the $B_{PK}$. We just cannot have everything. That is why transformer design is so tricky and rewarding at the same time.

## 10.4   Rule of Thumb for Quick Selection of Flyback Transformer Cores

We saw that for ferrites the typical energy density in the core material is

$$\frac{E_c}{V_e} \approx 18 \text{ J/m}^3$$

By adding an air gap, the energy increases as per

$$\frac{E}{E_c} = z$$

We have seen that a practical value of $z$ is about 10. So

$$\boxed{\frac{E}{V_e} \approx 180} \text{ J/m}^3 (\mu = 2000, \ B_{SAT} = 0.3 \text{ T, practical gapped core})$$

Using CGS units for volume, this is $180 \times 10^{-6} = 1.8 \times 10^{-4} \text{J/cm}^3$. However, we must remember that this is *not* the same parameter that most magnetics vendors provide when they draw *Hanna* curves. What they use is $LI^2/V_e$, whereas we are working with the actual physical energy density term $\frac{1}{2}LI^2/V_e$. However, the conversion is clear. So, our rule of thumb can also be written in a form that can be better compared with a vendor's datasheet.

$$\frac{LI^2}{V_e} \approx 3.6 \text{ J/cm}^3$$

We thus see that our rule of thumb is actually very close to the official guideline of 3.5 J/cm$^3$ provided for material-77 from Fair-Rite, on the basis of the Hanna curves.

Now, relating the peak energy $\frac{1}{2} \times LI^2$ to the input watts ($V_{IN} \times I_{SW\_CENTER} \times D$) for a flyback, we get a rather complicated expression

$$P_o = \frac{8r \times \left(E/V_e\right) \times f \times V_e}{(2+r)^2} \text{ W}$$

where $E$ is the total energy stored in the entire gapped structure (in joules), $V_e$ is the volume in m$^3$, and $f$ is the frequency in Hz.

Note, however, that both $D$ and $V_{IN}$ have got themselves cancelled out in this relationship.

For ferrites we just saw that the maximum energy density $E/V_e$ is about 180 J/m$^3$, so setting $r = 0.4$, we get a very simple relationship for predicting a ballpark figure for the power capability of a typical ferrite-based flyback transformer (assuming 100 percent efficiency):

$$P_o \approx 100 \times f \times V_e \text{ W}$$

This is based on our assumptions: $r = 0.4$, $B_{SAT} = 0.3T$, 90 Vac, and $\mu = 2000$. As expected, if we increase frequency (within reason), we can usually proportionally increase the power capability of a given core. However, we will eventually run into high-frequency effects like excessive core losses and proximity/eddy current losses.

## 10.5   Worked Example (Part 2)

We know that the EF25 has a volume of $52.5 \times 57.5 = 3019$ mm$^3$. So, at a frequency of 100 kHz,

$$P_o \approx 100 \times 10^5 \times 3019 \times 10^{-9} = 30.2 \text{ W.}$$

This means that the chosen core (EF 25) is almost certainly ok up to about 30 W output (*no holdup time designed in*). Thereafter, if we choose copper wire with a thickness just enough to give us a current density of about 400 cmils/A, we will probably never run into either a situation where we cannot accommodate the required copper in the available window area (for most commercial transformer cores), or where we end up with an overheated transformer. A discussion about current density will therefore follow next.

**Note:** Transformers commonly used in power supplies for information technology equipment are classified either as Class A (less than 65°C temperature rise as measured by thermocouples inserted into windings under rated maximum load), or Class B (similarly measured to be less than 85°C rise). Note that whereas Class A transformers just require individually approved materials, for Class B transformers *all* the materials have to be evaluated and approved *together*, thus forming a designated "insulation system."

**Note:** If we increase $r$ to 2 (critical conduction), we can actually almost double the throughput power capability of a given core. This happens because the term $8r/(2 + r)^2$ changes from 0.56 at $r = 0.4$ to 1 at $r = 2$—an increase of almost 100 percent. How does this correlate to our understanding that the size of the core is related to $I_{PK}^2$? If we decrease $L$, we increase $r$ and $I_{PK}$. However, energy is proportional to $LI_{PK}^2$. So, since $L$ is proportional to $N^2$, energy comes down despite the fact that the peak current has gone up. We can also think in terms of ampere-turns. This is related to the energy in the core. Though we increase the (peak) amperes, the turns have to be reduced at a faster rate so that the net ampere-turns actually decrease. But remember, we may have to oversize other components on the board. That is why an optimum of $r = 0.4$ is usually suggested.

**Note:** Vendors of some popular off-line integrated switcher ICs have often used the above logic to mask a major limitation of their devices, perhaps unknowingly. The problem is that they can only offer a family of devices with a certain limited set of discrete choices for current limit, as opposed to controllers where we can set virtually any current limit that we like. Since core size is essentially determined by the current limit, we do actually end up with a corresponding (and equally limited) number of discrete core sizes *for all applications* (within the range). This means that if, for example, the vendor has a 2-A (current limit) device and the next higher device is a 3-A device, then we are forced into using the latter device even for an application where the peak current is, say, only 2.2 A. We always expect and hope that the core size required for an application with a peak current of 2.2 A would be smaller than a *similar* application with a peak current of 3 A. "Similar" implies that we have set the same frequency, $r$ and $V_{OR}$ for each application, so we are not guilty of "comparing apples to oranges." Under these conditions, we would then expect the required $L$ to vary as $1/I_{PK}$, and so

$E$ would vary as $I_{PK}^2/I_{PK} = I_{PK}$. Therefore we expect the core size for the 2.2-A application to be 27 percent smaller than that required for a 3-A application (with an equivalent cost reduction). But since in off-line applications we always need to size the core according to current limit, not peak operating load current, *our inability to set the current limit at around 2.2 A means we cannot reduce the core size either*. The only way we can *seem* to reduce it is by changing our basic assumptions on $r$ and $V_{OR}$. Therefore, we *can* make the 2.2-A application have a smaller transformer, but all we may have done to achieve this is to *have increased* r *and/or* $V_{OR}$, and thereby simply *transferred the problem* over into the capacitors (larger rms currents). We should be cognizant about all these subtleties in transformer design and in component selection.

## 10.6    Circular mils (cmils)

We know that 1000 mils = 1 in = 25.4 mm. So 1 mil is also 25.4 $\mu$m ($\mu$m is also called a *micron*).

A circular mil (cmil) is a term used to define conductor cross-sectional areas using an arithmetic shortcut, in which the area of a round wire is taken as $d^2$ (diameter$^2$) in units of cmils, rather than the "correct" form $\pi d^2/4$ in units of sq. mils. Thus *1 cmil is equivalent to $\pi/4$ sq mils*. The rationale behind introducing this "shortcut" can be explained with the help of the following example. When we consider the case where we stack wires on top of each other to form a winding arrangement, the net area occupied by a wire is in effect the entire square cross-sectional area bounding that wire strand, as shown in Fig. 10.1. The rest is wasted space. The area allocated to each turn is thus d$^2$. So in this case, the

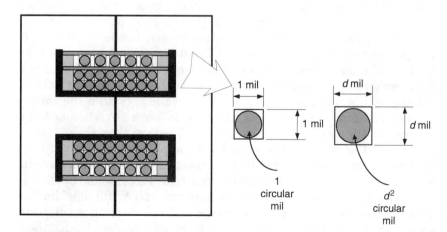

**Figure 10.1** Circular mil explained.

TABLE 10.1   Conversions for Diameter/Length, Area, and Current Density.

| Length | mil | In | ft | mm |
|---|---|---|---|---|
| mil | 1 | 0.001 | 0.000083 | 0.0254 |
| in | 1000 | 1 | 0.083 | 25.4 |
| ft | 12000 | 12 | 1 | 304.8 |
| mm | 39.37 | 0.03937 | 0.0033 | 1 |
| **Area** | **cmil** | **sq. mil** | **sq. in** | **sq. cm** |
| cmil | 1 | $\pi/4 = 0.7854$ | $7.854 \times 10^{-7}$ | $5.067 \times 10^{-6}$ |
| sq. mil | $4/\pi = 1.274$ | 1 | $10^{-6}$ | $6.452 \times 10^{-6}$ |
| sq. mm | 1973 | 1550 | 0.00155 | 0.01 |
| **Current density*** | **cmils/A** | **A/cm$^2$** | | **A/sq. in** |
| **x** cmils/A | x | 197353/x | | 1273000/x |
| **y** A/cm$^2$ | 197353/y | y | | 6.45 y |
| *440 cmils/A $\cong$ 440 A/cm$^2$, 1000 cm/A $\cong$ 1270 A/sq. in | | | | |

cross-sectional area expressed in cmils is numerically the same as the sq. mils occupied. For almost all other considerations this "shortcut" can become rather confusing, and we should be cautious.

See the conversions in Table 10.1.

**Note:** Some turns do settle into the spaces *between* turns on the layer below them, but that is unpredictable and so we are ignoring it here. Besides, we would also usually have interlayer insulation to prevent that from happening.

As a result, we have several important conversions in Table 10.1 (also connected to their corresponding metric units).

## 10.7   Current Carrying Capacity of Wires

The following two expressions are often used for calculating the diameter in mils for AWG (American Wire Gauge, also called the Brown & Sharpe or simply B&S wire gauge). The first form is easier to remember but the second is accurate and is the one we have used for generating the tables

$$d_{\text{mils}} = \frac{1000}{\pi} \times 10^{-AWG/20}$$

or

$$d_{\text{mils}} = 5 \times 92^{(36 - AWG)}\!/\!_{39}$$

For the range of wire diameters in common use for power supplies, there is almost no difference between the above two expressions. But we must remember that either way, what we get here is the diameter of the *bare* copper wire, *excluding* any insulation/coating.

Note: It is possible to buy half-integral AWG sizes too for critical applications. However many power supply companies just stock either even or odd number AWG sizes for logistical ease and lower inventory costs.

There are several issues related to a suitable choice of AWG in switching converter transformers and inductors as we now consider:

In inductors, (single winding) the current is relatively smooth and has fairly low high-frequency content. So, technically speaking, we could for example use a single strand of say AWG 10 (very thick wire) for a large current. There are, however, some DFM (design for manufacturability) issues and, therefore, since thick wire is harder to wind, it will usually be replaced by several strands twisted together.

In transformers we can have one winding suddenly stop carrying current, while one or more windings freewheel. As far as the core is concerned, it does not "know" the difference since all it demands is that we "allow" the total ampere-turns (summed over all the windings on the given core) to remain continuous (no sudden step changes). But as far as each particular winding is concerned, the amperes (or ampere-turns) of that winding can certainly jump discontinuously. This gives us the commonly seen trapezoidal/rectangular current waveforms seen in the windings of a transformer. This also means that we now have a large high-frequency Fourier content, and therefore we have to be concerned about *skin depth*, which is something that doesn't bother us much in single-winding inductors because of the much lower high-frequency content. It means that as we increase the diameter of the wire beyond a certain point (twice the skin depth), the high-frequency current stays restricted in an annular surface region of the wire. Though the cross-sectional area "available" for conduction does increase thereafter, it no longer varies as $d^2$ but as $d$. So, it is a situation of rapidly diminishing returns. What we need to do for higher currents is to use several strands of insulated wire, each strand-diameter equal to roughly twice the skin depth. Standard Litz wire used for radio-frequency applications is not suited for commercial switching power supplies, mainly because of the wasted space due to the silk/textile insulation covering. Instead, we can either make in-house, or order directly from several vendors, multi-strand twisted/braided bundles of standard AWG magnet wire (also increasingly and perhaps loosely now referred to as Litz wire).

## 10.8 Skin Depth

*Skin depth* is defined as the distance from the surface where the current density has fallen by a factor $1/e$ from the value at the surface (continuing to fall exponentially as we go deeper). But the high-frequency resistance (and the loss) is the same as if the entire current was distributed *uniformly* up to a depth equal to the skin depth, falling *abruptly* to zero thereafter. We can integrate any exponential curve to see that this is indeed true. This description of skin depth as an annular region of uniform current density, equal to the density we actually have on the surface of the inductor, leads to much easier computations.

Skin depth in mils for copper is typically presented as $2837/f^{1/2}$ and this is the equation we too have used for generating the design aids which follow. Note that in this equation $f$ (in Hz) is rather loosely taken to be the switching frequency, and the diameter of the wire then set to *twice* this value. However, in reality we should consider *all the harmonics* that go into making up the rectangular/trapezoidal current waveform. Recommendations on what the diameter of the wire really should be thus vary somewhat in literature.

A more complete form of skin depth of copper expressed in mm is

$$Skin\_depth = \frac{66.1 \times [1 + 0.0042(T - 20)]}{\sqrt{f}} \ \text{mm}$$

where $T$ is the temperature of the winding in °C, and $f$ is in Hz. Note that the 0.0042 comes from the fact that the resistance of any copper trace or winding increases by 4.2 percent every 10°C rise in temperature.

**Example 10.1** A 90 to 270 Vac (worldwide input) flyback with a turns ratio of 20 needs a transformer. Suggest suitable primary and secondary wire gauges for the following cases
a) Single output of 5 V@5 A (25 W)
b) Single output of 5 V@8 A (40 W)
c) Single output of 5 V@16 A (80 W)
d) Dual outputs of 5 V@10 A and 12 V@2.5 (50 + 30 = 80 W)

Assume the efficiency is 70 percent in all cases and the switching frequency is 70 kHz.

As seen earlier in Chap. 6, for Case (a) i.e., 5 V@5 A, the switch current ramp is 0.64 A at 90 Vac (using the *flat-top* approximation). For Case (b), i.e., 5 V@8 A it is $8 \times 0.64/5 = 1.02$ A. For Case (c) i.e., 5 V@16 A it is $16 \times 0.64/5 = 2.05$ A. Case (d) is identical to Case (c) as far as the primary side is concerned. From Fig. 10.2 we see that AWG 24 is the right choice for 70 kHz. No thickness greater than this is acceptable for the transformer. From Table 10.2, we see that at the most popular industry choice of current density (400 cmils/A), AWG 24 can handle a little over 1 A instantaneous current. Note that in all preceding cases the

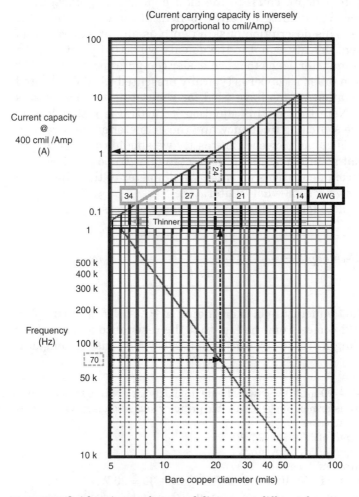

**Figure 10.2** Quick-estimate of area and diameter at different frequencies.

duty cycle is 0.44 at 90 Vac, as calculated earlier, since we are keeping the turns ratio fixed and the $V_{OR}$ is clearly 100. Therefore the average input current is

1. *0.64 A × 0.44 = 0.28 A*. From Table 10.2, a single strand of AWG 29 is adequate for the primary.
2. *1.02 A × 0.44 = 0.45 A*. From Table 10.2, a single strand of AWG 27 is adequate for the primary.
3. *2.05 A × 0.44 = 0.9 A*. From Table 10.2, a single strand of AWG 24 is adequate for the primary.
4. Same as (c) for primary.

For the secondary windings we can just take the average load current passing through each. Therefore

1. For a load current of 5 A, we can take five strands of AWG 24 twisted together. At 70 kHz the maximum allowed thickness is from Fig. 10.2 about 21 mils (twice the skin depth). If we take standard copper foil of $t$ mils thickness, and if the width of the foil is $w$ mils, the available square mils is $t \times w$ mils$^2$. If we are aiming for something close to, or slightly better than 400 cmils/A (i.e., $400 \times 0.785 = 314$ mils$^2$/A), then we need a foil of width greater than

$$w \geq \frac{0.785 \times \text{cmils}/_A \times I_O}{t} = \frac{0.785 \times 400 \times 5}{20} = 78.5 \text{ mils}$$

where we have used a standard 20-mil foil thickness. This is

$$w \geq 0.078 \text{ in}$$

or

$$w \geq 2 \text{ mm}$$

To this foil width we typically add $4 + 4 = 8$ mm margin tape as per safety requirements for universal input, off-line power supplies. Also, if we have a bobbin of thickness 1.5 mm, we should add another 3 mm to this width. So, what we are demanding is that

$$D \geq w + 11 = 13 \text{ mm}$$

Clearly, at this power level we have possible, but not necessarily practicable values for implementing a foil winding. We could have a wider copper foil, but the current density would be too low to justify it. So we would rather stay with $4 \times$ AWG 24 wire here.

**TABLE 10.2   AWG versus Current Carrying Capacity for Different cmil/A.**

| AWG | Current @ 1000 cmil/A or 197A/cm² | Current @ 900 cmil/A or 219A/cm² | Current @ 800 cmil/A or 247A/cm² | Current @ 700 cmil/A or 282A/cm² | Current @ 600 cmil/A or 329A/cm² | Current @ 500 cmil/A or 395A/cm² | Current @ 400 cmil/A or 493A/cm² | Current @ 250 cmil/A or 789A/cm² | Current @ 200 cmil/A or 987A/cm² |
|---|---|---|---|---|---|---|---|---|---|
| 10 | 10.383 | 11.537 | 12.979 | 14.833 | 17.305 | 20.766 | 25.958 | 41.532 | 51.915 |
| 11 | 8.2341 | 9.149 | 10.293 | 11.763 | 13.724 | 16.468 | 20.585 | 32.936 | 41.171 |
| 12 | 6.5299 | 7.2555 | 8.1624 | 9.3285 | 10.883 | 13.06 | 16.325 | 26.12 | 32.65 |
| 13 | 5.1785 | 5.7539 | 6.4731 | 7.3978 | 8.6308 | 10.357 | 12.946 | 20.714 | 25.892 |
| 14 | 4.1067 | 4.563 | 5.1334 | 5.8667 | 6.8445 | 8.2134 | 10.267 | 16.427 | 20.534 |
| 15 | 3.2568 | 3.6186 | 4.071 | 4.6525 | 5.428 | 6.5136 | 8.142 | 13.027 | 16.284 |
| 16 | 2.5827 | 2.8697 | 3.2284 | 3.6896 | 4.3046 | 5.1655 | 6.4569 | 10.331 | 12.914 |
| 17 | 2.0482 | 2.2758 | 2.5603 | 2.926 | 3.4137 | 4.0964 | 5.1205 | 8.1928 | 10.241 |
| 18 | 1.6243 | 1.8048 | 2.0304 | 2.3204 | 2.7072 | 3.2486 | 4.0608 | 6.4972 | 8.1215 |
| 19 | 1.2881 | 1.4313 | 1.6102 | 1.8402 | 2.1469 | 2.5763 | 3.2203 | 5.1525 | 6.4407 |
| 20 | 1.0215 | 1.135 | 1.2769 | 1.4593 | 1.7026 | 2.0431 | 2.5538 | 4.0861 | 5.1077 |
| 21 | 0.8101 | 0.9001 | 1.0126 | 1.1573 | 1.3502 | 1.6202 | 2.0253 | 3.2405 | 4.0506 |
| 22 | 0.6424 | 0.7138 | 0.8031 | 0.9178 | 1.0707 | 1.2849 | 1.6061 | 2.5698 | 3.2122 |
| 23 | 0.5095 | 0.5661 | 0.6369 | 0.7278 | 0.8491 | 1.019 | 1.2737 | 2.0379 | 2.5474 |
| 24 | 0.404 | 0.4489 | 0.5051 | 0.5772 | 0.6734 | 0.8081 | 1.0101 | 1.6162 | 2.0202 |
| 25 | 0.3204 | 0.356 | 0.4005 | 0.4577 | 0.534 | 0.6408 | 0.801 | 1.2817 | 1.6021 |
| 26 | 0.2541 | 0.2823 | 0.3176 | 0.363 | 0.4235 | 0.5082 | 0.6353 | 1.0164 | 1.2705 |
| 27 | 0.2015 | 0.2239 | 0.2519 | 0.2879 | 0.3359 | 0.403 | 0.5038 | 0.8061 | 1.0076 |

| | | | | | | | | | |
|---|---|---|---|---|---|---|---|---|---|
| 28 | 0.1598 | 0.1776 | 0.1998 | 0.2283 | 0.2663 | 0.3196 | 0.3995 | 0.6392 | 0.799 |
| 29 | 0.1267 | 0.1408 | 0.1584 | 0.181 | 0.2112 | 0.2535 | 0.3168 | 0.5069 | 0.6337 |
| 30 | 0.1005 | 0.1117 | 0.1256 | 0.1436 | 0.1675 | 0.201 | 0.2513 | 0.402 | 0.5025 |
| 31 | 0.0797 | 0.0886 | 0.0996 | 0.1139 | 0.1328 | 0.1594 | 0.1993 | 0.3188 | 0.3985 |
| 32 | 0.0632 | 0.0702 | 0.079 | 0.0903 | 0.1053 | 0.1264 | 0.158 | 0.2528 | 0.316 |
| 33 | 0.0501 | 0.0557 | 0.0627 | 0.0716 | 0.0835 | 0.1003 | 0.1253 | 0.2005 | 0.2506 |
| 34 | 0.0398 | 0.0442 | 0.0497 | 0.0568 | 0.0663 | 0.0795 | 0.0994 | 0.159 | 0.1988 |
| 35 | 0.0315 | 0.035 | 0.0394 | 0.045 | 0.0525 | 0.063 | 0.0788 | 0.1261 | 0.1576 |
| 36 | 0.025 | 0.0278 | 0.0313 | 0.0357 | 0.0417 | 0.05 | 0.0625 | 0.1 | 0.125 |
| 37 | 0.0198 | 0.022 | 0.0248 | 0.0283 | 0.033 | 0.0397 | 0.0496 | 0.0793 | 0.0991 |
| 38 | 0.0157 | 0.0175 | 0.0197 | 0.0225 | 0.0262 | 0.0314 | 0.0393 | 0.0629 | 0.0786 |
| 39 | 0.0125 | 0.0139 | 0.0156 | 0.0178 | 0.0208 | 0.0249 | 0.0312 | 0.0499 | 0.0623 |
| 40 | 0.0099 | 0.011 | 0.0124 | 0.0141 | 0.0165 | 0.0198 | 0.0247 | 0.0396 | 0.0494 |
| 41 | 0.0078 | 0.0087 | 0.0098 | 0.0112 | 0.0131 | 0.0157 | 0.0196 | 0.0314 | 0.0392 |
| 42 | 0.0062 | 0.0069 | 0.0078 | 0.0089 | 0.0104 | 0.0124 | 0.0155 | 0.0249 | 0.0311 |
| 43 | 0.0049 | 0.0055 | 0.0062 | 0.007 | 0.0082 | 0.0099 | 0.0123 | 0.0197 | 0.0247 |
| 44 | 0.0039 | 0.0043 | 0.0049 | 0.0056 | 0.0065 | 0.0078 | 0.0098 | 0.0156 | 0.0196 |
| 45 | 0.0031 | 0.0034 | 0.0039 | 0.0044 | 0.0052 | 0.0062 | 0.0078 | 0.0124 | 0.0155 |

2. For a load current of 8 A, we can take eight strands of AWG 24 twisted together. For a foil winding we need a foil of width greater than

$$w \geq \frac{0.785 \times \text{cmils}/_A \times I_O}{t} = \frac{0.785 \times 400 \times 8}{20} = 126 \text{ mils}$$

where we used a standard 20 mil foil thickness. This is

$$w \geq 3.2 \text{ mm}$$

So we are demanding that

$$D \geq w + 11 = 14.2 \text{ mm}$$

We see that we would rather stay with $8 \times$ AWG 24 wire here.

3. For a load current of 16 A, we can take 16 strands of AWG 24 twisted together. For a foil winding we need a foil of width greater than

$$w \geq \frac{0.785 \times \text{cmils}/_A \times I_O}{t} = \frac{0.785 \times 400 \times 16}{20} = 251 \text{ mils}$$

where we used a standard 20 mil foil thickness. This is

$$w \geq 6.4 \text{ mm}$$

So, we are demanding that

$$D \geq w + 11 = 17.4 \text{ mm}$$

This may be a better choice than bulky $16 \times$ AWG 24 wire. But if we do prefer to use wire instead of foil (possibly for ease of production), we can consider using two bundles of $8 \times$ AWG 24 wound bifilar. That would flatten the winding structure, improve the stacking, and also help reduce eddy current/proximity loss terms, which we have not accounted for so far.

4. For a load current of 10 A, we can take 10 strands of AWG 24 twisted together for the 5 V output and three strands of AWG 25 for the 12 V output. We could also consider a foil for the 5 V on lines of the previous discussions.

## 10.9 A Feel for Wire Gauges

The standard known as "ASTM B-258" specifies that AWG is based on geometric interpolation between gauge 0000, which is 0.46 in exactly, and gauge 36 which is 0.005 in exactly. We must realize that ASTM B-258 also specifies rounding rules which seem to be ignored by makers of most tables (though for convenience we will do the same too!). Actually, gauges up to 44 are to be specified with up to four significant figures, but no closer than 0.0001 in. Gauges from 44 to 56 are to be rounded to the nearest 0.00001 in.

Note that $92^{1/39} = 1.123$ and is very close to $2^{1/6} = 1.122$, so the *diameter is approximately halved for every six gauges*. In a similar manner we get the following quick rules for the diameter of bare (unclad) copper wires

- Diameter of gauge no. 36 is 5 mils
- Diameter changes approximately by a factor of 2 every 6 gauges
- Diameter changes approximately by a factor of 3 every 10 gauges
- Diameter changes approximately by a factor of 4 every 12 gauges
- Diameter changes approximately by a factor of 5 every 14 gauges
- Diameter changes approximately by a factor of 10 every 20 gauges
- Diameter changes approximately by a factor of 100 every 40 gauges

Using the definition of AWG we also see that the diameter of gauge no. 10 is 10380 cmils. Take this to be almost 10000 cmils. Then we have the following rules for area (resistance per unit length varies in the same manner, though inversely)

- The area of gauge no. 10 is 10000 cmils
- Area changes approximately by a factor of 2 every 3 gauges
- Area changes approximately by a factor of 10 every 10 gauges

We can also remember that

- Diameter increases by 12 percent every decrease in wire gauge number
- Area increases 26 percent every decrease in wire gauge number
- Resistance/length increases 26 percent every increase in wire gauge number

We now give some examples on using these trends, which provide acceptable accuracy for most purposes.

**Example 10.2** What is the area of no. 27 wire?
No. 10→10000 cmils ⇒ No. 20→1000 cmils ⇒ No. 30→100 cmils ⇒ No. 27→200 cmils

**Example 10.3** What is the area of no. 28 wire?
No. 10→10000 cmils ⇒ No. 13 → 5000 cmils ⇒ No. 16 →2500 cmils ⇒ No. 19 → 1250 cmils ⇒ No. 22 → 625 cmils ⇒ No. 25 → 312 cmils ⇒ No. 28 → 156 cmils

**Example 10.4**  What is the area of no. 26 wire?

No. 10→10000 cmils ⇒ No. 20 → 1000 cmils ⇒ No. 23 → 500 cmils ⇒ No. 26 → <u>250 cmils</u>

## 10.10  Diameter of Coated Wire

Typical magnet wire used in converter magnetics is coated with polyurethane plus polyamide type of insulation. It is most commonly available in *single* or *heavy* (double) insulation depending on the number of protective coatings applied on the bare copper conductor. The *insulation build* is by definition twice the thickness of the coating deposited. On the high-voltage (primary) side, it is unusual not to use double insulation, whereas on the secondary side we can usually use single insulation wire. Note that from the viewpoint of safety regulations the coating does not correspond to anything other than a functional insulation. It is really a reliability issue, especially related to the scratches the wire may receive in production during the winding process and the resulting possibility of flashovers while in operation later.

A well-known vendor is PD Wire and Cable (a unit of Phelps Dodge Corporation), available at *http://www.pdwcg.com*. It makes the popular Nyleze and Thermaleze brands. These are solderable/tinnable varieties meant for easy production flow.

We need to know the build thickness because only then can we accurately predict the number of turns we can lay on a certain bobbin width. A closed form equation closely approximating typical overall wire thicknesses for AWG's falling between 14 and 29 is

$$\boxed{d = d_{Cu} + 10^{\alpha - AWG/44.8}}\ \text{mils} \ \textit{(AWG 14 through 29)}$$

where $d_{Cu}$ is the diameter of the bare copper in mils, $\alpha = 0.518$ for single build, and $\alpha = 0.818$ for double build. For AWG's between 30 and 60 we should use

$$\boxed{d = d_{ref} \times \beta \times \left(\frac{d_{Cu}}{d_{ref}}\right)^{\gamma}}\ \text{mils} \ \textit{(AWG 30 through 60)}$$

where $d_{ref}$ is an arbitrarily defined diameter used to force the dimensions to be right. Here it is set to be the diameter of AWG 40 wire. $\beta = 1.12$ for single-build wire and it is 1.24 for double build. $\gamma = 0.96$ for single-build wire and it is 0.926 for double build. On this basis we have generated Fig. 10.3.

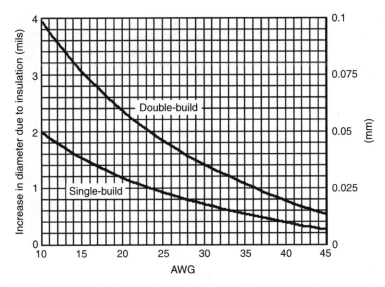

**Figure 10.3** How insulation coating changes bare copper diameter.

## 10.11  SWG Comparison

In the "rest of the world" the Standard wire gauge (SWG, also called Imperial or British wire gauge) is more commonly used. In Table 10.3 we have provided (bare) copper wire diameters for SWG too. In Fig. 10.4 we have also provided a simple graphical way of quickly picking the closest SWG or AWG equivalent as required. For example, we have shown how an AWG of 32 gives us the closest equivalent—an SWG of 36 (and vice versa).

For diameters of SWG the following quick rules apply within a limited range

- Diameter of no. 19 is 40 mils (1 mm).

- From no. 19 through no. 23, diameter decreases 4 mils each number.

- From no. 23 through no. 26, diameter decreases 2 mils each number.

- The closest match with AWG is that AWG no. 24 (or no. 25) is almost exactly the same diameter as SWG no. 25 (or SWG no. 26) (see Table 10.3). This gives us a convenient "bridge point" to go from one to the other system.

TABLE 10.3   Bare Copper Wire Diameters for SWG and AWG.

| Gauge No. | AWG in | AWG mm | SWG in | SWG mm | Gauge No. | AWG in | AWG mm | SWG in | SWG mm |
|---|---|---|---|---|---|---|---|---|---|
| 0 | 0.3249 | 8.25 | 0.324 | 8.23 | 23 | 0.0226 | 0.574 | 0.024 | 0.61 |
| 1 | 0.2893 | 7.35 | 0.3 | 7.62 | 24 | 0.0201 | 0.511 | 0.022 | 0.559 |
| 2 | 0.2576 | 6.54 | 0.276 | 7.01 | 25 | 0.0179 | 0.455 | 0.02 | 0.508 |
| 3 | 0.2294 | 5.83 | 0.252 | 6.4 | 26 | 0.0159 | 0.404 | 0.018 | 0.457 |
| 4 | 0.2043 | 5.19 | 0.232 | 5.89 | 27 | 0.0142 | 0.361 | 0.0164 | 0.417 |
| 5 | 0.1819 | 4.62 | 0.212 | 5.38 | 28 | 0.0126 | 0.32 | 0.0148 | 0.376 |
| 6 | 0.162 | 4.11 | 0.192 | 4.88 | 29 | 0.0113 | 0.287 | 0.0136 | 0.345 |
| 7 | 0.1443 | 3.67 | 0.176 | 4.47 | 30 | 0.01 | 0.254 | 0.0124 | 0.315 |
| 8 | 0.1285 | 3.26 | 0.16 | 4.06 | 31 | 0.0089 | 0.226 | 0.0116 | 0.295 |
| 9 | 0.1144 | 2.91 | 0.144 | 3.66 | 32 | 0.008 | 0.203 | 0.0108 | 0.274 |
| 10 | 0.1019 | 2.59 | 0.128 | 3.25 | 33 | 0.0071 | 0.18 | 0.01 | 0.254 |
| 11 | 0.0907 | 2.3 | 0.116 | 2.95 | 34 | 0.0063 | 0.16 | 0.0092 | 0.234 |
| 12 | 0.0808 | 2.05 | 0.104 | 2.64 | 35 | 0.0056 | 0.142 | 0.0084 | 0.213 |
| 13 | 0.072 | 1.83 | 0.092 | 2.34 | 36 | 0.005 | 0.127 | 0.0076 | 0.193 |
| 14 | 0.0641 | 1.63 | 0.08 | 2.03 | 37 | 0.0045 | 0.114 | 0.0068 | 0.173 |
| 15 | 0.0571 | 1.45 | 0.072 | 1.83 | 38 | 0.004 | 0.102 | 0.006 | 0.152 |
| 16 | 0.0508 | 1.29 | 0.064 | 1.63 | 39 | 0.0035 | 0.089 | 0.0052 | 0.132 |
| 17 | 0.0453 | 1.15 | 0.056 | 1.42 | 40 | 0.0031 | 0.079 | 0.0048 | 0.122 |
| 18 | 0.0403 | 1.02 | 0.048 | 1.22 | 41 | 0.0028 | 0.071 | 0.0044 | 0.112 |
| 19 | 0.0359 | 0.912 | 0.04 | 1.02 | 42 | 0.0025 | 0.064 | 0.004 | 0.102 |
| 20 | 0.032 | 0.813 | 0.036 | 0.914 | 43 | 0.0022 | 0.056 | 0.0036 | 0.091 |
| 21 | 0.0285 | 0.724 | 0.032 | 0.813 | 44 | 0.002 | 0.051 | 0.0032 | 0.081 |
| 22 | 0.0253 | 0.643 | 0.028 | 0.711 | 45 | 0.0018 | 0.046 | 0.0028 | 0.071 |

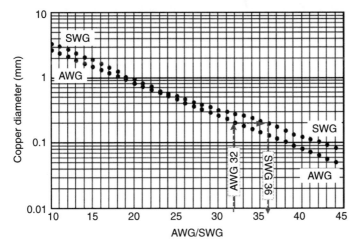

**Figure 10.4**  Graphical comparison of SWG and AWG.

# 11

# Forward Converter Magnetics Design

## 11.1 Introduction

Unless otherwise stated, the reader can assume that we are using the MKS (i.e., SI) system of units.

From Chap. 7 we have understood *transformer action*. When we come to the forward converter, there are actually two magnetic components to consider. First, we realize that its transformer is really one, unlike a flyback where the "transformer" is actually a multiwinding inductor. We also have the output choke to consider, but its primary purpose is energy storage, and we can handle it the same way as we handle any inductor.

The transformer of a forward converter has entirely different design and selection criteria. Let us understand the differences.

## 11.2 The Transformer and Choke Compared

A forward converter choke is usually always operated in continuous conduction mode (CCM) like any other energy storage magnetic component. Its current ripple ratio $r$ is also thus set to an optimum value of around 0.4. A forward converter transformer is, however, always operated in discontinuous conduction mode (DCM) as explained in Chap. 6.

For transformer design, we can use the voltage-dependent equation given in Chap. 7. We take $B_{AC}$ to be equal to half the *total swing* $\Delta B$, so we can also set $B_{AC} = B_{PK}/2$ for a DCM condition. In a typical off-line forward converter, the duty cycle $D$ is normally set around 0.3 to 0.35 at low-line, and that is virtually preordained from holdup time and input capacitance considerations (see worked example in Chap. 5). The applied voltage $V_{AVG}$ in the voltage-dependent equation during the on-time is $V_{IN}$ which is the rectified input rail. From the well-known input to output transfer function of a forward converter

$$V_O = V_{IN} \times \frac{D}{n}$$

we get a simple but noteworthy relationship

$$V_O \propto D \times V_{IN} = \text{constant}$$

Since $V_{IN}$ is the voltage across the transformer when the switch is *on*, and $D$ is by definition $t_{ON} \times f$, the above relationship means that the *voltseconds across the transformer of a forward converter is a constant* for any input voltage. For example, as we increase input voltage, the applied voltage certainly increases, but to ensure regulation, the controller commands the time for which this voltage is applied to decrease. It is just a happy coincidence that this decrease is by exactly the same factor. Therefore, the product of applied volts and its duration remains a constant. Note that this only happens because though the transformer is in DCM, the duty cycle is a CCM duty cycle as dictated by the output choke. Now, looking carefully at the voltage-dependent equation again, as applied to the transformer during the on-time, we see that we have the same $D \times V_{IN}$ term in its numerator. So for a forward converter transformer we can actually use *either the $V_{INMIN}$ or the $V_{INMAX}$* to design the transformer. From either of the two equations below, we will get the same result for $N$ (primary number of turns)

$$B_{AC} = \frac{B_{PK}}{2} = \frac{V_{INMIN} D_{MAX}}{2 \times NA_e f}$$

$$B_{AC} = \frac{B_{PK}}{2} = \frac{V_{INMAX} D_{MIN}}{2 \times NA_e f}$$

Here we can set $B_{PK}$ equal to $B_{SAT}$ as a worst case, or we can deliberately keep it less so as to reduce core loss. We also note that since core loss depends on $\Delta I$ (or $\Delta B$), *the core loss in the forward converter transformer is also independent of line voltage. The same is true for the peak current.* Of course, this assumes that the simple CCM duty cycle equation applies.

Note that the duty cycle equation can be written as

$$D = \frac{V_O}{V_{INR}}$$

where $V_{INR}$ is the *reflected input voltage*, equal to $V_{IN}/n$. Therefore, for the output choke, the voltage-dependent equation reads

$$B_{AC} = \frac{V_{ON}D}{2 \times NA_e f} = \frac{(V_{INR} - V_O)D}{2 \times NA_e f}$$

or equivalently

$$B_{AC} = \frac{V_{OFF}(1-D)}{2 \times NA_e f} = \frac{(V_O)(1-D)}{2 \times NA_e f}$$

So, here we see that though the following equation still holds

$$V_O \propto D \times V_{IN} = \text{constant}$$

the $D \times V_{IN}$ here is not what appears in the voltage-dependent equation (as applied to the choke).

From the second (equivalent) equation above we can see that $B_{AC}$ (or $\Delta I$) increases if $D$ decreases, because $V_O$ is a constant. We know that for all topologies, if $V_{IN}$ increases, $D$ decreases, and vice versa. *Therefore, the core losses (and peak currents) are higher at high-line for the forward converter choke.*

Thus, the design of the output choke should be done at high-line, just the same way as we design an inductor for a buck converter application. As mentioned, the basic design of the forward converter transformer can be done at either high-line or low-line, provided of course we start by *setting its duty cycle as per holdup time considerations at low-line.*

Analyzing the schematic in Fig. 11.1 we see

On the secondary side we effectively have a buck converter with an effective input voltage equal to $V_{INR}$. The center of the inductor current is the load current as for the buck.

On the primary side the forward converter "thinks" (almost entirely so) that it itself is just a buck converter with an output voltage of $V_{OR}$ and a load current of $I_{OR}$.

The difference is the magnetization current $I_M$ which adds to the reflected inductor current $I_{LR}$. This forms the primary winding (and switch) current. This current is not engaged in the transfer process, therefore $I_S = n(I_P - I_M)$.

$I_M$ is simply freewheeled back through the tertiary winding (T) as $I_T$, and the associated magnetization energy of the transformer returns to the input bulk capacitor during the switch off-time. This is clearly a circulating current and some of its energy is dissipated in the tertiary diode and tertiary winding resistance.

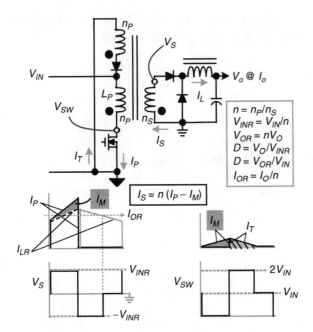

$$I_S = n (I_P - I_M)$$

$n = n_P/n_S$
$V_{INR} = V_{IN}/n$
$V_{OR} = nV_O$
$D = V_O/V_{INR}$
$D = V_{OR}/V_{IN}$
$I_{OR} = I_O/n$

**Figure 11.1**  Understanding the forward converter.

When the switch is *on*, the voltage across the transformer is $V_{IN}$. When the switch is *off*, the tertiary winding will reflect an equal and opposite voltage across the primary winding (provided it has the same number of turns as the primary winding). Thus the peak voltage on the drain of the switch is $2 \times V_{IN}$. This lasts only till the core has de-energized, and that is when $I_T$ falls to zero.

The magnetization ramp is determined by the primary inductance of the transformer, the applied input voltage, number of primary turns and the like. But the ramp it "rides on" (see Fig. 11.1) is being reflected from the secondary side. So this pedestal is determined completely by the load current (of course) and the inductance of the output choke.

The inductance of the choke should be so chosen that (as for any buck converter) $r$ is about 0.4 at $V_{INMAX}$. Note that we can use all the standard buck converter equations, including the selection criteria for inductors, if we just imagine that the effective input applied to the output step-down stage of the forward converter is $V_{INR}$.

Therefore, we can mentally visualize the forward converter as a buck converter, with an input dc voltage rail of $V_{IN}/n$ (created by preceding transformer action). Besides isolation, the only major difference is that in a buck converter we can, in principle, go up to 100 percent duty cycle, but in a forward converter, we need to allow the *transformer* to reset too (besides the output inductor). The (magnitude of) the slopes of the ramp-up and ramp-down of the magnetization current are the same ($=V_{IN}/L_P$). So, the worst case is at 50 percent duty cycle where the magnetization current is in *critical conduction* (boundary between continuous and discontinuous conduction modes). If we exceed this, in every cycle there will be a net increase in the magnetization current. Since the feedback loop connects only to the output choke, it will have no way of realizing and correcting this build-up of transformer magnetization current. Therefore, we cannot, in principle, even sustain a steady, stable continuous conduction mode condition in the transformer. The only option is to ensure that it resets itself by allowing enough time for the magnetization current to return to zero. Knowing what its slopes are, this means restricting the maximum duty cycle to less than 50 percent for a forward converter, as opposed to a buck converter.

We reiterate that as far as the core is concerned, it "thinks" that the only current flowing around it is the magnetization current. It is only from the temperature of the copper that we may actually "discover" what the actual primary and secondary currents are. The windings are found to be hot, though the core may be running quite cool. But unlike the choke, which also sees a high current, the primary and secondary windings have a *chopped* current waveform. So the high-frequency losses (ac resistance) can be very high because of the high-frequency harmonic content. Reducing the copper losses in the windings of a forward converter is thus very important, because the forward converter is in all other respects ideally suited for a high-power design. There is no reason why we should have the efficiency reduced by our inability to understand *proximity effects*.

## 11.3   Introducing the Proximity Effect

Whereas the skin depth considerations that we discussed in Chap. 10 still apply, here we must go further. Skin depth represents the case of a single wire actually. We didn't consider the fact that the field from the nearby windings may be affecting the current distribution significantly. So, the annular area on the surface of the wire that we had visualized as being fully available for the high-frequency current to flow through, is actually not. We may in fact need to further reduce the wire diameter to utilize it more effectively. This is the rationale and purpose behind our study of the proximity effect.

The power throughput of a forward converter transformer depends on two main criteria

- How much copper we can squeeze into the available window space? We know that, because of transformer action, we can theoretically keep increasing the load current almost without limit. Finally we get limited by various parasitics, of which the winding resistances play a major part. Since the copper gets very hot, it must be sized appropriately. Therefore, the amount of window space available becomes crucial in determining how much power we can get out of the transformer. However, *the utilization of the available window space* is also a key factor in optimizing the size of a forward converter transformer, and we need to be able to *get the most out of least copper*.

- How can we "get more out of the copper?" Realizing that the losses in a transformer are not just due to the dc winding resistance, but their *ac resistance* too, we can reduce the latter by understanding the proximity effect and optimizing the winding arrangement.

**Note:** We must account for space that is "wasted" because of the bobbin, insulation, and the like. More complicated winding arrangements (several splits) may ultimately provide even lesser copper space because of the three layers of tape required between every primary to secondary (*safety extra low voltage*, SELV) interface, and possibly one to two Faraday shields per interface too. This can also significantly increase the cost due to the additional processing steps, lead-out terminations, and the like.

In a forward converter transformer, the primary and secondary windings conduct simultaneously. Therefore their fields combine to vary in a certain way through the transformer. Adjacent turns produce local fields that affect the current distribution in each other. Therefore our earlier understanding of skin effect now needs to be modified. Let us define the terms more clearly now.

## 11.4  More about Skin Depth

In a stand-alone conductor, the current distribution falls as $e^{-1}$ inside the conductor. Ignoring the sinusoidal time variation here, its variation in space has the form

$$J = J_o e^{-x/\delta}$$

Looking at Fig. 11.2, we have arbitrarily set the surface current density $J_o$ to unity and the skin depth $\delta$ to 2. We can eyeball the curve, and also confirm by simple mathematics that by the properties of the exponential function, the area under the entire exponential curve is equal to the area of the hatched portion. This implies that we can, in

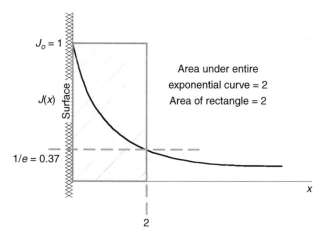

**Figure 11.2**  Understanding current density and skin depth.

effect, replace the entire exponential current distribution with a constant (uniform) current density, equal to the actual current density on the surface but now confined to a distance $\delta$ under the surface. This is by definition the skin depth.

This leads to the concept of ac resistance. In Fig. 11.3 we have shown a round conductor, and we can see that if dc current is passed, it will spread out uniformly through the entire cross-sectional area, i.e., $4^2$. But if a time-varying (sinusoidal) current tries to make it through the conductor, it is confined to an annular area on the surface, as shown. The skin depth is assumed unity here. So, as far as the ac is concerned, the area available to it is less than if it was dc. Since resistance is inversely proportional to area, the ac resistance must be related to the

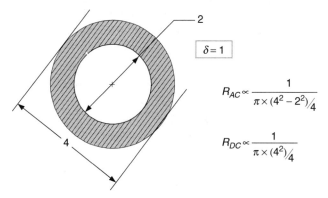

**Figure 11.3**  How $R_{AC}$ and $R_{DC}$ are calculated for a wire.

dc resistance by

$$F_r \equiv \frac{R_{AC}}{R_{DC}} = \frac{4^2}{4^2 - 2^2} = 1.3$$

*where we have also introduced the term* $F_R$. Clearly, $F_R$ can be as low as unity, but as we can see, it can be very high if the diameter is increased.

**Note:** For large wire diameters, the annular region is basically the circumference of a circle; so $R_{AC} \propto 1/d$, whereas we know that $R_{DC} \propto 1/d^2$. So, $F_R \propto d^2/d = d$. But in a copper foil, increasing the thickness of the foil will *not* affect $R_{AC}$ at all. Since here too $R_{DC} \propto 1/d^2$, we get for foil: $F_R \propto d^2$. So, as we increase the thickness—*d* (or *h*)—a foil will get resistive more quickly at high frequencies than round conductors (or bundles thereof). The only way to reduce the ac resistance of foils is to increase the *width* of the foil. This would normally mean a larger core size. However, one way out is to look for special EER cores, like the EER35, for achieving high output currents when using secondary foil windings. These are long (stretched out) versions of comparably sized cores, so it should not affect the cost as much as increasing the core volume normally will.

Now if we put similar current-carrying conductors next to each other, they will affect even the availability of the annular region displayed in Fig. 11.3. This can lead to a very steep increase in $F_R$.

## 11.5  Dowell's Equations

Dowell successfully reduced a very complex three-dimensional field problem into a manageable and accurate one-dimensional calculation. But first we must define a "portion" as that is what Dowell's equations apply to. As we look at how the magnetomotive force (mmf, see Chap. 7) varies through different winding arrangements in Fig. 11.4, we realize that the key to reducing eddy current losses is to reduce the

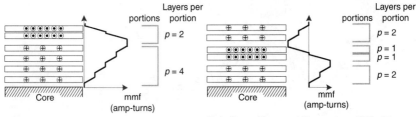

**Figure 11.4**  Magnetomotive force and "portions" for two winding arrangements.

local fields, and we can do that by reducing the peak value of the (local) mmf. So split windings will always help. In principle, additional levels of primary-secondary interleaving will help further, except for the added cost and complexity that may result. Therefore, most practical medium power designs will use only a split primary (sections in series) and a sandwiched secondary, as shown. A portion is then defined as the layers between a maxima and a zero of mmf (to know how the mmf varies from layer to layer, count the number of current arrows in Fig. 11.4).

**Note:** By a split winding we have reduced the peak mmf and field by a factor of 2. This translates into four times less energy in this leakage field and four times less leakage inductance too.

According to Dowell, the $F_R$ of a portion with an integral number of layers ($p$ being an integer) is

$$F_R(p, X) = A(X) + \frac{p^2 - 1}{3} B(X)$$

where $p$ is the number of layers in that portion, $X$ is $h/\delta$, $h$ being the thickness of the *equivalent foil*, and

$$A(X) = X \frac{e^{2X} - e^{-2X} + 2\sin(2X)}{e^{2X} + e^{-2X} - 2\cos(2X)}$$

$$B(X) = 2X \frac{e^X - e^{-X} - 2\sin(X)}{e^X + e^{-X} + 2\cos(X)}$$

For half-integral layers ($p$ being a half-integer) we can use

$$F_R(p, X) = \left(1 - \frac{0.5}{p}\right) \bullet [F_R(p - 0.5, X)] + \frac{C(X)}{4 \bullet p} + \frac{(p - 0.5) \bullet B(X)}{2}$$

where

$$C(X) = X \frac{e^X - e^{-X} + 2\sin(X)}{e^X + e^{-X} - 2\cos(X)}$$

We have plotted these out in Fig. 11.5. The equations presented above certainly apply to actual foils too, but they also can be made to apply to *wires spread out in a layer* (assuming *no space left between successive windings*), through the equivalent foil transformation which is described later. *Note that in related literature the same curves are presented with* X *on the horizontal axis. But we have preferred to use the number of layers in the portion p on this axis, with* X *as the parameter for the set of curves.* This way we can present the curves for many more layers, and that comes in handy when we use bundled/Litz wire as we will see.

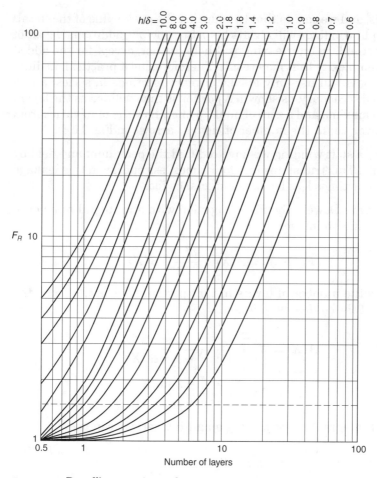

**Figure 11.5** Dowell's curves in an alternative representation.

These curves talk in terms of $\delta$, which depends on frequency; thus clearly *the curves are applicable to a sine wave only*. The actual switching current waveform is however usually a unidirectional (*unipolar*) rectangular/trapezoidal waveform. So, we must use Fourier analysis to split the waveform into harmonic components (each of amplitude $|c_n|$), evaluate the ac resistance of each, and then find the effective $F_R$ ($F_{R\_eff}$) for the composite waveform. Taking the dc level of the current waveform as the *zeroth harmonic* ($c_o$), we get the *effective winding resistance* to be

$$R_{AC\_eff} \times I_{RMS}^2 = R_{DC} \times \sum_{n=0}^{40} |c_n|^2 F_{Rn}$$

where $I_{RMS}$ is the usual rms of the current waveform ($\cong I_{SW} \times D^{1/2}$). Simplifying

$$F_{R\_eff} \equiv \frac{R_{AC\_eff}}{R_{DC}} = \frac{\sum_{n=0}^{40} |c_n|^2 F_{Rn}}{I_{RMS}^2}$$

or

$$R_{AC\_eff} = F_{R\_eff} \times R_{DC}$$

where $F_{Rn}$ is the $F_R$ of the $n$th harmonic, and by definition, $F_{R0} = 1$. We are summing only up to the 40th harmonic.

Optimization starts by varying $h$ (or equivalently the $X$) and seeing where we get the least $R_{AC\_eff}$. But as we saw, for a foil, $R_{DC}$ too will vary in the process as per $1/h$. So, we need to find the minima of the function $F_R/X$ where $X$ is now $h/\delta_1$, that is, *referred to the fundamental frequency* (the switching frequency). This function is plotted out in Fig. 11.6 from a math file, and we see that we get an optimum value for $X$ corresponding to the number of layers in the portion. We have set both the rise time and fall times to a typical 0.5 percent of the switching-cycle time period. The duty cycle is assumed to be 50 percent.

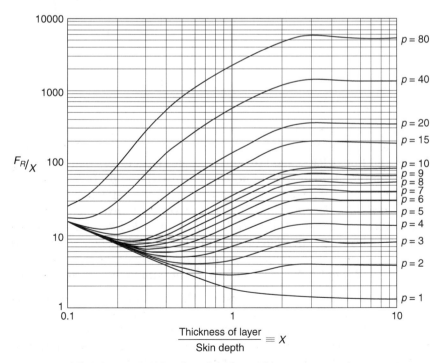

**Figure 11.6**  AC resistance as a function of thickness for a square waveform.

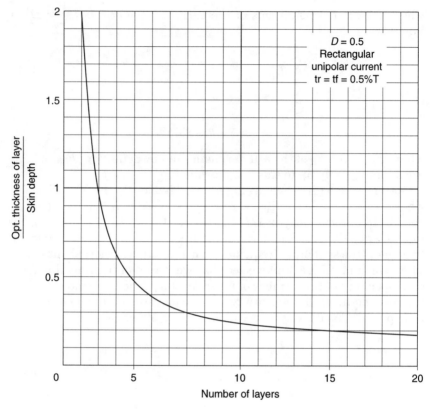

**Figure 11.7** Optimum layer thickness for a square waveform (first view).

In Figs. 11.7 and 11.8 we have collated the *optimum foil thickness* $(X/\delta)$ values thus generated. We have two design curves (different zoom levels and linear/log scales) to help us either with layers using a single conductor (small $p$) or with bundled/Litz wire (large $p$). In Fig. 11.9 we have the results of a mathematical iteration for the effective $F_R$ we can expect *if we use the optimum* $X/\delta$. We can see that *anything more than five to six layers is not going to reduce the* $F_R$ *much*. Even for very large number of layers per portion (as for bundled/Litz wire), *we are not going to be able to reduce the copper loss nuch lower than that corresponding to an* $F_R$ *of 2*. That is *the best achievable*, proximity effects considered.

**Note:** According to Bruce Carsten, the optimum foil thickness is roughly proportional to $D^{1/2}$ (where $D$ is the duty cycle). He also recommends that for *bipolar* current waveforms (e.g., bridge topologies), we should *halve* the unipolar current foil thickness so obtained. However, as per the author of this book, do not expect to be able to lower the $F_R$ much from the numbers indicated in Fig. 11.9.

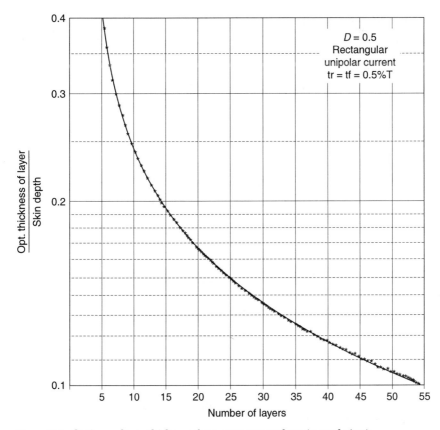

**Figure 11.8**  Optimum layer thickness for a square waveform (second view).

From the figures we can roughly see that $h/\delta$ seems roughly inversely proportional to the number of layers. This implies that the *total thickness of the portion, that is,* p × h *tends to become virtually constant, irrespective of the number of layers*, and more so when there are a large number of layers in the portion. In Fig. 11.10 we have plotted this out more accurately. We see that it does vary somewhat within the accuracy of the program used to generate these results. This curve will help us a great deal, especially when we start a design based on bundled/Litz wire. Knowing the core selected and the available window, we know beforehand how thick the primary portion can (or should) be. Therefore from this curve we can then estimate the thickness of the wire with which to start the iterative design procedure.

A good curve-fit to our mathematically generated results is

$$h = \delta \times \left[ \frac{2}{p^{1.2}} + 0.095 \right]$$

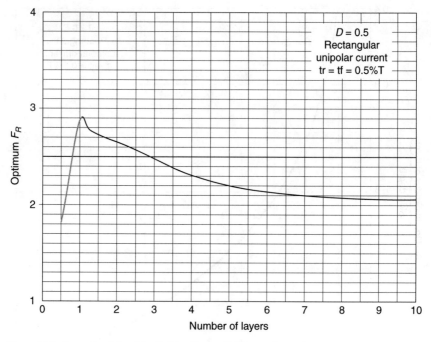

**Figure 11.9**  Lowest achievable $F_R$ if optimum thickness is used.

where $\delta$ is the skin depth at the switching frequency and $p$ the number of layers in the portion.

**Example 11.1**  We have six layers in the primary, and it is split into two sections. What is the optimum foil thickness if the switching frequency is 100 kHz?

Using the skin depth equation for a typical 60°C temperature rise, at 100 kHz we get a $\delta$ of 0.24 mm. Using the above equation for three layers per portion we get

$$h = 0.24 \times \left[ \frac{2}{3^{1.2}} + 0.095 \right] = 0.15 \text{ mm}$$

Note that $h \times p \cong 2\,\delta$. This is the required *thickness of the portion*. We can see how the proximity effect is affecting us. If we had ignored this effect, we would have normally chosen each layer to be $2\,\delta$ thick, giving us a total primary thickness of $12\,\delta$. Now we require the total thickness of the primary—both sections considered—to be only $2 \times 2\,\delta = 4\,\delta$. We can also, thus, start allocating space to the windings and see if the available window suffices.

We can also see from Fig. 11.9 that with this thickness, the effective $F_R$ for the entire primary will be about 2.5. So if we know the dc resistance we can easily compute the ac resistance and thus the losses in the primary winding. We can repeat the same process for the secondary and thus get the total copper loss in the transformer.

In the next section we show how to apply these results, which are basically for a foil, to the case of a wire winding.

## 11.6  The Equivalent Foil Transformation

To be able to apply Dowell's equations and the results of the preceding section, the round wire is to be mentally replaced by an equivalent square wire of the same area. Thus if we have found out $h$ from the previous curves or equations, the equivalent wire diameter is obtained by

$$d \Leftarrow \frac{2}{\sqrt{\pi}} \bullet h$$

We therefore just *need to multiply h obtained previously by 1.13 to get d*. So in the previous example we get an equivalent wire diameter of 0.17 mm, or AWG no. 33. Of course, we know that this gives us the lowest losses, but we still don't know yet if that is good enough, considering the output power and efficiency of the converter. We will come to this again later.

For now we can see from Fig. 11.9 that the best $F_R$ with three layers is going to be 2.5. Why not use bundled wire? The rule for generating this transformation is also shown in Fig. 11.11. We see that (for example) a bundle of 16 strands can be considered to be stacked as 4×4, and so where each wire was, we now get four effective layers. Finally, we merge them and we get a portion with $4 \times 3 = 12$ layers. From Fig. 11.9 we can see that the $F_R$ can be reduced to about 2, *provided we choose an equivalent foil thickness of*

$$h = 0.24 \times \left[ \frac{2}{12^{1.2}} + 0.095 \right] = 0.047 \text{ mm.}$$

The equivalent optimum wire diameter is 0.047 × 1.13 = 0.053 mm. This gives us AWG no. 43 for each strand.

When we come to bundled or Litz wire, if we have $n$ strands in the bundle, we should replace these with equivalent square wire strands, and then stack them together in a square. Ultimately *we get* $n^{1/2}$ *layers from each layer of wire* (see Fig. 11.11). Note that we are not implying anything about the total number of primary or secondary turns. From Dowell's point of view, that is in fact of no concern. What we are interested in are the number of layers of foil (or equivalent foils) and their respective portions.

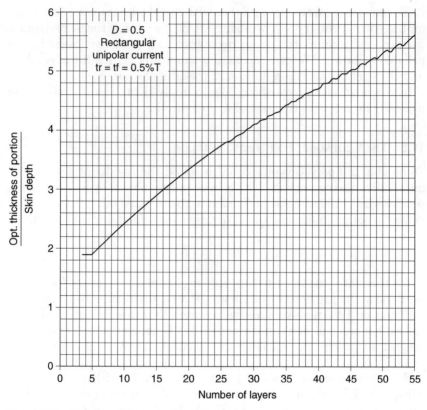

**Figure 11.10**  Optimum thickness of a portion.

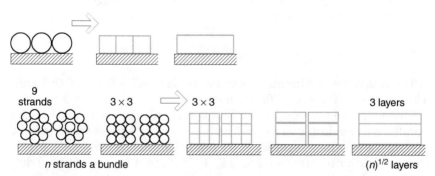

**Figure 11.11**  How single-strand wire transforms to a single layer of foil (top) and how wire with $n$ strands/bundle can be replaced by $\sqrt{n}$ layers of foil (bottom) for applying Dowell's curves.

## 11.7  Some Useful Equations for Quick Selection of Forward Converter Cores

For a transformer, its throughput power capability can be traced back to (a) the effective area $A_e$ of the core, which determines how many turns we need (using the voltage-dependent equation) and (b) the available window area, which determines how much copper we can actually squeeze in. Therefore, we do expect $P_o \propto A_e \times A_w$, where $A_w$ is the window available. The product $A_e A_w$ is called the *area product* (AP) of the core. However, we also know that the entire "available" window area will not have copper in it. We need space for the wire insulation, bobbin, tape layers and the like, and in addition, round copper wire will have wasted air spaces between windings, however closely packed it is. Therefore in general we introduce a window utilization factor $K_u$, which expresses how much of $A_w$ will be actually occupied with copper. However, out of this, only a part of the area will be occupied by the primary winding. In addition, generalizing further, there is going to be a dependence on the topology too (all however still being forward-converter variants i.e., buck-derived topologies). The most commonly used equation for selecting a core is given by

$$AP = \frac{11.1 \times P_{IN}{}^{1.32}}{K \times \Delta B \times f} \; \text{cm}^4$$

where $K = K_t \times K_u \times K_p$ ($K_u$ is window utilization factor $A_w'/A_w \approx 0.4$, $K_p$ is primary area factor $A_P/A_w'$, $K_t$ is topology factor). $\Delta B$ is the flux swing in tesla (in discontinuous conduction mode for unipolar current this equals $B_{PK}$ and is typically set to 0.15 to 0.2 T for ferrites). $f$ is in Hz. The $K$ factors are given in Table 11.1.

The reader can refer to the conversion tables in Chap. 10 for the following.

*Empirically, it is seen that the maximum current density is 400 to 450 A/cm² (about 450 cmils/A) for a core with* AP = 1 cm², *and a 30°C temperature rise attributable to the copper losses.*

**TABLE 11.1  Selecting the K-Factors for Different Topologies for Use in the Area Product Equation.**

|  | **K** | **K$_t$** | **K$_u$** | **K$_p$** |
|---|---|---|---|---|
| *Forward converter* | 0.141 | 0.71 | 0.4 | 0.5 |
| *Bridge/half-bridge* | 0.165 | 1.0 | 0.4 | 0.41 |
| *Full wave center tap* | 0.141 | 1.41 | 0.40 | 0.25 |

So, the current density for a 30°C rise is written as

$$J = 420 \times AP^{-0.24} \; \text{A/cm}^2$$

This was the relationship that was used to calculate the foregoing $AP$ relationship. But if the core losses are almost the same as the copper losses (not necessarily always true), and we do not want to exceed 30°C, we will have to allocate only 15°C for the copper losses. In that case the current density should be reduced to about 300 A/cm² (about 650 cmils/A).

$$J = 297 \times AP^{-0.24} \; \text{A/cm}^2$$

Note that

$$AP \propto \frac{1}{J}$$

So, the area product should be increased accordingly.

*Smaller cores have a large relative surface area for cooling.* That is because $V_e$ varies as $l^3$ but surface area goes only as $l^2$. So, the thermal resistance $R_{th}$ of large cores is higher than for small cores. The empirical relationship for IEC qualified safety transformers (for $\approx 40°C$ rise in temperature) is stated as

$$R_{th} = \frac{61}{V_e^{0.54}} \; °\text{C/W}$$

The empirical relationship between $V_e$ and $AP$ is

$$V_e = 5.7 \times AP^{0.68}$$

where $V_e$ is in cm³.

So, we also get the following relationship

$$R_{th} = \frac{24}{AP^{0.37}} \; °\text{C/W}$$

We can also rewrite the $AP$ relationship in terms of the easier to find $V_e$ parameter, to aid a quick first selection of core. This is

$$V_e = \frac{29.3 \times P_{IN}^{0.9}}{(K \times \Delta B \times f)^{0.68}} \ \Bigg| \ \text{cm}^3$$

**Example 11.2**   For a 200 W forward converter, what core should we select? The switching frequency is 100 kHz and the expected efficiency is 80 percent.

$$V_e = \frac{29.3 \times {200/0.8}^{0.9}}{(0.141 \times 0.15 \times 10^5)^{0.68}} = 23.1 \text{ cm}^3$$

From Table 11.2, we see that the EE42/21/20 (two halves making 42/42/20 size) will suffice. One size smaller (the EE42/42/15) may also suffice depending on the application (considering how many secondaries we have, for example), how good our design is with respect to winding arrangement, proximity and core losses, and the like.

**TABLE 11.2   Dimensions and Effective Parameters for Various Popular Core Sizes.**

| Core | $V_e$ (cm$^3$) | $A_e$ (cm$^2$) | $l_e$ (mm) | $2 \times D$ (mm)* |
|---|---|---|---|---|
| E 13/7/4 | 0.384 | 0.13 | 29.6 | 9 |
| E 16/8/5 | 0.753 | 0.201 | 37.6 | 11.4 |
| E 20/10/6 | 1.5 | 0.335 | 44.9 | 14 |
| E 25/13/7 | 3.02 | 0.525 | 57.5 | 17.4 |
| E 30/15/7 | 4 | 0.6 | 67 | 19.4 |
| E 32/16/9 | 6.18 | 0.83 | 74 | 22.4 |
| E 42/21/15 | 17.6 | 1.82 | 97 | 29.6 |
| E 42/21/20 | 23.1 | 2.36 | 98 | 29.6 |
| E 55/28/21 | 43.7 | 3.54 | 123 | 37 |
| E 55/28/25 | 52 | 4.2 | 123 | 37 |
| E 65/32/27 | 78.2 | 5.32 | 147 | 44.4 |
| ETD 29 | 5.47 | 0.76 | 72 | 22.0 |
| ETD 34 | 7.64 | 0.97 | 78.6 | 23.6 |
| ETD 39 | 11.50 | 1.25 | 92.2 | 28.4 |
| ETD 44 | 17.80 | 1.73 | 103 | 32.2 |
| ETD 49 | 24.0 | 2.11 | 114 | 35.4 |
| ETD 54 | 35.50 | 2.80 | 127 | 40.4 |
| ETD 59 | 51.50 | 3.68 | 139 | 45.0 |

*$2 \times D$ is as per the nomenclature presented in Chap. 10 using E-cores (so $2 \times D$ is twice the length of any limb).

**Tip.** The *mean length per turn* (MLT) of ETD cores is smaller than E-cores because of the round central limb. This lowers the resistance for a given number of turns.

## 11.8  Stacking Wires and Bundles

When we stack round wires next to each other, it is trivial to calculate how many turns can be accommodated per inch in a layer (transverse, $t$ direction) and how many can be stacked on top of each other, i.e., height-wise ($h$ direction). With insulation between each layer, both are expected to be the same. We just need to know the diameter with the coating. That is provided in Chap. 10. In the $t$ direction we will need to allow for any margin tape requirement.

When we take bunched wire (bundles twisted together, but *preferably braided*) the picture is not so obvious. Some relevant empirical results are presented in Table 11.3 and have been verified to be providing good stacking predictions. For example, if the diameter of each strand of a four-strand bundle is $d$, then the bundle can be considered to be looking like an effective "diameter" of $2.45 \times d$ when wound side by side (i.e., along a layer). But in the vertical direction, since the bundles on top tend to slip slightly into the spaces of the preceding layer, the effective diameter of each bundle looks more like $2.31 \times d$. We can then very clearly estimate how the available window space is filling up with the bundles and whether the transformer suffices.

**Note:** In the lab we sometimes just take several AWG strands and twist them into a bundle. This is not going to help reduce proximity/eddy current losses much since the same strand(s) tends to remain on the surface. What we really want to do is braid them so that they take turns on the surface and inside. It is therefore better to directly order bunched bundles from vendors who have automatic machines to do the braiding. A quick option in the lab is to twist together subbundles made out of a lesser number of strands.

**Tip.** For a chosen diameter, the $F_R$ increases significantly if the number of layers increases. Further, even if a few turns are left over and wound

TABLE 11.3  **Stacking of Bundled Wire: Transverse and Vertical (height).**

| | Strands/bundle | | | | | | |
|---|---|---|---|---|---|---|---|
| | 4 | 5 | 6 | 7 | 8 | 9 | 10 |
| $n_t$ | 2.45 | 2.94 | 2.98 | 3.11 | 3.61 | 3.89 | 4.34 |
| $n_h$ | 2.31 | 2.69 | 2.93 | 2.93 | 3.16 | 3.16 | 3.36 |

on the last layer, from Dowell's point of view, that counts as a complete layer. Therefore, if just a few turns are left over, *it is better not to create an additional layer.* We can do this either by omitting these extra turns entirely (i.e., reduce $n_P$ slightly and incur slightly higher flux swing and core losses) or by reducing the diameter of the wire slightly (i.e., keep the number of turns constant but incur slightly higher dc resistance). Both ways are usually preferable to increasing the $F_R$.

**Tip.**  As we can see from Table 11.3, a bundle with seven strands has an $n_h$ equal to that of the bundle with six strands. This is a correct indication of the fact that a seven-strand bundle is the most effective, i.e., it has more copper within a given space. One strand tends to stay in the middle with six strands distributed evenly around its circumference. The strands can take turns, but this basic arrangement remains consistent at any cross-sectional point along its length.

## 11.9  Core Loss Calculations

Core loss depends on the flux swing $\Delta B$ and the frequency $f$ (and also temperature but we are ignoring this here). Note, however, that by convention core loss is usually always quoted in terms of $B_{AC}$ instead of $\Delta B$. This is *half* the actual swing as indicated earlier. Unfortunately, there are also several units in which core loss is expressed in related literature, and the designer can get really confused. Therefore, they will all be touched upon here, including their relative transformations.

In general

$$Core\ Loss = \text{constant}_1 \times B^{\text{constant}_2} \times f^{\text{constant}_3} \times V_e$$

We write the loss *per unit volume* as

$$P_{CORE} = \frac{Core\ Loss}{V_e}$$

In Table 11.4 we have indicated the three main systems of units used. In Table 11.5 we have the values for the constants and therefore core loss can be calculated for any application. We of course need to know the $\Delta B$ from the magnetics equations.

Note that the table is only a rough initial guide. For example, the core loss coefficients are not really constants, but vary with temperature. *Refer to vendors' datasheets for more accurate information including frequency and* $B_{SAT}$.

**TABLE 11.4   Converting Between Different Systems of Core Loss**

|          | Constant | Exponent of $B$ | Exponent of $f$ | $B$ | $f$ | $V_e$ | $P_{CORE}$ |
|----------|----------|-----------------|-----------------|-----|-----|-------|------------|
| System A | Cc<br>$= \dfrac{C \times 10^{4 \times p}}{10^3}$ | Cb<br>$= p$ | Cf<br>$= d$ | T | Hz | cm$^3$ | W/cm$^3$ |
| System B | C<br>$= \dfrac{Cc \times 10^3}{10^{4 \times Cb}}$ | p<br>$= Cb$ | d<br>$= Cf$ | G | Hz | cm$^3$ | mW/cm$^3$ |
| System C | Kp<br>$= \dfrac{C}{10^3}$ | n<br>$= p$ | m<br>$= d$ | G | Hz | cm$^3$ | W/cm$^3$ |

**TABLE 11.5   Typical Core Loss Coefficients of Common Materials.**

| Material | $C$ | $d$ | $p$ | $\mu$ | $\approx B_{SAT}$ (gauss) | $\approx$ max freq. (Hz) |
|----------|-----|-----|-----|-------|---------------------------|--------------------------|
| Powdered Iron 8 | 4.3E-10 | 1.13 | 2.41 | 35 | 12500 | 1E+8 |
| Powdered Iron 18 | 6.4E-10 | 1.18 | 2.27 | 55 | 10300 | 1E+8 |
| Powdered Iron 26 | 7E-10 | 1.36 | 2.03 | 75 | 13800 | 2E+6 |
| Powdered Iron 52 | 9.1E-10 | 1.26 | 2.11 | 75 | 14000 | 1E+7 |
| Kool Mu 60 | 2.5E-11 | 1.5 | 2 | 60 | 10000 | 5E+5 |
| Kool Mu 75 | 2.5E-11 | 1.5 | 2 | 75 | 10000 | 5E+5 |
| Kool Mu 90 | 2.5E-11 | 1.5 | 2 | 90 | 10000 | 5E+5 |
| Kool Mu 125 | 2.5E-11 | 1.5 | 2 | 125 | 10000 | 5E+5 |
| MolyPermalloy 60 | 7E-12 | 1.41 | 2.24 | 60 | 6500 | 5E+6 |
| MolyPermalloy 125 | 1.8E-11 | 1.33 | 2.31 | 125 | 7500 | 3E+6 |
| MolyPermalloy 200 | 3.2E-12 | 1.58 | 2.29 | 200 | 7700 | 1E+6 |
| MolyPermalloy 300 | 3.7E-12 | 1.58 | 2.26 | 300 | 7700 | 4E+5 |
| MolyPermalloy 550 | 4.3E-12 | 1.59 | 2.36 | 550 | 7700 | 2E+5 |
| HighFlux 14 | 1.1E-10 | 1.26 | 2.52 | 14 | 15000 | 1E+7 |
| HighFlux 26 | 5.4E-11 | 1.25 | 2.55 | 26 | 15000 | 6E+6 |
| HighFlux 60 | 2.6E-11 | 1.23 | 2.56 | 60 | 15000 | 3E+6 |
| HighFlux 125 | 1.1E-11 | 1.33 | 2.59 | 125 | 15000 | 1E+6 |
| HighFlux 160 | 3.7E-12 | 1.41 | 2.56 | 160 | 15000 | 8E+5 |
| Ferrite Magnetics F | 1.8E-14 | 1.62 | 2.57 | 3000 | 3000 | 1.3E+6 |
| Ferrite Magnetics K | 2.2E-18 | 2 | 3.1 | 1500 | 3000 | 2E+6 |
| Ferrite Magnetics P | 2.9E-17 | 2.06 | 2.7 | 2500 | 3000 | 1.2E+6 |
| Ferrite Magnetics R | 1.1E-16 | 1.98 | 2.63 | 2300 | 3000 | 1.5E+6 |
| Ferrite Philips 3C80 | 6.4E-12 | 1.3 | 2.32 | 2000 | 3000 | 1E+6 |
| Ferrite Philips 3C81 | 6.8E-14 | 1.6 | 2.5 | 2700 | 3000 | 1E+6 |
| Ferrite Philips 3C85 | 2.2E-14 | 1.8 | 2.2 | 2000 | 3000 | 1E+6 |
| Ferrite Philips 3F3 | 1.3E-16 | 2 | 2.5 | 1800 | 3000 | 1E+6 |
| Ferrite TDK PC30 | 2.2E-14 | 1.7 | 2.4 | 2500 | 3000 | 1E+6 |
| Ferrite TDK PC40 | 4.5E-14 | 1.55 | 2.5 | 2300 | 3000 | 1E+6 |
| Ferrite FairRite 77 | 1.7E-12 | 1.5 | 2.3 | 2000 | 3000 | 1E+6 |

Note: Philips Ferrites is now Ferroxcube. Powdered iron grades are from Micrometals. High Flux, Kool Mu are registered trademarks of Magnetics Inc.

**Note:** *Optimum results* (in terms of overall losses) are said to be attained if

$$\frac{Core\ Loss}{Copper\ Loss} = \frac{2}{\text{exponent of } B}$$

But treat this only as a general guideline. For example, in most off-the-shelf inductors for dc-dc converters, the copper loss is over 90 percent of the total loss. Besides, we may not have enough window available to wind that much of copper when dealing with E-cores.

# 12

# PCBs and Layout

## 12.1  Introduction

When it comes to switching regulators, it is not enough to concern oneself with just the basic routing/connectivity and related mechanical/production issues. Both the power supply designer and the CAD person need to be aware that the design of a switching power converter is only as good as its layout.

The overall area of printed circuit board (PCB) design is an extremely wide one, embracing several test/mechanical/production issues and applicable compliance/regulatory issues. Most of the issues discussed in this chapter revolve around simply assuring basic functionality. Though luckily, as the beleaguered switcher designer will be happy to know, in general, all the electrical aspects involved are related and point in the same general "direction." For example, an ideal layout, that is, one that helps the IC function properly, also leads to reduced electromagnetic emissions and vice-versa. However, there are some exceptions to this helpful trend, particularly the practice of indiscriminate copper-filling, and this will be touched upon too.

## 12.2  Trace Analysis

We must first learn to identify the troublesome or *critical trace sections* of any topology. The following rule is simple and applies to all topologies.

During a crossover transition the current flow in some trace sections has to suddenly come to a stop, and in certain others it has to start equally suddenly. These are the *critical traces* for any switcher PCB layout. These should be identified.

In Fig. 12.1 we have trace analysis for the buck. We have omitted most of the control traces like feedback, bootstrap, and enable, as these

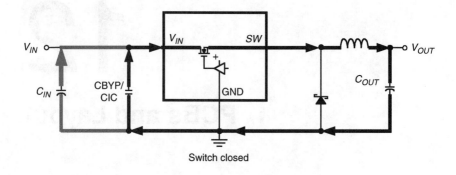

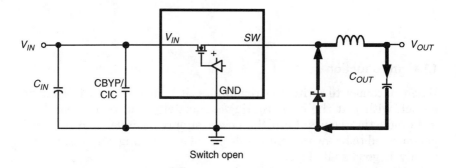

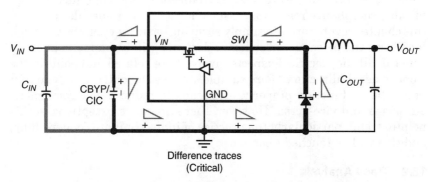

**Figure 12.1**  Trace analysis of a buck converter to identify critical traces.

hardly carry any current and are *not* critical from the perspective of current/power flow. But do note that the routing of the feedback trace (which will be taken up later) is certainly important too, but from the viewpoint of noise pickup.

The bold traces in the schematic show the power flow. We have shown the traces which are passing current during the switch on-time, followed by the traces conducting during the off-time, and finally in the lower-most schematic we work out the difference between the two previous schematics. These *difference traces* are the ones in which the current must abruptly either turn *on* or turn *off* during a crossover transition. These are the critical traces, by definition.

All traces have a nonzero inductance. So, if current is passing through them, they have a certain amount of stored energy $1/2 \times LI^2$. When the current in such a trace is commanded by the switch to stop flowing, its stored energy cannot disappear immediately. So, it "complains loudly" in the form of a voltage spike (shown with gray triangles). Similarly, the same traces generate a voltage spike in the opposite direction, when we try to force current suddenly through them. The amplitude of the spikes can be significant as per the standard equation $V = LdI/dt$. This means that even if $L$ is small, if the $dI/dt$ is very high so will be the spike. The $dI/dt$ depends on how fast the FET/BJT switches from on-state to off-state (and the other way around).

The problem with the spikes is that not only can they cause EMI and output ripple, but they easily ingress into the control sections of the IC through their connecting pins, causing the IC to misbehave. The noise rejection of switcher ICs is never specified in datasheets. However, a "good device" is one that is fairly tolerant to these noise spikes. But we can really test its patience if we do not reduce the inductance of the identified critical traces. This aspect almost completely separates a constantly griping switcher IC customer from a satisfied repeat-order one. It's usually just the layout, not the IC.

We do note that ICs with BJT switches always tend to switch more slowly than ICs which use FET switches. So it is no surprise that there are less "layout issues" (and apparent IC misbehavior complaints) from customers who are using the slower devices.

We need to highlight some more details from Fig. 12.1. CIC is a bypass capacitor specifically meant to smoothen the supply rails (which are ultimately going to the control sections of the IC). If CIC is very close to the IC it provides most of the high-frequency content demanded (mainly by the edges) of the switch current waveform. CIN is the input bulk capacitor and it provides most of the remaining current waveform. It is being constantly dc-refreshed from a distant bulk capacitor, in this case belonging to the dc bench power supply.

We can see that the way to reduce the length, and thereby the inductance of the critical traces, is to bring their *associated components close to the IC*. That will automatically reduce the corresponding trace lengths. It will also reduce the high-frequency *current loop* and thereby reduce EMI too. For example, the position of CIC is critical in Fig. 12.1 (for this topology). This should thus be as close as possible to the IC. If it is correctly placed, then CIN can usually be an inch or two away. Since

TABLE 12.1  Critical Components in PCB Layout for Integrated Switchers.

| | CIN (input bulk cap) | CIC (bypass cap) | COUT (output cap) | Inductor | Diode |
|---|---|---|---|---|---|
| **Buck** | Critical | Critical | Not critical | Not critical | Critical |
| **Buck-boost** | Critical | Critical | Critical | Not critical | Critical |
| **Boost** | Not critical | Critical | Critical | Not critical | Critical |

it is not very critical it is shown in bold gray lines in the schematic. If the decoupling capacitor is omitted, (and it can usually be left out completely if we have a BJT-based switcher IC), then CIN must be brought very close to the IC. If CIN is more than even 0.5 in away from the IC, it is known to have caused problems. Similarly, for a FET-based switcher even a pair of vias intervening between the IC and the decoupling capacitor has been known to cause problems. In Table 12.1 we have summarized which are the critical components for the three topologies, based on similar trace analysis.

## 12.3  Miscellaneous Points to Note

1. The often-repeated thumb rule is that *every inch of trace length has an inductance of about 20 nH.*

2. The change in current in the critical trace sections of a typical buck converter is about 1.2 times the load current ($I_o$) during the switch turn-off transition and is about 0.8 times the load current during the switch turn-on transition (for $r = 0.4$). The worst case is at *maximum* input voltage for this topology.

3. The change in current in the critical trace sections of a typical buck-boost and boost converter is about 1.2 times $I_o/(1-D)$ during the switch turn-off transition and is about 0.8 times $I_o/(1-D)$ during the switch turn-on transition. That is assuming $r$ is set to about 0.4. So the worst case is thus at *minimum* input voltage (highest $D$).

4. The transition time is about 30 ns for high-speed FET switchers and is about 75 ns for the slower bipolar switchers. This also means that the voltage spikes in the high-speed family will be more than twice that in the slower family, for a comparable layout and load. Therefore, layout becomes all the more critical in high-speed switchers. For 1 in of trace switching, say, 1 A of instantaneous current in a transition time of 30 ns gives 0.7 V. For 3 A and 2 in of trace the induced voltage tries to be 4 V!

5. None of these problems can be easily corrected, or even "band-aided," once the layout is bad. So the important thing is to get the layout "right" to start with.

6. The curious designer may well ask why is it that these current changes are a problem with the parasitic trace inductances and not with the main inductor of the buck converter. That is because all inductors try to resist any sudden current change. But since the main inductor has a much larger inductance (and energy storage) as compared to the parasitic trace inductances, it therefore ends up "dominating". The parasitic (trace) inductances have the necessary degree of freedom available, and the current in these inductors does reverse. But not before "complaining."

7. For the buck topology the output capacitor current is smooth (because the inductor is in series with it). In a boost topology the situation is reversed, i.e., the input capacitor current is smooth, and the current into the output capacitor is pulsed. In a buck-boost or flyback both the input and the output capacitor currents are pulsed.

In some "not-so-bad" buck converter layouts where the catch diode was not appropriately placed, the converter was successfully "band-aided" by a small RC (resistor-capacitor) snubber. This consists typically of a resistor (low inductive type preferred) of value 10 to100 $\Omega$ in series with a capacitor, which should be a high-frequency type (e.g., ceramic) of value 470 pf to 2.2 nF. Larger capacitance than this would lead to unacceptably higher dissipation ($=1/2 \times C \times V^2 \times$ f), chiefly in the resistor, and would serve no additional purpose. However note that this RC snubber needs to be placed very close to and across the switching pin and ground pin of the IC, with short leads/traces. Sometimes it is felt that this is across the diode, because on the schematic there is no way to tell the difference. However, the purpose of this RC snubber is to absorb the voltage spikes of the trace inductances and therefore its position must be such that it bypasses the critical or ac trace sections of the output side.

**Note:** Traditionally, one purpose of introducing an RC snubber was to reduce the d$V$/d$t$ as the switch crossed over, because the peak of the diode reverse recovery current decreases if we slow the rate of change of voltage across it. However, most low voltage integrated switchers recommend only a Schottky diode for trouble-free performance. One reason is that the wide variations possible in the reverse recovery currents of ultrafast diodes may sometimes "fool" the switcher's current limit circuitry into thinking that current limit has been reached, forcing it into premature pulse termination. A Schottky diode is preferred, not only from the viewpoint of efficiency, but because it has no reverse recovery current in principle. However, we should remember that its higher body capacitance can cause it to mimic recovery current effects somewhat.

8. The traces to the critical components should be short, reasonably wide, and should not go through any vias. The inductance of a via is

given by

$$L = \frac{h}{5}\left(1 + \ln\frac{4h}{d}\right) \text{ nH}$$

where $h$ is the height of the via in mm (equal to the thickness of the board) and $d$ is the diameter in mm. If vias have to be used for some reason, *several vias in parallel* will yield better results than a single via. And *larger via diameters* would help further.

9. A first approximation for the inductance of a conductor having length $l$ and diameter $d$ is

$$L = 2l \bullet \left(\ln\frac{4l}{d} - 0.75\right) \text{ nH}$$

where $l$ and $d$ are in centimeters. Note that the equation for a PCB trace is not much different from that of a wire.

$$L = 2l \bullet \left(\ln\frac{2l}{w} - 0.5 + 0.2235\frac{w}{l}\right) \text{ nH}$$

where w is the width of the trace. For PCB traces, $L$ hardly depends on the thickness of the copper (1-oz or 2-oz board). We all know that halving the length of a trace, halves its inductance. But few realize that we have to roughly decrease the width (or diameter) by a factor of 10 to get the inductance to halve (this is related to the mutual inductance of the several thin parallel strips constituting the trace). As a corollary, making traces as wide as possible may not be a very good idea. It doesn't reduce the inductance much and it can also become a nice antenna if there is a pulsating voltage across it (e.g., if it is the switching node). So at some point we must learn to resist the tendency of indiscriminate copper-filling. See Fig. 12.2 for a plot based on the inductance equations.

10. The current handling capability of traces is related to the temperature rise we permit. Though U. S. Military Standards (MIL-STDs) recommend maximum 10°C rise, we can easily go up in commercial designs to two or three times that. In Fig. 12.3 we have the MIL-STD 275E curves. Note that they can be applied to both 1-oz copper board or 2-oz copper board in the way the x axis is presented. For a moderate temperature rise (less than 30°C) and currents less than 5 A we can use the following rule of thumb:

- Use at least 12 mils width of copper per amp for 1-oz board
- Use at least 7 mils width of copper per amp for 2-oz board

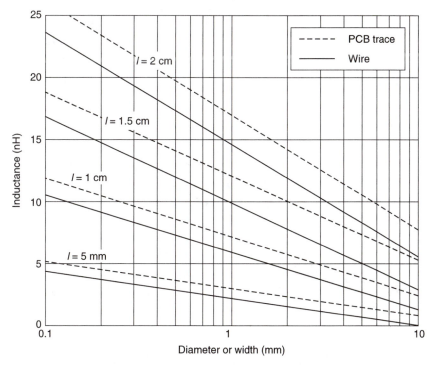

**Figure 12.2** Trace and wire inductances.

11. Commercial PCBs are often referred to as *1*-oz or *2*-oz for example. This refers to the *weight of copper in ounces per square foot* deposited on the copper clad laminate. Note that 1 oz is equivalent to 1.4 mils copper thickness (or 35 $\mu$m). Similarly 2 oz is 2.8 mils or 75 $\mu$m.

## 12.4   Routing the Feedback Trace

The only critical signal trace is the feedback trace. This is more liable to pick up noise if the pin it connects to is a high-impedance node. ICs with *fixed output voltage options* like 5 V and 3.3 V require no external voltage divider, because the divider is internal to the IC. Therefore, for such parts the feedback trace is not going to a high-impedance input pin, and their feedback trace is relatively immune to noise pickup. For the "adjustable voltage parts" greater care must be taken in the way we route this trace. We should keep the feedback trace short *if possible* so as to minimize pickup but it should certainly be kept away from noise sources (e.g., switch, diode, and even inductor). We must not compromise the positions of the input decoupling capacitor and catch diode in

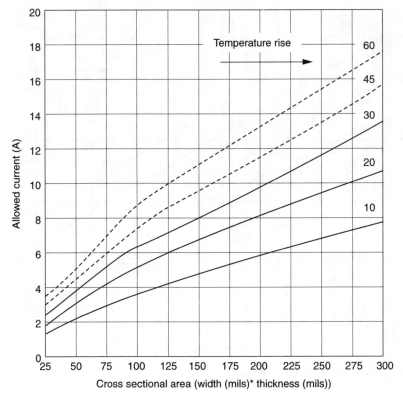

**Figure 12.3** MIL-STD-215E curves for temperature rise of PCB traces.

the process. If necessary, we can use a via to drop into the *ground plane* and route the feedback trace through the ground plane. This literally keeps its surroundings "quiet". In multilayer boards we should be very cautious that a noisy power trace does not run even a few millimeters parallel to the feedback trace, directly above or below. Any noisy traces must pass it at right angles.

## 12.5   The Ground Plane

With double-sided boards, it is a common practice to almost completely fill one side with ground. There are people who usually, rightly so, consider this a panacea for most problems. Every signal has a return and at low frequencies, the return tries to take the shortest path (straight line) in an attempt to minimize the dc resistance. But as its harmonics get higher, the return path tries to readjust itself in an effort to reduce the growing ac impedance. It does this by trying to get directly under the signal path since this leads to field cancellation and reduction of trace inductance. So, by keeping a large ground plane, we basically allow enough room for all the returns to choose their respective preferred paths. The ground plane is also a good heatsink in effect, by virtue of its large conducting surface area (see Chap. 13). It therefore helps in thermal management since heat from various components can get coupled into it by conduction through the PCB material. The ground plane can also capacitively link to noisy traces above it, causing a general reduction in noise and EMI.

The ground plane can also, however, end up radiating on its own, if caution is not exercised. One way this can happen is to have *too much* capacitive coupling from noisy traces into this plane, especially if this plane is of rather thin copper gauge, since it would then not be very effective in quashing the voltage gradients being created across it. By its sheer area, it can thus become a good antenna too.

To reiterate, at low frequencies, the return current flowing in the ground plane tries to return by the path of least resistance, i.e., a straight line running across the ground plane between the two points. But at higher frequencies (higher harmonics) it tries to image itself directly under the forward trace as it tries to cancel the mutual field, minimize the loop area, and lower the total inductance, all at the same time. This is still the path of least impedance that it tries to choose. But if the ground plane is *partitioned, or cut into* in odd ways, either to create thermal islands or to route other traces, the current flow patterns in it can become "unnatural". With thoughtless cuts in the ground plane, we can get effects quite similar to a slot antenna. We must learn to allow nature to help us. It always does!

We also note that generally, a ground plane should supplement other recommendations. For example, it does not necessarily substitute for the correct placement of the critical components in particular.

Sometimes we forget that board parasitics are also known to inadvertently *help* us in some cases, such as in preventing unwanted interactions between different circuit blocks—the reason why we put decoupling in the first place. So designers should weigh everything carefully before they decide on their board configuration and layout. One example where "poor layout" helps is when we try to put several dc-dc converter stages in parallel across the same input rails. They can end up interacting and oscillating if their input traces are very short and wide. We should either use deliberate decoupling (in the form of an input LC section for each converter stage), or introduce decoupling by creating high frequency separation through the use of thin and long traces coming from the common input rails. The decoupling capacitors should be very close to their respective stages.

In multilayer boards we should try to keep the ground plane as the internal plane just below the component side (though in this position it won't help much from a thermal viewpoint). By the fact that the distance from the ground plane to the components is now much less (assuming that the overall board thickness is still the standard 1.6 mm), the length of the interconnecting vias is much smaller and the coupling, in general, is better. Therefore, noise gets suppressed much better. Current sense sections also seem to work better. EMI is better and the layout becomes rather forgiving to bad layout practices. So, if the engineer is rather nonchalant, or just too busy to sit with the CAD guy through all this, this is the safest bet. Never mind the added cost. But if the cost is already inherent in the design, as for example if the converter section is merely being added on an already mandated multilayer system card, this becomes a no-brainer.

## 12.6   Some Manufacturing Issues

Any 1-oz double-sided board will pass through an electroless copper plating process stage (before solder mask is applied) to create the vias (also called PTH for *plated through hole*), and so it may end up effectively as being considered closer to a 1.4-oz copper board. Therefore it is a good idea to check this out with the PCB manufacturer before even starting the layout. This will help determine how wide the copper traces actually need to be for a particular current-carrying capacity and temperature rise. Also, note that even a single-sided board passes through a *hot-air solder-level* finishing stage (HASL) after solder mask, where a thin tin-lead layer is deposited on the "unmasked" (no solder mask) copper areas. This does increase the effective thickness of these traces, but doesn't help as much as copper plating since tin-lead solder has 10 times higher resistivity than copper.

We are also getting so used to seeing standard *green boards* (FR-4) that we may forget that there are cheaper alternatives available. In power supplies where cents count and the PCB could cost several dollars, we cannot afford to forget that cheaper materials and board configurations could cut the overall PCB cost by almost a factor of 4. The designer may, however, have to struggle since all board laminate materials are not equal and they have limitations that need to be overcome by careful layout.

A strategy once used by the author was to move as many control components as possible onto *one* side of a double-sided PTH daughter board *made of FR-4*, while placing *all* of the bulkier components which were *not capable* of being handled by pick and place surface mount technology (SMT) machines on to a cheaper single-sided CEM-1 board. This was considered to be a major cost saver for the company, not only in terms of component cost but also production and test costs. Note that both primary and secondary-side SMD components were on the same control board, which implies that we had to be using surface mount optocouplers. The designer should pay heed to mandatory clearances and "creepages" when selecting such optocouplers. Just because they are "safety approved," does not imply that they meet the separations as required of our application. See Chap. 17 to understand these terms.

We need a small discussion about what are FR-4 and CEM-1. FR stands for *fire resistant* or *fire retardant*. That is not the same issue as *flammability* (or resistance to thereof), which is the main concern of safety agencies. All FR-4 laminates, and even the cheaper CEM laminates (composite epoxy material), are all typically available with the highest flammability rating of safety agencies (called *94V−0*, or just *V−0*). So there should be no problem using even cheaper materials from that point of view.

→ *CEM-1.*   This consists of a core of a cellulose paper core and a woven glass cloth surface, combined with an epoxy resin binder. It can be recognized by its whitish to cream color (opaque). It has good mechanical strength, is easily punched, and drills well too. It is very cheap. It is not suitable for plated through-holes (PTH or vias). Useful for single-sided boards.

*CEM-2.*   Considered virtually identical to CEM-1.

→ *CEM-3.*   Similar to CEM-1 except that the core is of nonwoven glass. CEM-3 is considered virtually interchangeable with FR-4. It is recognizable by its natural light brown color (opaque). PTH is possible with this material. In Japan, for example, CEM-3 has a very large market share.

*CEM-4.*   Considered virtually identical to CEM-3, particularly from an electrical standpoint.

*FR-2.*   This is a paper-base laminate with phenolic resin binder. It has high moisture resistance and is flame retardant.

*FR-3.*   This is a paper-base laminate with epoxy resin binder. It has high flexural strength and is flame retardant.

→ *FR-4.*   This is a woven glass cloth with epoxy resin binder. Same as CEM-3 except that it is fire retardant.

Some general comments and observations are as follows:

FR-4 is used for its inherent stability. Its characteristics remain constant over long periods of time, even under adverse environmental conditions. However, there are different grades and specifications even within FR-4, with special versions for chemical resistance and for use in multilayer designs. Different grades for higher temperatures are also available. The rating we just cannot exceed is the $T_g$ (*glass transition temperature*) of the PCB laminate.

An important reason to use FR-4 along with SMD components is that almost all manufacturers of such components spend a great deal of effort matching their CTE or TCE (coefficient of thermal expansion) to that of standard FR-4 boards. If the thermal expansions are not well matched, components may start to crack under extended periods of thermal cycling, and also solder joints will develop fatigue and start to give.

For boards devoted to through-hole components, we can consider using either CEM-1 or CEM-3 to reduce costs. But note that whenever we use through-hole components, that board will pass through a wave soldering process. We should however avoid immersing SMD components in hot solder as will happen during wave soldering if they are on the underside of the board. Generally, *if SMD components have got to be on the same board as through-hole components, for various reasons we should try to restrict them to the component side only* (where the power components are placed). Just good manufacturing practices!

If a double-sided SMD control board is being used, we may like to avoid placing components on both sides of the board; because, if the components lie only on one side, they are just placed on top of the board by the pick and place machines and pass through the reflow soldering. They thus stay in place with the help of gravity (and solder paste). However, if we have components on both sides, we will have to ensure that the components on one side do not fall off. Adhesives will become necessary. All this adds to the manufacturing cost.

A possible concern with the use of single-sided boards for power components is that under a vibration test, a single-sided board's copper traces can come undone rather easily. Double-sided boards offer much greater *holding capability*, especially when the heavy component's leads pass through plated through-holes. Here the solder seeps into the PTH creating a very firm bond. So with a single-sided board we should at least try to leave large copper areas around the legs of heavier components, like the transformer. Gluing the component to the board with RTV/hot-melt will also probably be required. *RTV* stands for *room temperature vulcanizing* (silicone glue).

Single-sided boards will require jumpers. Most of the time these may have to be hand-inserted, which can add to the cost depending on the geographical area the production is being done. Jumpers have a higher inductance than traces. We must therefore study the layout to see where currents *change direction* during a given switch transition.

As mentioned, such traces are considered critical and no jumpers should ever be placed in these sections.

The transition temperature $T_g$ of the PCB material is also important. We don't want to use a low temperature board material and have to provide more expensive better cooling systems. The standard FR-4 has a $T_g$ of approximately 130°C. The CEM boards may be rated 20 to 30°C less than FR-4 depending on their manufacturer.

One of the most important ratings of a PCB laminate is its CTI (*comparative tracking index*). We should confirm that the CTI is from the specific vendor as it has a bearing on the required "creepage" in off-line power supplies. See Chap. 17.

## 12.7 PCB Vendors and Gerber Files

We must understand that our PCB should match the ability of the vendor if we want to reduce costs. For example, every vendor has his or her own list of *preferred drill sizes*. We should get that information beforehand.

Too often we also don't understand the information we need to provide the vendor. Often vendors just don't ask, and simply use their default values. So, one vendor's' boards may behave very differently from another's, simply because their default values were different. It will take us a while to figure out why the new build is not working properly anymore. We need to avoid this situation by specifying clearly all that is important to us. One of the most common problems is that the copper thickness may have changed in going from one PCB vendor to another, say from 1 oz to $1/2$ oz. This can cause lots of problems with the performance of any switcher circuit. Visually, it is virtually impossible to check this out. Usually, such an error can be traced to the power supply engineer not having specified clearly what copper thickness is required for his or her board, or simply not knowing the copper thickness originally used on the first tested and approved prototype.

The Gerber File format is the industry standard format for files used to generate artwork necessary for circuit board imaging. The preferred Gerber format is RS274X, which embeds the apertures within the specific files. The apertures assign specific values to design data (such as specific pad size and trace width) and these values make up a $D$-code list.

When files are not saved as RS274X, a text file with values must be included because the values must be hand-entered by CAM operators. This slows down the process and increases the margin for human error as well as lead-time and cost.

Here is a sample of a request to a PCB vendor:

Please build a quantity of 200 pieces of this six-layer board. We need 1-oz base copper weight, 0.062-in thick FR-4 material, with green solder mask, and white legend.

File name: FLYBACK_70_REVA.ZIP includes:

FLYBACK70_REVA.CMP Layer 1 (component side)
FLYBACK70_REVA.GND Layer 2 (ground layer)
FLYBACK70_REVA.IN1 Layer 3 (internal signal)
FLYBACK70_REVA.IN2 Layer 4 (internal signal)
FLYBACK70_REVA.VCC Layer 5 (power layer)
FLYBACK70_REVA.SLD Layer 6 (solder side)
FLYBACK70_REVA.CSM (component-side solder mask)
FLYBACK70_REVA.SSM (solder-side solder mask)
FLYBACK70_REVA.TSK (component-side silkscreen/nomenclature)
FLYBACK70_REVA.BSK (solder-side silkscreen)
FLYBACK70_REVA.DRL (Excellon drill file)
FLYBACK70_REVA.TOL (drill tool description)
FLYBACK70_REVA.APT (aperture information)
FLYBACK70_REVA.DWG (drawing or print)

# 13

# Thermal Management

## 13.1 Introduction

Natural convection starts with the simple statement that "hot air rises." For example, a power semiconductor mounted on a metal plate gets exposed to the natural movement of air, and this transfers heat from the device to the surroundings. Perhaps we know intuitively that using larger metal plates helps this process and this lowers the temperature. But we probably also know that very large plates don't help much beyond a certain point. We are also perhaps aware that higher dissipations actually help the cooling process, though that certainly doesn't mean we should try to increase our dissipation! It's not the temperature rise but the temperature rise *per watt* that is lesser under high-dissipation conditions. Clearly, nature is always willing to provide a helping hand, provided we understand its inner workings and foster its beneficial effects. In thermal management, the way to go about this is to understand the equations of convection clearly.

Thermal management is not a black art. The rules are actually quite simple. We noticed that one of the major problems is *not* that there was a lack of available rules, empirical or otherwise, but that they were being cast in so many diverse forms that engineers just don't know how they compared and which one to choose. Our purpose here is not to declare which rule is the most accurate, but to show the different forms each rule can be written in. Further, if they are all converted into the same format, they become amenable to a simple point comparison. Then we will see how amazingly close the apparently different equations really are, in their respective predictions of temperature. Finally, we can pick either a "conservative estimate" or a more practical estimate.

## 13.2   Thermal Measurements and Efficiency Estimates

One problem in thermal management is *measurement*. Measuring a temperature rise and relating it to the dissipation is one of the most difficult bench studies an engineer can undertake. Some issues to be aware of are as follows:

*Where temperature should be measured.* Heatsink manufacturers often describe the measurement procedures they use, but these are unfortunately almost impractical in an actual switching converter scenario. For example, we really cannot be expected to drill a small hole through the heatsink to allow a thermocouple to be inserted to make contact with the tab. Alternatively, some say that the best point for judging the junction temperature of a 3-terminal FET, for example, is the central lead, just at the point where it exits the plastic. But actually, this gives measurements which are slightly lower than with other methods. Some say the tab of a mounted transistor is the best point, and the thermocouple should be glued on (centered) just below the mounting hole on the plastic body at a point closest to the tab. This may in fact be necessary if the tab is the drain of the FET, and is therefore at a high (swinging) voltage, which could render measurements noisy and unreliable. One thing is clear—the location which gives us the highest steady temperature measurement is likely to be the most accurate. The author personally prefers to loosen the mounting screw slightly, insert the thermocouple under the tab, put in a lot of thermal grease to fill the air space, and then tighten the screw again. Since the thermocouple is in galvanic contact with the tab, this actually works only if the tab is "quiet" as is the case with some high-voltage integrated switcher ICs. Note that in previous methods the thermocouple will always need to be glued on to the plastic part of the device. Thermally conductive adhesives are available, but some take several hours to cure. A quick method, and one that is internally allowed by some major power supply vendors, is to use *super glue* (cyanoacrylate). This has been found to be ok, provided the thermocouple is pressed firmly against the tab for the minute or so it takes for the glue to harden. Moisture (a little dab of water) can hasten the curing process for this glue, but we can also use a formal accelerator spray.

**Note:** Thermocouples are commonly either (a) T type: these are of copperconstantin and work over the range of −270 to 600°C with a swing of 25 mV, providing 40.6 µV/°C at 25°C, or (b) J type: these are of iron-constantin and work over the range of −270 to 1000°C with a swing of 60 mV, providing 51.70 µV/°C at 25°C. The ends are usually spot-welded together, but soldering often works too for quick results. But the solder should be 60 percent lead and 40 percent tin as it has a higher eutectic temperature than the more commonly used 40 percent lead and 60 percent tin. Clearly we need a high-temperature iron too.

*Variable ambient conditions.*   In a typical room, unbeknownst to us, small movements of air (draughts) keep taking place, like someone walking past, or an overhead air-conditioning vent suddenly coming into play, and the like. The measurements therefore rarely turn out to be reproducible (within an acceptable couple of degrees of accuracy). An enclosed box may be built in an effort to solve the problem, but now local pockets of hot air can form inside. To keep a proper average ambient temperature inside, we need some movement of air. If a fan is put in it may be too much. We wanted to create natural convection, not forced air convection. So round and round we go, trying to create the right environment—too much air, now too less. . .

*What is the dissipation?*   Even if we do get the temperature, how do we relate it to the dissipation? A FET for example has an Rds that is a function of temperature. So, is its dissipation. So, we land up with a cyclical argument. We want to know the temperature at a certain dissipation, but the dissipation itself depends on temperature. Nature ultimately attempts to achieve a stable solution, but for us the math may be rather hard to figure out. One thing we can do is to replace the FET with a diode in the same package. We can pass dc current through the diode and monitor the voltage across it by a Kelvin measurement using a multimeter. We should increase our current slowly (and wait!) so that we get the same temperature we got with the switching FET in place. We thus now know the dissipation correctly, and we also have the corresponding temperature reading. Having characterized the heatsink, we can put in the FET again and try to validate our design estimates against the measured dissipation.

*Measuring conduction loss.*   We go back to the bench and try to measure the forward voltage drop across the FET (or diode), for example. This is in our actual switching application, and we are thus trying to validate our estimate of conduction loss, and possibly create a model for purposes of optimization. Clearly, a multimeter will not work here. So we use an oscilloscope and try to "see" the forward voltage across the FET. We are surprised to find that it seems too high, or too low, or even goes negative. We make sure our probes are well compensated by connecting the probe to the test signal on the front panel of the scope and adjusting the trimpot on the probe by the small, insulated screwdriver provided in the accompanying probe kit. Now our test signal looks nice and square, but when we look at the FET again, we still get absurd results. The problem here is that we have adjusted the vertical scale of the scope to a few mV per division as we try to zoom in on the small forward drop. So when the FET turns *off*, the voltage is high, and out of scale. The invisible part of the signal

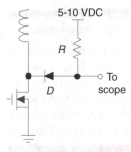

**Figure 13.1**   Measuring the forward voltage drop.

we are trying to measure therefore actually ends up overdriving the
internal amplifiers of the scope, and when the FET turns *on* again
in the next cycle, the scope just can't faithfully reproduce the signal
correctly (for a certain undefined recovery time). The solution to this
is to clamp the voltage to a lower level of about 5 to 10 V as shown in
Fig. 13.1. We would need to characterize the *VI* characteristics of the
diode beforehand since we are now going to be reading the forward
drop across the FET plus the diode drop. The diode should be carefully selected to be a fast small-signal PN type diode. *R* will also need
to be adjusted, typically to around 1 kΩ to 2 kΩ, to bias the diode
correctly. The dc voltage may also need to be further adjusted to keep
the signal limited to the screen.

*Measuring switching losses.*   Here we look at the overlap between
the voltage waveform and the current waveform. We have to remember that active current probes are not so fast and typically have
a delay. Therefore, switching loss measurements can be erroneous.
Passive probes (which have just a coil inside) may work much better for this measurement. Further, even the scope probes used for
voltage measurement have a few nanoseconds of delay, and this can
vary from probe to probe, especially if their lengths are different. We
should check for these "propagation delays" and correct our readings
accordingly. On some scopes we can apply this correction on-screen
too. Scopes with *general purpose interface bus* (GPIB) output can be
used to drop the data into an Excel spreadsheet, in which we can
apply the necessary offsets, $\Delta t$ and $\Delta v$, and then do the usual *VI*
crossover calculation automatically.

## 13.3   The Equations of Natural Convection

If we know the temperature of the heatsink (which could just be the
copper in the immediate vicinity of a surface-mounted device) and also
the dissipation, we can estimate the junction temperature by using

$$Rth_{JA} = Rth_{JH} + Rth_{HA}$$

where $Rth$ is the thermal resistance in °C/W, $J$ refers to the junction, $H$ is the heatsink, and $A$ stands for ambient. If $P$ is the dissipation in watts,

$$T_J = P \times (Rth_{JH} + Rth_{HA}) + T_A$$

Note that $Rth_{JH}$ can be further split

$$Rth_{JH} = Rth_{JC} + Rth_{CH}$$

where $C$ is for case. But the only parameter we probably have full control over is $Rth_{HA}$. Hence, this chapter focuses on understanding this critical link. *Henceforth,* Rth *just refers to* $Rth_{HA}$.

There are abundant equations or empirical formulas available for the purpose—though, as we will see, they actually fall under only two or three umbrellas, and these too are very close. But that is not obvious at first sight. One source of confusion is that the area used in these equations often refers to the area of the plate (*one* side of it) even though we know that *both* sides are exposed to natural convection. Some equations recognize this explicitly and use twice the value for area. For the units, we may find the area expressed in square inches, or square meters, or square centimeters, and the like. Further, some may assume that the area is square in shape and use the length of each side instead of the area. What we have to be careful about is that *even the most minor change in the manner of expression of a given equation can render it virtually unrecognizable.* That is what we will strive to show in this chapter too.

In this chapter, $A$ refers to the area of *one* side of a plate, assuming *both* its sides are exposed to cooling. The total exposed area is being called $\underline{A}$ (so $\underline{A} = 2\ A$).

## 13.4 Historical Definitions

We take the simplest case of a square plate made from a very good thermally conducting material, dissipating $P$ watts. After some time we will find that the plate stabilizes at a certain temperature rise of $\Delta T$ over the ambient.

We expect that the temperature rise will be proportional to the dissipation. The proportionality constant is called the *thermal resistance* $Rth$ in °C/W. So,

$$Rth = \frac{\Delta T}{P}$$

Similarly, we expect that the thermal resistance will vary inversely with the area

$THERMAL\ RESISTANCE$ $Rth \propto \dfrac{1}{A} \propto \dfrac{1}{h}$

The inverse of the proportionality constant above is $h$ in W/°C per unit area, and is called by various names, like *convection coefficient* or *heat transfer coefficient*.

$h \equiv CONVECTION\ OR\ HEAT\ TRANSFER\ COEF.$ $Rth = \dfrac{1}{h\underline{A}} = \dfrac{1}{2hA}$

Finally, we have the basic equation set

$$P = h \times \underline{A} \times \Delta T = 2 \times h \times A \times \Delta T = \frac{\Delta T}{Rth}$$

Explicitly,

$$Rth = \frac{\text{temperature rise}}{\text{watts}}$$

$$h = \frac{\text{watts}}{\text{total exposed area} \times \text{temperature rise}}$$

and

$$h\underline{A} = \frac{1}{Rth}$$

The reason we kept using the word "expect" above is that historically speaking these relations were presumed to be true, and it was thought that $Rth$ and $h$ were merely proportionality *constants*. But later it was realized that this was not so. The above classical equations were still maintained for the sake of consistency, but what changed was that no longer were $h$ or $Rth$ considered constants. They were now "allowed" to depend on area, dissipation, and the like, the intention being to indirectly factor in the observed deviations from the "expected" results.

### 13.5   Available Equations

As a first approximation, $h$ is often stated (at sea level) to be

$$h = 0.006 \text{ W/in}^2 - {}^\circ\text{C}$$

If area is expressed in meters this becomes

$$h = 0.006 \times (39.37)^2 = 9.3 \text{ W/m}^2 - {}^\circ\text{C}$$

(since there are 39.37 in. in a meter).

We now know that in reality $h$ can easily vary by even a factor of 4 from the commonly assumed value above.

So, in the literature we can find the following generalized empirical equation for $h$, and this becomes our *standard equation no. 1*:

$$h = 0.00221 \times \left(\frac{\Delta T}{L}\right)^{0.25} \text{ watt/in}^2 - {}^\circ\text{C (standard equation no. 1)}$$

where $L$ is the length along the direction of natural convection (vertical). In the case of the simple square plate, $L = A^{0.5}$, so we can write this as

$$h = 0.00221 \times \Delta T^{0.25} \times A^{-0.125} \text{ watt/in}^2 - {}^\circ\text{C (standard equation no. 1)}$$

Also observe that the above equation uses $A$, which is actually half the area exposed to cooling. So, equivalently, we can rewrite it in terms of the actual area involved in the cooling process:

$$h = 0.00221 \times \Delta T^{0.25} \times \left(\frac{A}{2}\right)^{-0.125} \text{ watt/in}^2 - {}^\circ\text{C}$$

$$h = 0.00241 \times \Delta T^{0.25} \times \underline{A}^{-0.125} \text{ watt/in}^2 - {}^\circ\text{C}$$

True.

These are all available and published forms of the same equation for $h$.

**Note:** The above equation predicts that $h$ has a certain specified dependency on the exposed area of the plate and also on its temperature differential with respect to ambient. This dependency (i.e., $A^{-0.125}$) implies that the cooling efficiency *per unit area* (i.e., $h$) of *large* plates is *worse* than that of small plates. However, if this sounds surprising, we note that the overall/total cooling efficiency of a plate is $h \times A$, which depends on $A^{+0.875}$. So, thermal resistance goes as $1/A^{+0.875}$ and is clearly lower for a large plate than for a small plate as we would expect. Compare this to the "ideal" $1/A$ variation which was, classically speaking, expected for thermal resistance.

In the literature we often find the following "standard" formula (area in sq. inches), hereafter referred to as our *standard equation no. 2*:

$$Rth = 80 \times P^{-0.15} \times A^{-0.70} \quad \text{(A in in}^2\text{)}(standard\ equation\ no.\ 2)$$

We notice that the first equation is written in terms of $h$ and the second in terms of $Rth$. How do we compare them?

We will now do some more manipulations on these equations to bring them to a comparable format.

## 13.6  Manipulating the Equations

We have provided tables for the purpose but let us do one such manipulation to get comfortable with the process.

1. We can rewrite our standard equation no. 1 in terms of dissipation instead of temperature rise:

$$h = 0.00221 \times \left[\frac{P}{h \times A \times 2}\right]^{0.25} \times A^{-0.125}$$

So,

$$h = 0.00654 \times P^{0.2} \times A^{-0.3} \text{ watt/in}^2 - {}^\circ\text{C}$$

2. We can also write it in terms of the total exposed area:

$$h = 0.008 \times P^{0.2} \times \underline{A}^{-0.3} \text{ watt/in}^2 - {}^\circ\text{C}$$

3. We can also now try to see what this will look like in MKS (SI) units. The conversion is not obvious and so we proceed as follows:

   Take an imaginary plate of size 39.37 in × 39.37 in, or 1 m × 1 m. Clearly, the thermal resistance of the plate is in °C/watt and is therefore independent of the units used to measure area, and must remain unchanged by any change in the system of units used. This means that $1/(h \times \underline{A})$ is independent of units, and so is $h \times \underline{A}$. Therefore let us assume a similar form for $h$ in MKS units:

$$h = C \times \Delta T^{0.25} \times A^{-0.125} \text{ watt/m}^2 - {}^\circ\text{C}$$

Equating,

$$h \times A = C \times \Delta T^{0.25} \times A_{m^2}^{-0.125} \times A_{m^2} = 0.00221 \times \Delta T^{0.25} \times A_{in^2}^{-0.125} \times A_{in^2}$$

$$C \times A_{m^2}^{0.875} = 0.00221 \times A_{in^2}^{0.875}$$

$$C = (39.37^2)^{0.875} \times 0.00221 = 1.37$$

So, finally in MKS units

$$h = 1.37 \times \Delta T^{0.25} \times A^{-0.125} \text{ watt/m}^2 - {}^\circ\text{C}$$

4. In terms of the total exposed area:

$$h = 1.49 \times \Delta T^{0.25} \times \underline{A}^{-0.125} \text{ watt/m}^2 - {}^\circ\text{C}$$

5. We can also express $h$ in terms of $P$ instead of temperature as before:

$$h = 1.12 \times P^{0.2} \times A^{-0.3} \text{ watt/m}^2 - {}^\circ\text{C}$$

6. In terms of the total exposed area:

$$h = 1.38 \times P^{0.2} \times \underline{A}^{-0.3} \text{ watt/m}^2 - {}^\circ\text{C}$$

7. We can also recast standard equation no. 1 in terms of thermal resistance instead of $h$. We get several different forms:

$$Rth = \frac{1}{2hA} = 76.5 \times P^{-0.20} \times A^{-0.70} \text{ (area in in}^2)$$

8. Or in terms of the total exposed area:

$$Rth = \frac{1}{h\underline{A}} = 124.3 \times P^{-0.20} \times \underline{A}^{-0.70} \text{ (area in in}^2)$$

9. In MKS units

$$Rth = \frac{1}{2hA} = 0.45 \times P^{-0.20} \times A^{-0.70} \text{ (area in m}^2)$$

10. Or in terms of the total exposed area:

$$Rth = \frac{1}{h\underline{A}} = 0.72 \times P^{-0.20} \times \underline{A}^{-0.70} \text{ (area in m}^2)$$

## 13.7 Comparing the Two Standard Equations

Our standard equation no. 2 is

$$Rth = 80 \times P^{-0.15} \times A^{-0.70} \text{ (area in in}^2)$$

The result of our manipulations on standard equation no. 1 gives us

$$Rth = \frac{1}{2hA} = 76.5 \times P^{-0.20} \times A^{-0.70} \text{ (area in in}^2)$$

And we thus see that the two equations, one initially expressed in terms of $h$ and the other in terms of $Rth$ are not very different at all, if brought to a like form as we have done.

## 13.8   *h* from Thermodynamic Theory

Without going too deep into thermodynamics, let us do a quick check on the equations derived theoretically. We have the dimensionless Nusselt number $Nu$, which is the ratio of the convection heat transfer to the conduction heat transfer. We also have the dimensionless Grashof number $Gr$, which is the ratio of buoyant flow to viscous flow. Under natural convection (laminar flow) we have the following defining equations in MKS units

$$Nu = 3.5 + 0.5 \times Gr^{\frac{1}{4}}$$

where

$$Gr = \frac{g \times \frac{1}{Tamb+273} \times \Delta T \times L^3}{v^2}$$

where $g = 9.8$ (acceleration due to gravity in m/s$^2$), and $v = 15.9 \times 10^{-6}$ (kinematic viscosity in m$^2$/s). At an ambient temperature $T_{amb} = 40°C$ it can be shown that this simplifies to

$$Nu = 3.5 + 52.7 \times \Delta T^{0.25} \times L^{0.75}$$

The coefficient of cooling is by definition

$$h = \frac{Nu \times K_{AIR}}{L}$$

where $K_{AIR}$ is the thermal conductivity of air (0.026 watt/m$-$°C). So we get our *standard equation no. 3*:

$$h = 0.091 + 1.371 \times \left(\frac{\Delta T}{L}\right)^{0.25} \text{watt/m}^2 - °C \text{ (standard equation no. 3)}$$

or

$$h = 0.091 + 1.371 \times \Delta T^{0.25} \times A^{-0.125} \text{ watt/m}^2 - °C$$

*(standard equation no. 3)*

   Comparing this to the previously given empirical equations, we find that this equation too is surprisingly close, especially to the comparable form of our standard equation no. 1.

Unfortunately, though this theoretical form may be more accurate because of the constant term in its equation, for that very reason it is more difficult to manipulate into the forms the previous equations could be easily manipulated into. So we won't even try here. We would rather manipulate the previous empirical equations into a form similar to the theoretical equation, and then compare them (as we do toward the end).

### 13.9   Working with the Tables of the Standard Equations

In Table 13.1 we have the complete procedure for manipulating an equation given in terms of $h$ into all the other forms. We have four cases each time:

*Case 1.* "Area" used in the equation is half of the exposed area (in inch$^2$)

*Case 2.* "Area" used in the equation is the full exposed area (in inch$^2$)

*Case 3.* "Area" used in the equation is half of the exposed area (in meter$^2$)

*Case 4.* "Area" used in the equation is the full exposed area (in meter$^2$)

In the same table we have shown in gray the numbers thus generated from our standard equation no. 1. At the bottom of the same table we show *how to generate all the forms from an equation given in terms of* Rth (*such as from our standard equation no. 2*). *In one step we can go from an equation in terms of* Rth *to the same expressed in terms of* h, *that is, to the very beginning of the table, and then we can work our way down as before, generating all the other forms.*

In Table 13.2 we have compared the numerical results in each form for our two standard equations. Sure enough, we have seen many of these (or something very close) in related literature, though we probably hadn't realized until now that they were all the same equation.

We can provide a simple equation for estimating the copper area on a PCB. This is not a plate but a copper island on a PCB, and only one side is exposed to cooling. This is not the same as using the area of one side of a plate, both sides of which are exposed to cooling. Here we use the equation which uses the entire exposed area. For this, the standard equation no. 1 gives us

$$Rth = \frac{124.2}{P^{0.20} \times Area^{0.70}} \, °\text{C/W (area in inch}^2)$$

**TABLE 13.1  Conversion Table for Comparing Equations of Natural Convection.**

$$h = \alpha \times \frac{\Delta T^{\beta}}{Area^{\gamma}}$$

| | | | |
|---|---|---|---|
| Case 1<br>inches, half area | $\alpha_1 \equiv \alpha =?$<br>0.00221 | $\beta_1 \equiv \beta =?$<br>0.25 | $\gamma_1 \equiv \gamma =?$<br>0.125 |
| Case 2<br>inches, full area | $\alpha_2 = \alpha \times 2^{\gamma}$<br>0.00241 | $\beta_2 = \beta$<br>0.25 | $\gamma_2 = \gamma$<br>0.125 |
| Case 3<br>meters, half area | $\alpha_3 = \alpha \times 39.37^{2-2\gamma}$<br>1.37 | $\beta_3 = \beta$<br>0.25 | $\gamma_3 = \gamma$<br>0.125 |
| Case 4<br>meters, full area | $\alpha_4 = \alpha \times 39.37^{2-2\gamma} \times 2^{\gamma}$<br>1.49 | $\beta_4 = \beta$<br>0.25 | $\gamma_4 = \gamma$<br>0.125 |

$$h = x \times \frac{P^{y}}{Area^{z}}$$

| | | | |
|---|---|---|---|
| Case 1<br>inches, half area | $x_1 = \left(\dfrac{\alpha_1}{2^{\beta}}\right)^{1/(\beta + 1)}$<br>0.00653 | $y_1 = \dfrac{\beta}{\beta + 1}$<br>0.20 | $z_1 = \dfrac{\beta + \gamma}{\beta + 1}$<br>0.30 |
| Case 2<br>inches, full area | $x_2 = (\alpha_2)^{1/(\beta + 1)}$<br>0.00805 | $y_2 = \dfrac{\beta}{\beta + 1}$<br>0.20 | $z_2 = \dfrac{\beta + \gamma}{\beta + 1}$<br>0.30 |
| Case 3<br>meters, half area | $x_3 = \left(\dfrac{\alpha_3}{2^{\beta}}\right)^{1/(\beta + 1)}$<br>1.12 | $y_3 = \dfrac{\beta}{\beta + 1}$<br>0.20 | $z_3 = \dfrac{\beta + \gamma}{\beta + 1}$<br>0.30 |
| Case 4<br>meters, full area | $x_4 = (\alpha_4)^{1/(\beta + 1)}$<br>1.38 | $y_4 = \dfrac{\beta}{\beta + 1}$<br>0.20 | $z_4 = \dfrac{\beta + \gamma}{\beta + 1}$<br>0.30 |

$$Rth = \frac{C\alpha}{\Delta T^{C\beta} \times Area^{C\gamma}}$$

| | | | |
|---|---|---|---|
| Case 1<br>inches, half area | $C\alpha_1 = \dfrac{1}{2 \times \alpha_1}$<br>226.2 | $C\beta_1 = \beta$<br>0.25 | $C\gamma_1 = 1 - \gamma$<br>0.875 |
| Case 2<br>inches, full area | $C\alpha_2 = \dfrac{1}{\alpha_2}$<br>415 | $C\beta_2 = \beta$<br>0.25 | $C\gamma_2 = 1 - \gamma$<br>0.875 |
| Case 3<br>meters, half area | $C\alpha_3 = \dfrac{1}{2 \times \alpha_3}$<br>0.365 | $C\beta_3 = \beta$<br>0.25 | $C\gamma_3 = 1 - \gamma$<br>0.875 |
| Case 4<br>meters, full area | $C\alpha_4 = \dfrac{1}{\alpha_4}$<br>0.67 | $C\beta_4 = \beta$<br>0.25 | $C\gamma_4 = 1 - \gamma$<br>0.875 |

*(Continued)*

**TABLE 13.1 Conversion Table for Comparing Equations of Natural Convection (*Continued*).**

$$Rth = \frac{Cx}{P^{Cy} \times Area^{Cz}}$$

| Case 1 | $Cx_1 = \dfrac{1}{2 \times x_1}$ | $Cy_1 = y_1$ | $Cz_1 = 1 - z_1$ |
|---|---|---|---|
| inches, half area | 76.5 | 0.20 | 0.70 |
| Case 2 | $Cx_2 = \dfrac{1}{x_2}$ | $Cy_2 = y_2$ | $Cz_2 = 1 - z_2$ |
| inches, full area | 124.2 | 0.20 | 0.70 |
| Case 3 | $Cx_3 = \dfrac{1}{2 \times x_3}$ | $Cy_3 = y_3$ | $Cz_3 = 1 - z_3$ |
| meters, half area | 0.45 | 0.20 | 0.70 |
| Case 4 | $Cx_4 = \dfrac{1}{x_4}$ | $Cy_4 = y_4$ | $Cz_4 = 1 - z_4$ |
| meters, full area | 0.72 | 0.20 | 0.70 |

Direct

⇓⇓⇓

$$Cx = \frac{1}{2} \times \left( \frac{2^\beta}{\alpha} \right)^{\frac{1}{\beta+1}} \qquad Cy = \frac{\beta}{\beta+1} \qquad Cz = 1 - \frac{\beta+\gamma}{\beta+1}$$

(Cases 1 and 3 only)

Direct

⇑⇑⇑

$$\alpha = \frac{1}{2} \times \frac{1}{Cx^{1/(1-Cy)}} \qquad \beta = \frac{Cy}{1-Cy} \qquad \gamma = 1 - \frac{Cz}{1-Cy}$$

(Cases 1 and 3 only)

Solving for $A$, we get

$$Area = \left( \frac{124.2}{P^{0.20} \times Rth} \right)^{1/0.70}$$

$$\boxed{Area = 981 \times Rth^{-1.43} \times P^{-0.29}} \quad (\text{area in inch}^2)$$

**Example 13.1** We have a dissipation of 0.45 W from an SMT device, and we want to restrict the temperature of the PCB to a maximum of 100°C to avoid getting too close to the glass transition of the board (which is around 120°C for FR-4). The worst-case ambient is 55°C. Let us find the amount of copper which should be made available to the device.

The required $Rth$ is

$$Rth = \frac{°C}{W} = \frac{100 - 55}{0.45} = 100 \ °C/W$$

So from our equation (based on standard equation no. 1 we get

$$\text{Area} = 981 \times 100^{-1.43} \times 0.45^{-0.29} = 1.707 \text{ in}^2$$

So we need a square copper area of side $1.707^{0.5} = 1.3$ in.

We also plot out the standard design equations no. 1 and no. 2 in Fig. 13.2. We see that standard equation no. 2 is always more conservative than standard equation no. 1, i.e., it calls for slightly larger areas,

**TABLE 13.2  Numerical Comparison of the Two Standard Equations of Natural Convection.**

$$h = \alpha \times \frac{\Delta T^{\beta}}{Area^{\gamma}} \equiv \alpha \times \frac{\Delta T^{\beta}}{L^{2\gamma}}$$

|  | Standard equation no. | $\alpha$ | $\beta$ | $\gamma$ |
|---|---|---|---|---|
| Case 1 | 1 | 0.00221 | 0.25 | 0.125 |
| inches, half area | 2 | 0.00288 | 0.18 | 0.18 |
| Case 2 | 1 | 0.0024 | 0.25 | 0.125 |
| inches, full area | 2 | 0.0033 | 0.18 | 0.18 |
| Case 3 | 1 | 1.37 | 0.25 | 0.125 |
| meters, half area | 2 | 1.22 | 0.18 | 0.18 |
| Case 4 | 1 | 1.49 | 0.25 | 0.125 |
| meters, full area | 2 | 1.38 | 0.18 | 0.18 |

$$h = x \bullet \frac{P_y}{Area^z} \equiv x \bullet \frac{P_y}{L^{2z}}$$

|  | Standard equation no. | $x$ | $y$ | $z$ |
|---|---|---|---|---|
| Case 1 | 1 | 0.0065 | 0.20 | 0.30 |
| inches, half area | 2 | 0.0063 | 0.15 | 0.30 |
| Case 2 | 1 | 0.0081 | 0.20 | 0.30 |
| inches, full area | 2 | 0.0077 | 0.15 | 0.30 |
| Case 3 | 1 | 1.12 | 0.20 | 0.30 |
| meters, half area | 2 | 1.07 | 0.15 | 0.30 |
| Case 4 | 1 | 1.38 | 0.20 | 0.30 |
| meters, full area | 2 | 1.32 | 0.15 | 0.30 |

*(Continued)*

TABLE 13.2  **Numerical Comparison of the Two Standard Equations of Natural Convection (Continued).**

$$Rth = \frac{C\alpha}{\Delta T^{C\beta} \times Area^{C\gamma}}$$

|  | Standard equation no. | $C_\alpha$ | $C_\beta$ | $C_\gamma$ |
|---|---|---|---|---|
| Case 1 | 1 | 226.2 | 0.25 | 0.875 |
| inches, half area | 2 | 173.4 | 0.18 | 0.82 |
| Case 2 | 1 | 414.9 | 0.25 | 0.875 |
| inches, full area | 2 | 306.8 | 0.18 | 0.82 |
| Case 3 | 1 | 0.37 | 0.25 | 0.875 |
| meters, half area | 2 | 0.41 | 0.18 | 0.82 |
| Case 4 | 1 | 0.67 | 0.25 | 0.875 |
| meters, full area | 2 | 0.72 | 0.18 | 0.82 |

$$Rth = \frac{Cx}{P^{Cy} \times Area^{Cz}}$$

|  | Standard equation no. | $C_x$ | $C_y$ | $C_z$ |
|---|---|---|---|---|
| Case 1 | 1 | 76.5 | 0.20 | 0.70 |
| inches, half area | 2 | 80 | 0.15 | 0.70 |
| Case 2 | 1 | 124.3 | 0.20 | 0.70 |
| inches, full area | 2 | 130.0 | 0.15 | 0.70 |
| Case 3 | 1 | 0.45 | 0.20 | 0.70 |
| meters, half area | 2 | 0.47 | 0.15 | 0.70 |
| Case 4 | 1 | 0.73 | 0.20 | 0.70 |
| meters, full area | 2 | 0.76 | 0.15 | 0.70 |

and, therefore, may be a "safer bet" in most cases. Though not compared, standard equation no. 3 is almost coincident with standard equation no. 1, though it predicts slightly lower temperatures.

## 13.10  PCBs for Heatsinking

Commercial PCBs are often referred to as *1 oz* or *2 oz*, for example. This refers to the weight of copper in ounces per square foot deposited on the copper-clad laminate. Actually, 1 oz is equivalent to 1.4 mils copper thickness (or 35 μm). Similarly, 2 oz is twice that (70 μm).

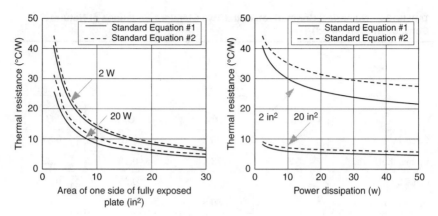

**Figure 13.2** Comparing the two standard design equations of natural convection.

PCB

Larger and larger areas of copper do not help, especially with thinner copper. A point of diminishing returns is reached for a square copper area of length (each side) 1 in. Some improvement continues till about 3 in, especially for 2-oz boards and better, but beyond that external heatsinks are required. A reasonable practical value attainable for the thermal resistance (from the case of the power device to the ambient) is about 30°C/W.

That is not to say that heat is lost only from the copper side. The usual laminate (board material) used for SMT applications is epoxy-glass FR4, which is a fairly good conductor of heat. So some of the heat from the device side does get to the other side where it contacts the air. Therefore, putting a copper plane on the other side (this need not even be electrically the same node, it could be the ground plane) also helps, but only by about 10 to 20 percent as compared to a copper plane on only one side. A much greater reduction of thermal resistance by about 50 to 70 percent can be produced if *thermal vias* are used to conduct heat to the other side. This *shunts* or bypasses the board material to get the heat to the other side where there is more air waiting to act.

The ground/tab can be a wide copper plane since it is "quiet" and will not radiate. It can then serve to carry heat away from the device to aid convection. If a double-sided board is used, several small vias sunk right next to the IC ground can be used to connect to a ground plane on the other side of the PCB. These vias not only help in the correct electrical implementation of grounding but also serve as thermal shunts. They are therefore called *thermal vias*. It is recommended that they be small (0.3 to 0.33 mm barrel diameter) so that the hole is essentially filled up during the plating process. Too large a hole can cause "solder wicking" during the reflow soldering process. The pitch (distance between the centers) of several such thermal vias in an area is typically 1 to 1.2 mm and a grid of thermal vias can be created right under the tab.

However, overestimating the amount of the copper plane for device cooling is a common mistake and can lead to excessive EMI too. The switching node is the biggest culprit. We should think twice about the copper really needed here. In some topologies the cathode of the catch diode (which is usually its tab/substrate) needs to be connected to the switching node (e.g., a buck). We should carefully estimate the dissipation before providing the required copper area for it. For example, we know that a typical Schottky diode has a forward voltage drop of 0.5 V. If the load current is 5 A and the duty cycle is 0.4, the dissipation is only $5 \times 0.5 \times (1 - 0.4) = 1.5$ W for the buck. For the boost and the buck-boost, the average diode current is fixed—it is the load current.

## 13.11 Natural Convection at an Altitude

At sea level, over 70 percent of heat is transferred by natural convection and the rest by radiation. Only at very high altitudes (70,000 ft and above) the ratio inverts and the heat lost by radiation could be 70 to 90 percent of the total, even though the radiated transfer is unchanged. So by about 10,000 ft the overall efficiency of cooling typically falls to 80 percent, at 20,000 ft it is only 60 percent, at 30,000 ft it is 50 percent.

Knowing that the coefficient of natural convection goes as $P^{1/2}$, where $P$ is the pressure of air, a good curve fit gives us the following useful relationship

$$\frac{Rth(feet)}{Rth(sealevel)} = [(-30 \times 10^{-6} \times feet) + 1]^{-0.5}$$

For example, we find that at 10,000 ft, all of the $Rths$ in Table 13.1 and Table 13.2 will increase by 19.5 percent.

## 13.12 Forced Air Cooling

Fans are rated for a certain cubic feet of minute $cfm$. The actual cooling, however, depends on the linear feet per minute $lfm$ to which the heatsink is subjected. Two parameters are needed to find the velocity in $lfm$: (1) the volume of air discharged from the fan in $cfm$ and (2) the cross-sectional area through which the cooling air passes in m$^2$. So $lfm = cfm/area$. But finally we should derate the calculated $lfm$ by 60 to 80 percent to account for backpressure.

At sea level the following formula gives a rough estimate of the required airflow:

$$cfm = \frac{1825}{\Delta T} \times P_{kW}$$

The $\Delta T$ is the differential between the inlet and outlet temperatures. It is typically set to about 10 to 15°C.

Note that if the inlet temperature, which is the room ambient, is 55°C, for example, then we need to add to this differential $\Delta T$ as the actual local ambient inside the power supply (when doing our *initial calculations*). However, ultimately we will be carrying out an actual temperature test by attaching thermocouples to all the components. We will thus certainly see an advantage in moving hotter components closer to the inlet during the design phase.

The linear speed is often expressed in terms of m/s. 1 m/s is equal to an lfm of 196.85. Roughly, *1 m/s is 200* lfm.

Some empirical results are as follows: at 30 W dissipation, an unblackened plate of $10 \times 10$ cm, has the following *Rth*: 3.9°C/W under natural cooling, 3.2°C/W with 1 m/s, 2.4°C/W with 2 m/s, and 1.2°C/W with 5 m/s. Provided the air flows parallel to the fins, with a speed greater than 0.5 m/s, the thermal resistance hardly depends on the power dissipation. That is because, on its own, even in static air, hot plates produce enough air movement around them to help in the heat transfer. Also note that blackening of plates has some effect under natural convection, but curves for forced convection depend very little on this aspect. Radiation is improved by blackening, but at sea level it is only a small part of the overall heat transfer. In general, black anodized heatsinks in typical forced air designs are a waste, and they should be replaced with uncoated aluminum.

Under steady state, roughly 2 mm-thick copper is almost exactly equivalent to 3 mm-thick aluminum. The only advantage of copper is its better thermal conductivity; thus it may be used to avoid thermal constriction effects when using very large areas.

The curve of thermal resistance to air flow falls off roughly exponentially, and so the improvement in thermal resistance in going from still air to 200 lfm is the same as from 200 to 1000 lfm. Velocities in excess of 1000 lfm (about 5 m/s) do not cause significant improvement.

Under forced convection the Nusselt number at sea level is

$$Nu_F = 0.664 \times \text{Re}^{1/2} \times \text{Pr}^{1/3} \quad (laminar\ flow)$$

$$Nu_F = 0.037 \times \text{Re}^{4/5} \times \text{Pr}^{1/3} \quad (turbulent\ flow)$$

Note that generally for natural convection, we can assume laminar flow. But under high dissipation the hot air tends to rise so fast that it breaks up into turbulence. This is actually very useful in reducing the thermal resistance (increasing the $h$). For forced air it is common to cut fingers on the sides of plate metal sinks and bend them alternately in and out. The purpose here is to actually create turbulent flow in the vicinity of the heatsink, thus lowering its thermal resistance. However, we do note from the formal analysis and equations which follow that turbulent flow provides better cooling (high $h$) under conditions of high lfm and/or large plates only. Laminar flow provides better cooling otherwise.

Above we have defined the Prandtl number $Pr$, which is the ratio of momentum diffusion to thermal diffusion. We can take its value at sea level to be 0.7. $Re$ is the dimensionless Reynolds's number, which is the ratio of momentum flow to viscous flow. If the plate has two dimensions $L1$ and $L2$ (so that $L1 \times L2 = A$), and $L1$ is the dimension along the flow of air, then $Re$ is

$$Re = \frac{lfm_{sealevel} \times L1_{meters}}{196.85 \times \nu}$$

where we already know $\nu = 15.9 \times 10^{-6}$ (the kinematic viscosity in m²/s). Thus we get the $h$ under forced convection:

$$h_F = \frac{Nu_F \times K_{AIR}}{L1_{meters}} \text{ watt/m}^2 - {}^\circ\text{C}$$

where $K_{AIR}$ is the thermal conductivity of air ($0.026$ W/m $- {}^\circ$C). Putting all the numbers together we simplify to get

$$\boxed{h_{FORCED} = 0.086 \times lfm^{0.8} \times L^{-0.2}}$$ (turbulent flow, $L$ in meters, sea level)

$$\boxed{h_{FORCED} = 0.273 \times lfm^{0.5} \times L^{-0.5}}$$ (laminar flow, L in meters, sea level)

At higher altitudes we need to increase the cfm calculated at sea level by the following factor so as to maintain the same effective cooling. This is because, a fan is a constant volume mover, not a constant mass mover, and at high altitudes the air density is much lower. Therefore the cfm has to be increased in inverse proportion to the pressure.

$$\boxed{\frac{cfm(feet)}{cfm(sealevel)} = \frac{1}{(-30 \times 10^{-6} \times feet) + 1}}$$

For example, at 10,000 ft the calculated cfm at sea level has to be increased by 43 percent to maintain the same $h_{FORCED}$.

## 13.13 Radiative Heat Transfer

Radiation does not depend on air and can take place even in vacuum since it is electromagnetic in nature. At high altitudes, radiative heat transfer can become a significant part of the overall heat transfer. The equation for $h$ is

$$h_{RAD} = \frac{\varepsilon \times (5.67 \times 10^{-8}) \times [(T_{HS} + 273)^4 - (T_{AMB} + 273)^4]}{T_{HS} - T_{AMB}} \; \text{watt/m}^2 - {}^\circ\text{C}$$

Note that at high altitudes, under forced air cooling, the cfm falls and so the inlet to outlet $\Delta T$ increases somewhat. Therefore $T_{AMB}$ goes up, and this affects $h_{RAD}$. So it ends up looking like radiation too is getting affected at higher altitudes. Luckily, this actually improves the situation somewhat—by about 2 percent every 10,000 ft in typical applications.

$\varepsilon$ is the emissivity of the surface. It is 1 for a perfect blackbody, but for polished metal surfaces we should take this as 0.1. If the surface is anodized, we can assume it to be about 0.9.

### 13.14   Miscellaneous Issues

A typical power supply specification will ask for meeting an altitude requirement of 10,000 ft (3000 m). They will usually not "relax" the ambient temperature up to about 6000 ft, after which they will allow us to reduce the upper ambient limit by about 1°C every 1000 ft higher.

A typical industry thumb rule for testing power supplies at sea level for a certain altitude requirement is to *add 1°C every 1000 ft to the upper limit of the maximum specified operating ambient.* So, if the power supply is designed for 55°C at sea level, we should test it at 65°C. However, this is not fully adequate nor do any temperature derating margins at sea level necessarily help. A key limiting factor is not the junction temperature but the temperature on the PCB where the device is mounted. We usually cannot exceed more than about 100 to 110°C on the PCB or it will start turning black.

We can sum over all of the $h$ calculated in this chapter as follows

$$h_{total} = h_{RAD} + \left(h_{FORCED}^3 + h_{NATURAL}^3\right)^{1/3}$$

Extruded heatsinks are certainly very useful under forced air cooling because then the efficiency of cooling depends on their surface area. But correlation of experimental data indicates that their cooling capabilities under natural convection conditions are a function of the volume of the space they occupy i.e., their "envelope" (ignoring the finer detail of their fin structure). That is because heat lost

TABLE 13.3  Typical Thermal
Resistance of Cores.

| Core sizes | °C/W |
|---|---|
| EC35/17/10 | 17.4 |
| EC41/19/12 | 15.5 |
| EE42/42/15 | 10.4 |
| EE42/42/20 | 10.0 |
| EE30/30/7 | 23.4 |
| EE25/25/7 | 30.0 |
| EE20/20/5 | 35.4 |
| EE42/54/20 | 8.3 |
| EE55/55/21 | 6.7 |
| EE55/55/25 | 6.2 |
| UU15/22/6 | 33.3 |
| UU20/32/7 | 24.2 |
| UU25/40/13 | 15.7 |
| UU30/50/16 | 10.2 |

from one fin is largely reacquired by the adjacent fins, and so there are very small deviations with regard to the "exoticness" of their actual shape. Typical values drawn from published curves are as follows: 0.1 in$^3$ will give about 30°C/W to 50°C/W, 0.5 in$^3$ will give about 15°C/W to 20°C/W, 1 in$^3$ will give about 10°C/W, 5 in$^3$ will give about 5°C/W, 100 in$^3$ will give about 0.5°C/W to 1°C/W. The above data are for one device mounted on the heatsink. Roughly, there will be a further 20 percent improvement in the thermal resistance if two devices share the dissipation and are mounted slightly apart.

For the common E-type magnetic cores (like the E-cores, ETD cores, and EFD cores), thermal resistance under natural convection can be approximated by

$$Rth \cong 53 \times V_e^{-0.54}$$

where $V_e$ is in cm$^3$. We have some typical thermal resistances based on actual empirical testing in Table 13.3.

When using the equation for extrusions, the volume to be used is the overall space they occupy (i.e., ignoring the details of their fin structure).

With extrusion heatsinks, if the fins are too close to each other, heat radiated from one is simply absorbed by the adjacent pins. So the overall emissivity is not as high as we may have thought.

If the fins are too close, they also impede the flow of air. Therefore, the recommended optimum fin spacing is about 0.25 in for natural convection, at 200 lfm it is about 0.15 in, and at 500 lfm it is about 0.1 in. This applies for heatsinks up to 3 in in length. We can increase the fin spacing by about 0.05 in for heatsinks as long as 6 in.

Here is a quick run-down on fans: ball-bearing fans are more expensive. They have a longer life when the temperature (as seen by the bearing system) is higher. But they can get noisier over time. If useful life of a fan was defined as ending when the fan became noisy, the ball-bearing fan would have a smaller life than the sleeve-bearing fan. Sleeve-bearing fans are less expensive, and quieter, and easily handle any mounting attitude (angle). Their life is as good as a ball-bearing fan provided temperatures are not very high. They can sustain multiple shocks (without impacting noise or life).

NOTES:

p 288 MEASURE TEMP ON A FET

$h \propto \frac{1}{A R_{TH}}$   HEAT TRANSFER COEFF.

PCB

— ~1 IN² CU IS ABOUT MAX BEFORE DIM RETURNS

VIA'S ADD 70% EFF TO 2ND LAYER, GROUND IS BEST BECAUSE OF NOISE.

- .3 TO .33mm HOLD FOR VIAS ON 1 TO 1.2mm GRID

> 100 TO 110°C PCB TURNS BLACK.

HEAT SINKS

∝ VOLUME INCLUDING FINS p 307

.25" SPACING IS OPTIMUM FOR FINS w/ NATURAL CONVEC

# 14

# Stabilizing Current Mode Converters

## 14.1 Background

Setting the current ripple ratio $r = 0.4$, as has been recommended at several places in this book, will usually stand up to scrutiny, provided we are using voltage mode control. But when we come to peak current mode control that criterion may not be enough. If we have current mode control *and* our duty cycle is over 50 percent, *and* we are in continuous conduction mode, then as we reduce the inductance we could enter into a strange type of instability. On the oscilloscope, this will manifest itself as an alternating pattern of one large pulse followed by one small pulse. This is a *steady state* of sorts, except that the pattern is repeating itself every two switching cycles instead of one. This is, therefore, alternatively called *subharmonic instability, alternate cycle instability, half frequency instability, period-doubling instability* and the like. They all mean the same thing. One of the first things to happen as a result is that the output voltage ripple suddenly increases. We will also find that the transient response has suddenly deteriorated, and significantly so. However, the author has seen cases where switcher ICs continued to operate acceptably in this quasi-steady-state mode for a long time before the customer even discovered it.

The familiar cure for subharmonic instability is *slope compensation*. The amount of slope compensation required and the value of the inductance are completely interlinked. So, for example, a poorly designed slope compensation circuit (i.e., with too little slope compensation) may demand that we use a higher inductance (than that which accrues from the usual recommendation of $r = 0.4$). This is clearly something we need to fix in the slope compensation circuit itself, rather than in component selection. On the other hand, we may even desire that $r$ be increased (by reducing inductance) so as to reduce the physical size of our magnetics (irrespective of the possible impact on the related power components). That is also understandable because reducing the size of the wound components is the main reason why we aim for higher and higher switching frequencies in the first place. But if inductance is decreased, we need a higher amount of slope compensation, or we will run into subharmonic instability. The amount of slope compensation required is also a function of the set output voltage and possibly the input voltage too.

In all cases, we should be clear that for a *given* amount of slope compensation (and constant input/output conditions) we get a *minimum inductance* requirement, which means that we can *increase* the inductance freely, but *not decrease it* (without first *increasing* the amount of slope compensation). Thus, if we *increase* the slope compensation, we can use a *lower* inductance. That may be our aim if we are trying to *increase* $r$ toward the optimum value of 0.4, or if we just want ever smaller magnetic components (possibly for marketing reasons!).

The designer is cautioned that there are other reasons why we may see the same oscillatory type of behavior, and these are not *subharmonic oscillations* in the sense that we are going to describe. In fact, in Chap. 1 we presented one such "other possible cause" that will mimic this very symptom, though its solution is clearly different (actually opposite, since we need to *decrease* the inductance).

The problem with subharmonic instability is that "playing" with the usual loop compensation components is not likely to help at all, simply because the root cause is not the usual "incorrect" placement of poles and zeroes. This particular instability is not even included in the usual small-signal models we use. And it applies to all topologies with peak current mode control. Understanding the equations behind this phenomenon helps us pick the smallest possible inductor in such a switching application (though not necessarily the most optimum one in terms of its overall impact).

We will see that we then need to keep the inductance greater than a certain minimum value (to be unspecified) and also that this *value of inductance is the highest at the lowest input voltage* (for the buck and the boost). Therefore, we should always design our slope compensation circuit at $V_{INMIN}$ for these topologies. For the buck-boost, we should do so roughly around the point where $V_{IN}$ is equal to $V_O$.

Designers familiar with the popular peak current mode controller, the 3842/3844 series, know that surprisingly, this peak current mode IC has no built-in slope compensation. For the 3844, we can understand that because subharmonic instability occurs only if $D$ is greater than 50 percent, and the 3844 being meant for forward converters is by design limited to a maximum of $D$ slightly less than 50 percent (though the

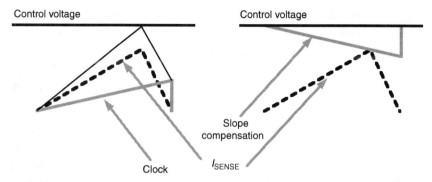

Control voltage

Control voltage

Slope
compensation

Clock        $I_{SENSE}$

**Figure 14.1** Applying a fixed ramp to the current sense is equivalent to slope compensation.

designer is cautioned that it is now known that subharmonic instability can show up at duty cycles even less than 50 percent). But what about the 3842? We all remember putting in that mysterious 47 pF (sometimes only 10 pF or 22 pF, or sometimes 100 pF) from the clock pin to the current sense pin. What it was really doing was mixing some of the clock ramp with the current sense signal in a certain ratio. In doing so, we may have said that we were introducing *a little voltage mode control*, because the ramp going to the PWM comparator is now partly current sense signal and partly fixed ramp. In fact, this is what we essentially do even in formal slope compensation. However, instead of looking at the situation as a ramp *added* to the current sense, we can equivalently look at it as the current ramp having remained the same, and instead, an *inverted clock ramp* being applied to the output of the error amplifier (see Fig. 14.1). The comparator is concerned only with the difference in voltage at its inputs. So, in fact, both these viewpoints are equivalent as far as the comparator is concerned.

This *inverted clock ramp* is the applied *slope compensation*. The composite signal is the *control voltage*. The control voltage is compared with the current sense signal and is thus the trip level of the current (or PWM) comparator. Once tripped, the switch turns *off* and the current ramps down. The control voltage can move "up" or "down" to change the $D$, as it is basically the error amplifier output. But it belongs to the slower voltage feedback loop, and thus it takes the controller a little while to "know" that the output needs to be corrected.

**Note:** The small capacitor mentioned above is also found to sometimes help with the 3844 and forward converters in general. But that is for a different reason. Here the current sense signal is a few mV at light loads, and the signal to noise ratio is very poor. So, we get a lot of jitter due to noise. By providing a small fixed ramp we mimic voltage mode control at very light loads and thus get cleaner and less noise-sensitive pulse patterns.

We also note that the way we implement slope compensation in the 3842 (and also in several older controllers and switchers) *affects the current limit too*; thus we will have to either decrease our sense resistor somewhat or derate the power available, otherwise we may not be able to guarantee full output power over the entire input range. But in more recent controllers the current limit circuit block is independent of the PWM comparator section, and so current limit remains "flat" and unaffected with changes in input voltage. If not, the overall converter design can become very tricky since whenever we adjust the slope compensation, we will need to double-check whether we are even going to be able to deliver maximum power (without hitting current limit), over the entire input range.

Recognizing that subharmonic instability formally occurs only if the duty cycle is greater than 50 percent, some controllers try to avoid impacting the current limit over the entire input voltage range by introducing this slope compensation ramp only for $D > 50$ percent.

Note that the applied slope compensation can eventually be expressed in terms of an equivalent amperes/second (even though internally it may be a voltage ramp, possibly expressed as volt/second).

This is also a good place to point out that as switching frequencies increase, the typical 50 to 150 ns blanking time requirements and oversensitivity to layout have taken a lot of the "shine" out of current mode control. The subharmonic oscillation problem with peak current mode control has not helped either. So, voltage mode controllers are certainly making a comeback. Using techniques like *input voltage feedforward*, where the up-slope of the internal ramp is made proportional to the input voltage, we replicate some of the advantages of current mode control in terms of its ability to react extremely fast (on a pulse by pulse basis) to changes in input voltage.

**Note:** In current mode control, the ramp to the PWM comparator is the current sense signal, and from $V = L \times dI/dt$ we know that the up-slope $dI/dt$ is proportional to the voltage applied during the on-time. This voltage is called $V_{ON}$ at several places in the book, but for a buck converter it is actually $V_{IN} - V_O$, not $V_{IN}$. Therefore, we should examine what the input voltage feedfoward really requires, and that could depend on topology.

## 14.2  Why Slope Compensation?

In Fig. 14.2 we have a peak current mode controller and we apply a small disturbance (in gray). We can see that this disturbs the regular pulse pattern. But at least we hope it subsides. If it subsides, we won't

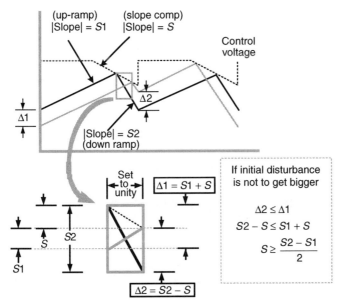

**Figure 14.2** How a load disturbance or noise can trigger instability.

get sustained oscillations, but *if the disturbance increases every cycle,* we will. On the other hand, we would also prefer that it subsides in a fairly short time.

The "control voltage" in the figure is the output of the error amplifier (with any applied slope compensation). Let us assume that we have a fixed slope compensation, arbitrarily fixed so far, of *magnitude S*. In the figure we have expanded the gray rectangular box, and shown by geometry that the only way we can ensure that the disturbance will not increase is if the amount of slope compensation is limited as per

$$S \geq \frac{S2 - S1}{2} \equiv \frac{1}{2} \ (difference\ between\ down-ramp\ and\ up-ramp)$$

We have several possibilities:

- In the figure we can see what happens if

$$S = S2$$

$S = S2$. The disturbance will clearly be "killed" in the very same cycle. Of course, to do this, we either have to increase the slope compensation (and we should not arbitrarily), or we have to increase our inductance significantly to have a small $r$.

- If, however,

$$S = \frac{S2 - S1}{2} \quad \text{or} \quad S2 = 2S + S1$$

then even though the inductance is the smallest possible, the disturbance will last forever.

- It is commonly stated in textbooks that the condition to avoid subharmonic instability is

$$S = \frac{S2}{2} \quad \text{or} \quad S2 = 2S$$

which, for a fixed slope compensation means that we are demanding a smaller down-slope $S2$ than in the previous case where the disturbance lasted forever.

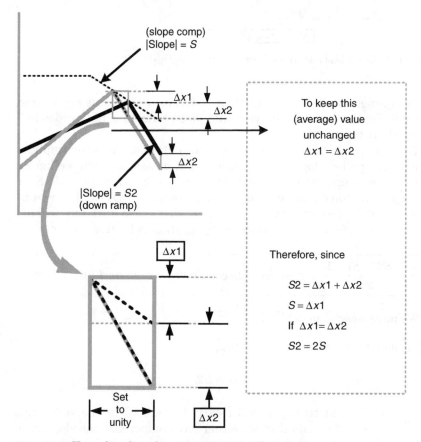

**Figure 14.3**  How a line disturbance can trigger oscillations.

But besides the fact that the disturbance may last forever, there is another problem as shown in Fig. 14.3. The output voltage is determined largely or in part (depending on the topology) by the average current in the inductor. If the average comes down in response to a small increase in input voltage, it may pose a problem. In a boost or buck-boost converter the load current is average diode current. So, for example, if the center of the inductor current comes down, the average diode current $I_{L\_AVG} \times (1 - D)$ will tend to fall too. Since $D$ decreases simultaneously, it could end up keeping the average load maintained. In a buck (and buck-derived topologies like the forward converter, and half-bridge) however, the average inductor current is the load current, and we have to keep it fixed. Any drop in the average will tend to reduce the output voltage.

In Fig. 14.3 we have explained what happens if there is a slight line disturbance. We are assuming that we are looking at the event even before the control voltage has had time to react. The control voltage (without slope compensation) is essentially the output of the error amplifier; thus we are in effect assuming that the voltage feedback loop is yet "unaware" that there is correction needed. In the figure we have actually "forced" both the waveforms to have the same average value (center of inductor current ramp) and then worked backward to see what geometric relationship needs to be met. Note that if the upper vertex, of the new waveform, is $\Delta x$ higher than the previous waveform, and its lower vertex is $\Delta x$ lower than the previous waveform, the geometric center of both waveforms will be the same. Thus we get one of the same conditions we got previously

$$S = \frac{S2}{2} \quad \text{or} \quad S2 = 2S$$

We notice that the up-slope didn't enter the picture. So, this type of temporary instability can, in fact, occur at *any duty cycle*. Usually, however, in the author's experience, this issue is ignored by engineers. Line transient testing is also rarely performed. It is load transient response and related issues that seem to occupy most engineers. But it is good to keep this issue in mind too.

Let us now discuss slope compensation in terms of a worked example.

**Example 14.1** We have a dc-dc converter displaying signs of subharmonic instability. Its output is 5 V and its input is 8 V. The applied slope compensation is 0.25 A/$\mu$s a typical value at switching frequencies of around 100 kHz). What are the possible choices for the inductance (ignoring switch and diode drops)?

This is a buck converter with a duty cycle $V_O/V_{IN}$ of greater than 50 percent. Clearly, it is a candidate for the subharmonic problem. Its down-slope is simply $V_O/L$ and its up-slope is $(V_{IN} - V_O)/L$. So,

- If $5/L = 0.25$, we get L = 20 μH. This gives us a disturbance that dies out immediately.

- The up-ramp has a slope of $(8 - 5)/L$. The down-slope is $5/L$. So,

$$S2 - S1 = \frac{5}{L} - \frac{8-5}{L} = \frac{2}{L}$$

- Therefore, if $2/L = 0.25$, we get $L = 4$μH. This disturbance will last forever.

- If $S2 = 2S$, we get $L = 10$ μH. This will damp out the oscillations after some unknown number of cycles. It is also the condition to avoid the line disturbance oscillation condition described earlier.

### 14.3   Generalized Rule for Avoiding Subharmonic Instability

As we said, our early small-signal models do not account for subharmonic instability. However, if we do a bench measurement for the Bode plot, and we zoom in on the area around half the switching frequency (with sufficient bandwidth), we will see that the gain plot which should have been continuing to fall smoothly past the crossover frequency has a "spike" at exactly half the switching frequency. This peaking is the physical manifestation of subharmonic instability. It has recently been modeled by theoretical analysis, and a quality factor $Q$ has been assigned to it, as for any resonance. In practical bench measurements we will see that if $Q$ is set low, say around 0.5, then the half frequency peak is not visible. If $Q$ is very large, this peak can be high enough to intersect the 0 dB line ($x$ axis of Bode plot), and thereafter almost immediately the Bode plot will change into a rather abnormal looking plot. That is because, though quasi-stable, we have entered the undesirable alternate-cycle pattern, and the control loop, as we know it, has gone. The converter will never recover without intervention. We note that a conservative $Q$ of around 0.5 leads to excessively high inductances, but a $Q$ of around 2 is just about right, as also indicated by several bench measurements the author performed. $Q = 2$ requires a fairly small inductance and is assuredly stable.

What do we mean by saying "set the $Q$ to 2?" The required relationship is provided below. It applies to all the topologies.

$$Q = \frac{1}{\pi\left[\left(1 - \frac{S}{S1}\right) \times (1 - D) - 0.5\right]}$$

Note that all the slopes are only their respective *magnitudes* here. Now we solve for $L$ and for each topology we get the following useful equations

$$L_{\mu H} = \frac{\frac{1}{\pi Q} + D - 0.5}{SlopeComp(\text{A}/\mu\text{s})} \times V_{IN} \quad (buck,\ set\ Q \approx 2)$$

$$L_{\mu H} = \frac{\frac{1}{\pi Q} + D - 0.5}{SlopeComp(\text{A}/\mu\text{s})} \times V_O \quad (boost,\ set\ Q \approx 2)$$

$$L_{\mu H} = \frac{\frac{1}{\pi Q} + D - 0.5}{SlopeComp(\text{A}/\mu\text{s})} \times (V_{IN} + V_O) \quad (buck\text{-}boost,\ set\ Q \approx 2)$$

In Fig. 14.4 we have plotted the minimum inductance required to avoid instability. We have set $Q = 2$ and the slope compensation value used is 0.25 A/µs. Since inductance is inversely proportional to slope compensation we can calculate the inductance for a general condition. Note that *for the curves to apply, we should be in a region of D > 50 percent and in CCM.* If the slope compensation is affecting current limit, we must ensure that peak power is still able to be delivered. This is

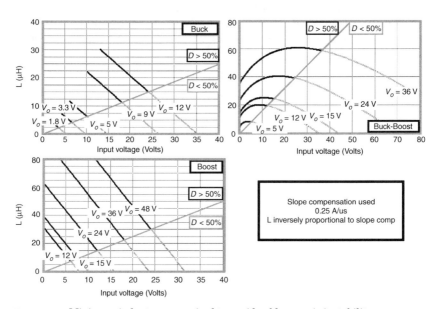

**Figure 14.4**  Minimum inductance required to avoid subharmonic instability.

particularly true for the buck converter, since it is the only topology in which the peak current increases with input voltage (decreasing D). The peak current equation for a buck is

$$I_{PK} = I_O \bullet \left[ 1 + \frac{r}{2} \right]$$

where

$$r = \frac{V_O + V_D}{I_O \bullet L \bullet f} \bullet (1 - D) \bullet 10^6$$

Therefore the peak current will change at a rate dependent upon voltage, load current, inductance, and the like. We have to make sure that the calculated peak current is not in excess of the effective current limit (with slope compensation included), or we will not get the output power we want. We must test this out at $V_{INMAX}$. But we remember that the inductor design for avoiding slope compensation is at $V_{INMIN}$. Therefore our general philosophy should be:

1. Find the minimum inductance required to keep the converter in CCM.

2. Find the minimum inductance required to meet peak power requirement (at $V_{INMAX}$).

3. Find the minimum inductance to avoid subharmonic instability.

Take the maximum of all three *minimum inductances*. Note that the last condition comes into the picture only if $D > 50$ percent. We may have to iterate a few times to arrive at the final and most optimum result.

# 15

# Practical EMI Filter Design

## 15.1 The CISPR 22 Standard

The applicable standard we usually follow for IT (information technology) equipment is CISPR 22, now called EN 550022 (EN stands for European norm). In Table 15.1 we have the *conducted emission limits*. OEM specifications usually ask us to meet the more stringent Class B limits. One rather unusual addition by the author in the table provided here is that dBμV has also been expressed as mV. Hopefully that is more comfortable to some designers who have trouble thinking logarithmically. Note that by definition, dB = 20 × log (voltage ratio). So, a dBμV implies a logarithmic comparison to a reference of 1 μV. We get the corresponding mV by using

$$(\text{mV}) = (10^{(\text{dB}\mu\text{V})/20}) \times 10^{-3}$$

The corresponding EMI requirement in the United States (FCC Part 15) does not have average limits, only quasi-peak limits, though it does accept certification to CISPR 22. For CISPR 22 (Class B) we can use the following equations for the 150 kHz to 500 kHz range

■ *The average limit* is (almost)

$$(\text{dB}\mu\text{V}_{AVG}) = -20 \times \log(f_{\text{MHz}}) + 40$$

■ *The quasi-peak limit CISPR curve (conducted emissions)* is (almost)

$$(\text{dB}\mu\text{V}_{QP}) = -20 \times \log(f_{\text{MHz}}) + 50$$

TABLE 15.1    Conducted Emissions Limits

| | CLASS A (industrial) | | | | | | | |
|---|---|---|---|---|---|---|---|---|
| | FCC Part 15 | | | | CISPR 22 | | | |
| Freq (MHz) | Quasi-peak | | Average | | Quasi-peak | | Average | |
| | $dB\mu V$ | mV | $dB\mu V$ | mV | $dB\mu V$ | mV | $dB\mu V$ | mV |
| 0.15–0.45 | NA† | NA | NA | NA | 79 | 9.0 | 66 | 2.0 |
| 0.45–0.5 | 60 | 1.0 | NA | NA | 79 | 9.0 | 66 | 2.0 |
| 0.5–1.705 | 60 | 1.0 | NA | NA | 73 | 4.5 | 60 | 1.0 |
| 1.705–30 | 69.5 | 3.0 | NA | NA | 73 | 4.5 | 60 | 1.0 |

| | CLASS B (residential) | | | | | | | |
|---|---|---|---|---|---|---|---|---|
| | FCC Part 15 | | | | CISPR 22 | | | |
| Freq (MHz) | Quasi-peak | | Average | | Quasi-peak | | Average | |
| | $dB\mu V$ | mV | $dB\mu V$ | mV | $dB\mu V$ | mV | $dB\mu V$ | mV |
| 0.15–0.45 - | NA | NA | NA | NA | 66-56.9* | 2.0-0.7* | 56-46.9* | 0.63-0.22* |
| 0.45–0.5 | 48 | 0.25 | NA | NA | 56.9-56* | 0.7-0.63* | 46.9-46* | 0.22-0.2* |
| 0.5–5 | 48 | 0.25 | NA | NA | 56 | 0.63 | 46 | 0.2 |
| 5–30 | 48 | 0.25 | NA | NA | 60 | 1.0 | 50 | 0.32 |

*This is a straight line on the standard $dB\mu$V versus log $f$ plot.
†NA stands for not applicable.

We can see that for Class B, the quasi-peak ($QP$) limit is always 10 dB higher.

## 15.2  The LISN

First note that "CM" or "cm" both stand for *common-mode* noise in this chapter, and "DM" or "dm" refer to *differential-mode* noise. We measure the conducted EMI emissions across a *line impedance stabilizing network* (LISN). The LISN provides the following load impedances to the noise generators (in the absence of any input filter).

- The CM load impedance is 25 Ω
- The DM load impedance is 100 Ω

As we flick the switch on the front panel of the LISN, we measure the following noise voltages (subscript $L$ stands for *line* and $N$ for *neutral*)

$$V_L = 25 \times I_{cm} + 50 \times I_{dm}$$

$$V_N = 25 \times I_{cm} - 50 \times I_{dm}$$

Since we have an ac line input, both lines are essentially symmetric at the input of the power supply, and so we don't usually see anything more than minor differences in the two signals above. Special LISNs are available (e.g., from Laplace Instruments) which can provide separated DM and CM components to help in trouble-shooting.

## 15.3  Fourier Series

For a function $f(x)$ with time period expressed as an angle $(2\pi)$ we can write

$$f(x) = \frac{1}{2}a_o + \sum_{n=1}^{\infty}(a_n \cos nx + b_n \sin nx)$$

$$a_n = \frac{1}{\pi}\int_0^{2\pi}[f(x)\cos nx]\,dx$$

$$b_n = \frac{1}{\pi}\int_0^{2\pi}[f(x)\sin nx]\,dx$$

*Alternatively,*

$$f(x) = \frac{1}{2}a_o + \sum_{n=1}^{\infty}c_n \cos(nx - \phi_n)$$

$$c_n^2 = a_n^2 + b_n^2$$

$$\tan\phi_n = \frac{b_n}{a_n}$$

Here the period is expressed as $2\pi$, but in power supplies we know that the period we are interested in is in units of time, not angle i.e., $T = 1/f$. The way to convert angle $\theta$ to $t$ is to use the equivalence

$$\frac{\theta}{2\pi} \rightarrow \frac{t}{T}$$

or

$$\theta \rightarrow 2\pi \times \frac{t}{T}$$

The first term $(\frac{1}{2}\times a_O)$ of the Fourier expansion really doesn't matter, and it can even be expressed differently. But it is simply the arithmetic average of the waveform (a pure dc). The sign of the $c_n$ is similarly not important, either from the viewpoint of the measured EMI spectrum or our EMI suppression methods.

## 15.4  The Trapezoid

If we take a rectangular wave with nonzero rise and fall times, we can show that (for the case of equal rise and fall times)

$$c_n = A \times \frac{2 \times (t_{ON})}{T} \times \left[ \frac{\sin \left\{ \frac{n \times \pi \times t_R}{T} \right\}}{\frac{n \times \pi \times t_R}{T}} \right] \times \left[ \frac{\sin \left\{ \frac{n \times \pi \times (t_{ON})}{T} \right\}}{\frac{n \times \pi \times (t_{ON})}{T}} \right]$$

where $t_{RISE} = t_{FALL} = t_R$, and $A$ is the amplitude (*actually peak-to-peak*). We are ignoring any signs as they are essentially irrelevant.

We have two "break points" of slope (in the log versus log plot). The first occurs at

$$\frac{n \times \pi \times t_{ON}}{T} = 1$$

that is,

$$n_1 = \frac{T}{\pi \times t_{ON}}$$

Since $n$ = frequency of harmonic/fundamental frequency, i.e., $n = f \times T$, we get the corresponding break frequency to be

$$\boxed{f_{BREAK\_1} = \frac{1}{\pi \times t_{ON}} = \frac{0.32}{t_{ON}}}$$

The second break point is at

$$n_2 = \frac{T}{\pi \times t_R}$$

i.e., the break frequency is

$$\boxed{f_{BREAK\_2} = \frac{0.32}{t_R}}$$

We know what to expect too, that after the second break point, the net roll-off will be at $20 + 20 = 40$ dB per decade. See Fig. 15.1.

Note that $n$ must be an integer to have any physical meaning. The first break point, therefore, may not be visible. What we will apparently perceive is that the envelope ramps down almost from the lowest frequency at the rate of 20 dB per decade.

Below the first break frequency, the envelope of the harmonics actually becomes flat. The first break point should therefore be calculated and the envelope should be *guillotined* (truncated) below this frequency. We can use the following equations to describe the $c_n$. Note that in these equations, the $c_n$ are no longer the actual coefficients of the Fourier

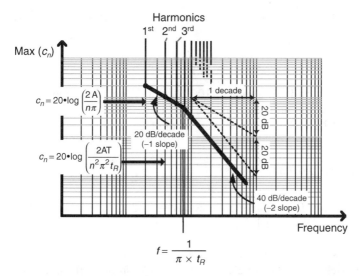

**Figure 15.1** Envelope of amplitudes of harmonic of rectangular wave-form of amplitude $A$ and $t_R = t_F$.

expansion, rather they represent the *envelope* (since that is all that matters from the standpoint of the EMI filter design).

$$c_n = 20 \log \left( \frac{2A}{n\pi} \right)$$

$$c_n = 20 \log \left( \frac{2A}{n^2 \pi^2 t_R f_{SW}} \right)$$

The first of the two equations above is valid between the first and second break points, and the second equation is valid for all frequencies higher than the second break point. Note that the switching frequency is $f_{SW} = 1/T$.

## 15.5   Practical DM Filter Design

Differential mode noise is caused by the voltage drop produced across the equivalent series resistance (ESR) of the input bulk capacitor by the high-frequency switching current.

The voltage across the ESR of the capacitor is

$$v = ESR \times I_{SW} \text{ volts}$$

If there was no filter present, the switching noise current received by the LISN is

$$I_{LISN} = \frac{v}{100} = \frac{ESR \times I_{SW}}{100} \text{ amperes}$$

since the LISN has an impedance of 100 Ω for DM noise. However, the analyzer measures the noise across one of the two effective series 50 Ω resistors in the LISN. So the measured level of noise is

$$\boxed{V_{LISN\_DM\_NOFILTER} = I_{LISN} \times 50 = \frac{ESR \times I_{SW}}{2}} \text{ volts}$$

We have assumed that $C_{BULK}$ is very large and that it has no ESL, and also that its ESR is much less than 100 Ω.

**Example 15.1**  What is the DM noise spectrum measured at the LISN for a 5 V@15 A flyback at an input of 265 Vac, with a transformer ratio of 20? We are using an aluminum electrolytic bulk capacitor whose datasheet states that it has a capacitance of 270 µF, a dissipation factor (tangent of loss angle) of tan δ = 0.15 measured at 120 Hz, and a frequency multiplier factor of 1.5 at 100 kHz.

First the ESR is to be computed at the 120 Hz test frequency. By definition

$$ESR_{120} = \frac{\tan \delta}{2\pi f \times C} = \frac{0.15 \times 10^6}{2 \times 3.142 \times 120 \times 270} = 0.74 \ \Omega$$

At a high frequency the ripple current is allowed to increase by the frequency multiplicative factor of 1.5, therefore since the heating ($I^2 \times$ ESR) must still be the same, it means that the ESR at a high frequency must be $1/(1.5)^2$ times the ESR at low frequency. Therefore for our purpose

$$ESR = \frac{1}{1.5^2} \times 0.74 = 0.33 \ \Omega$$

$$V_{LISN\_DM\_NOFILTER} = \frac{ESR \times I_{SW}}{2} = 0.17 \times {n_s}/{n_p} \times I_O = 0.13 \text{ V}$$

This is the amplitude of the measured signal. It is the $A$ in the corresponding Fourier series. In terms of dBµV its value is $20 \times \log(0.13/10^{-6}) = 102$ dBµV. We can thus predict that the spectrum is as shown in Fig. 15.2.

We have also shown how the spectrum relates to the CISPR 22 Class B quasi-peak emission limits (bold line). We are showing a sample case where the switching frequency is just below where the CISPR 22 limits

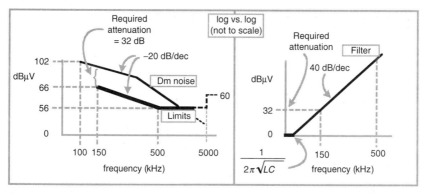

**Figure 15.2** Worked example for DM filter design.

start. Let us assume that it is 100 kHz here. For simplicity, we are also assuming that the common-mode noise is not a major contributor at these low frequencies. In practice, we should allocate some dB of margin for the CM noise (which will be estimated in the next section). We are also not explicitly accounting for any other required margins. But since we are comparing peak values (which are always higher than quasi-peak values) with quasi-peak limits, this automatically provides us with some natural headroom/margin.

The equation of the envelope at a frequency $f$ kHz is

$$\mathrm{dB}\mu\ \mathrm{V}(f) = -slope\{[\log(f) - \log(100)]\} + 102$$

Since the (magnitude of the) slope is 20 dB per decade, at 150 kHz we get

$$\mathrm{dB}\mu\ \mathrm{V}(f) = -20\{[\log(150) - \log(100)]\} + 102 = 98\ \mathrm{dB}$$

This is $98 - 66 = 32$ dB higher than it should be. So, we need to attenuate the noise by 32 dB. We, therefore, need to pick a low-pass $LC$ filter which provides this attenuation at 150 kHz. We can thus calculate its break frequency. For example, if we are using an $LC$ low-pass filter, it has an *attenuation characteristic of about 40 dB per decade* above its break frequency (i.e., $1/2\pi(LC)^{0.5}$). So, from Fig. 15.2 we can see that the equation it needs to satisfy is

$$32 = slope \times \{[\log(f) - \log(f_{BREAK})]\} = 40\times \{[\log(150) - \log(f_{BREAK})]\}$$

Solving

$$\log(f) = \log(150) - \frac{32}{40} = 1.38$$

$$f = 10^{1.38} = 24\ kHz$$

We therefore need a filter that has an $LC$ of

$$LC = \left(\frac{1}{2\pi \times 24000}\right)^2 = 4.4 \times 10^{-11} \text{ sec}^2$$

Therefore, if the net X-capacitance (line-to-line capacitance) $C$ is say 0.22 µF, we get the corresponding $L$ to be 200 µH.

Since the break point associated with the *rise and fall times* (crossover) didn't enter the picture, does this mean that it doesn't matter how fast we turn-on and turn-off the FET? Yes, from the DM-noise viewpoint it really doesn't matter much. However, there are parasitics that we ignored, chiefly the ESL and trace inductances. Since, unlike the ESR, these will produce frequency-dependent voltage spikes, it is in our interest not to keep the FET crossover times too small.

## 15.6   Practical CM Filter Design

We are assuming that the FET heatsink is tied to the chassis and that at one end of the parasitic capacitor $C_p$ between the (earthed) case and chassis we are applying a trapezoid (describing the drain waveform). This causes a CM noise current $I_{cm}$ to flow through the earth wire. We assume that the parasitic capacitances and/or X-caps cause this injected current to split *equally* between the $L$ and $N$ wires (that is CM noise by definition). So, we have $I_{cm}/2$ flowing in each of these two wires.

We already know the Fourier components of the trapezoid, and so we can treat them individually and determine the injected current due to each harmonic. As an example, we take the forward converter. Here the peak-to-peak amplitude of the drain to source waveform ($V_{max}$ or $A$) is twice the supply rail ($V_{IN}$).

$$c_n = A \times \frac{2 \times (t_{ON})}{T} \times \left[\frac{\sin\left\{\frac{n \times \pi \times t_R}{T}\right\}}{\frac{n \times \pi \times t_R}{T}}\right] \times \left[\frac{\sin\left\{\frac{n \times \pi \times (t_{ON})}{T}\right\}}{\frac{n \times \pi \times (t_{ON})}{T}}\right]$$

Since $\sin x / x \sim 1$ if $x$ is very small, we get

$$c_n \approx 2A \times \left[\frac{\sin\left\{\frac{n \times \pi \times (t_{ON})}{T}\right\}}{n \times \pi}\right]$$

So, assuming duty cycle is about 50 percent, the amplitude of the fundamental (first harmonic) is

$$c_1 \cong \frac{2A}{\pi} = \frac{4 \times V_{IN}}{\pi} \text{ volts}$$

We again realize that as for DM noise, from the viewpoint of the noise envelope and the required attenuation, it is the fundamental harmonic (only) which really counts. The current caused by this is

$$I_{cm} = \frac{V_{DS}}{25 - j\frac{T}{2\pi \times C_p}}$$

since the LISN presents an impedance of 25 $\Omega$ to $I_{cm}$. The voltage measured across the 50 $\Omega$ resistor of the LISN is due to $I_{cm}/2$ flowing through it. So,

$$V_{cm} = \frac{I_{cm}}{2} \times 50 = I_{cm} \times 25 \text{ volts}$$

Simplifying

$$V_{cm} = \frac{4 \times V_{IN}}{\pi - \frac{j}{50 \times C_p \times f_{SW}}} \text{ volts}$$

$$|V_{cm}| = \frac{200 \times V_{IN} \times C_p \times f_{SW}}{\sqrt{(50\pi \times C_p \times f_{SW})^2 + 1}} \text{ volts}$$

In terms of dBµV this is

$$V_{LISN\_DM\_NOFILTER} = 20\log\left(\frac{|V_{cm}|}{10^{-6}}\right) = 120 + 20\log\left(|V_{cm}|\right) \text{ dBµV}$$

So, if, for example, $V_{IN} = 100$ V ($A = 200$ V), $C_p = 200$ pF, $f_{SW} = 100$ kHz, we get $V_{cm} = 0.4$ V or 112 dBµV (for the first harmonic). We can follow a similar procedure as for the DM filter to calculate the $LC$ of the common-mode filter. Thereby we can calculate $L_{cm}$, and the net $Y$-capacitance (line to earth).

We can make the following key observations with regard to common-mode noise as borne out of an even more detailed computation

- The envelope is flat and fixed at 100 $V_{max}C_p f_{SW}$. At the break point described by $f = 1/(\pi t_{RISE})$, the envelope rolls-off at 20 dB per decade.

- The pedestal (flat part) does *not* depend on rise or fall times *contrary to popular perception*. So, the envelope does change, but not at the low-frequency end. For EMI purposes it is that end which is our starting point for the filter design. So, any subsequent roll-off is not going to affect the filter design.

Since the pedestal of the common-mode noise envelope is independent of the rise and fall times, does this mean that it doesn't matter how fast we turn-on and turn-off the FET? Yes, it doesn't. But see the last paragraph on DM filter design.

In Chap. 17 we have provided reasons why the amount of Y-capacitance we can use is restricted due to IEC safety regulations. Since $C$ is necessarily small, $L$ has to be much larger for CM filter stages. For DM filters, we usually put several X-caps in parallel (each of which is usually a maximum of 0.22 µF, that being based only on various manufacturing constraints). So the $L$ for a DM stage can in turn be made much smaller. The usual practice for low to medium power off-line converters is *not* to use any actual choke for the DM filtering. Instead, typically, we just use one (sometimes two) standard *common-mode chokes* in which the *leakage inductance* between the two windings provides the small amount of necessary DM filter inductance. Suppose $L_{cm}$ is the inductance of *each* winding of the CM choke, and $L_k$ the leakage inductance present in *each* limb of the CM choke, then effectively, the $LC$ filter inductance of the CM stage is $L_{cm}$, and the $LC$ filter inductance of the DM stage is $2 \times L_k$. If the line-to-line capacitance is $C_x$, then the effective $LC$ filter capacitance of the DM stage is $C_x$. If we have two Y-caps (each connected between one of the two line inputs and earth), and each has a capacitance $C_y$, then the effective $LC$ filter capacitance of the CM stage is $2 \times C_y$. We can thus relate our theoretical calculations above to a real-life filter.

We should be aware of some key factors impacting our filter design in the low-frequency region (150 kHz to 500 kHz).

- We know that the limit lines allow for progressively *higher emissions* below 500 kHz.

- But the sensitivity of the LISN *decreases* as the frequency falls off. This effectively allows us *more noise*. Roughly, we can say that the LISN impedance falls from 50 $\Omega$ at about 500 kHz to about 5 $\Omega$ at very low frequencies, at an approximate rate of 10 dB per decade below 500 kHz.

- However, we note that the EMI filter becomes less effective at low frequencies, since it is naturally a low pass type. Typically, its attenuation rolls off at the rate of 40 dB per decade.

Let us see what all this nets us. Suppose we have, by suitable design, achieved compliance at the lowest frequency. If the switching frequency is less than 150 kHz, it would mean that we have about 2 mV (66 dBµV) of noise emissions at 150 kHz (Class B). Now let us go in the *reverse direction*, i.e., from low to high frequency. This is what we find:

1. The LISN sensitivity *increases* (at the rate of ~10 dB per decade). So we would start getting higher and higher noise readings.

2. But the EMI filter starts becoming more and more effective, attenuating the signal at a typical rate of 40 dB per decade.
3. This swamps out the increasing LISN sensitivity, and so our measured noise actually *falls* at the rate of $40 - 10 = 30$ dB per decade.
4. But the limit lines are asking us to decrease the noise level at the rate of only 20 dB per decade.

Therefore the measured noise level continues to fall below the limit lines with an *increasing headroom* of $30 - 20 = 10$ dB per decade. This is why we try to go about first trying to achieve compliance at the lowest frequency, since then we usually have automatic compliance at higher frequencies (unless EMI spikes due to parasitic resonances, inadequate grounding and/or radiation problems are present).

# 16

# Things to Try

## 16.1  Introduction

Too often we grope within our minds to remember a circuit we may have seen somewhere, or even built ourselves, which at that time had seemed "neat." Too bad we threw it away while relocating. We could have used it now, but it is history. The author actually did manage to hold on to some of the most interesting circuits he either worked with, saw, or built, some of which are presented here with suitable comments. Most of them are well-tried circuits, not design ideas, and may even have been built in extremely large commercial volumes. But the reader must satisfy himself that there are no patent protections in force before using any of these, though most likely they are only some tricks of the trade.

## 16.2  Synchronizing two 3844 ICs

It is often believed that it is possible to synchronize two 3842 ICs but not two 3844 ICs because the latter includes a frequency doubler. The 3844 is intended for a forward converter in which we are not allowed to exceed 50 percent duty cycle. So the 3844 simply omits every alternate cycle generated by its clock to achieve an effective $D_{MAX}$ of (a little less than) 50 percent. Therefore if we synchronize the clocks of the two 3844 ICs, we cannot guarantee that the two ICs will synchronize *in-phase* or *out-of-phase*. The clock simply doesn't know which pulse we are omitting inside the IC, and there is no pin coming out from this low-cost eight-pin IC that gives us that information either. The idea in Fig. 16.1 was created by the author, and it solves the problem (assuming continuous conduction mode) in a very simple way—by exploiting the rule that synchronization is only possible if the "master" has a slightly *higher* frequency than the "slave." So we have a phase detector on the outputs of

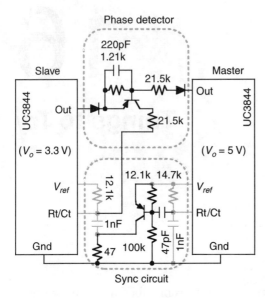

**Figure 16.1** Synchronizing two xx3844 controllers.

the 3844 ICs which basically interrupts the synchronization if it detects that the outputs are out of phase. It does this by injecting an additional charging current through the 21.5 kΩ resistor into the timing capacitor of the slave. Finally, after a few unsynchronized cycles, the outputs will get into phase again (by sheer chance), at which point synchronization would be immediately allowed again by turning *off* the upper transistor. The outputs would then lock in-phase again. In-phase synchronization is required for implementing the PFC-PWM synchronization scheme discussed in Chap. 5 if there are *two* PWM power-trains running from the output rail of the PFC preregulator. But if there is no PFC present, we should use *out-of-phase* synchronization to reduce the input capacitor ripple current. With a very simple modification of this circuit we can get the two 3844 ICs to synchronize out-of-phase.

## 16.3   A Self-Oscillating Low-Cost Standby/Auxiliary Power Supply

Most practical power supplies have a separate on-board low-power auxiliary power supply for various reasons. For example, we almost invariably need a current limit on *each* output, particularly for meeting safety regulations—specifically those limiting the maximum energy we can derive from outputs designated "SELV" (safety extra low voltage). We also need output overvoltage/undervoltage protection (OVP/UVP), input undervoltage lockout (UVLO), and the like. Most often, by customer specification, the overcurrent protection (OCP) needs to be *self-recovering* on removal of the overload. In contrast, if there is an overvoltage we will usually require the power supply to latch-off, requiring the mains input to be "recycled" before it tries to restore the output rail. In OCP we would normally sense each output, then pass the OR-ed information to the primary side through the *fault optocoupler* (i.e., the one placed in addition to the main *regulation optocoupler*). So, if there is an overcurrent on *any* output we usually get a situation where all outputs will collapse. Now, if *hiccup mode* is allowed under OCP then there may be no need for any auxiliary power supply, since the power supply can just keep resetting and trying. But if hiccup mode is not allowed, *we need to be able to continue various activities like monitoring the current during an overload, maintaining switching action,* and *maintaining all the fault logic even as the main outputs power down.* Therefore, now the supply rails going to the secondary- and primary-side logic sections cannot be allowed to collapse when the main outputs do. We, thus, need an auxiliary/standby power supply to provide the internal supply rails for functioning under abnormal conditions. But why make it self-oscillating? That is considered quite advantageous since such power supplies are not only cheaper, they are inherently self-protecting by their ability to change their frequency (naturally) under overload conditions. But self-oscillating power supplies always seem to have a poorer efficiency due to their sluggish turn-on and turn-off. The circuit in Fig. 16.2 therefore has several waveshaping zener diodes to make the gate drive "sharper." It provides power to both the primary-side and secondary-side logic, and also to the PWM controller IC. It can

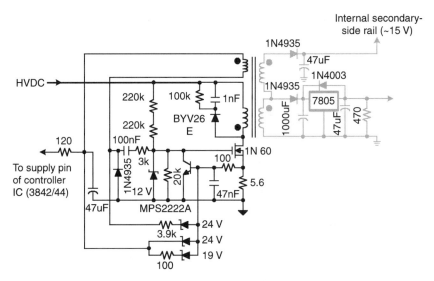

**Figure 16.2**  A self-oscillating auxiliary power supply for 90 to 270 Vac.

also provide a standby output rail of 5 V which is accessible outside the power supply box. Here we should remember that any rail coming out of the box could be subject to safety regulations if it becomes accessible to the user. However, if it comes from an LDO stage (like the 7805) it is considered inherently protected and is not required to have any separate current-limiting circuit.

## 16.4   An Adapter with Battery Charging Function

In Fig. 16.3, we see on the top left side the required characteristics. We have a CV (constant voltage) region, followed by a CP (constant power) region, and then a CC (constant current) region. Actually a constant

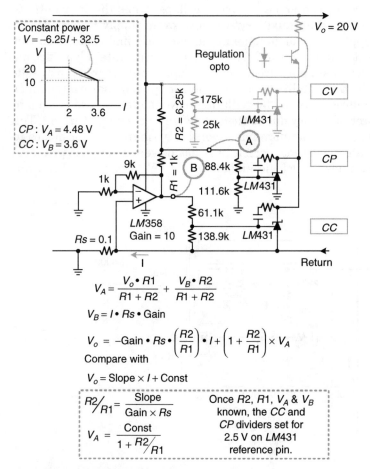

$$V_A = \frac{V_o \cdot R1}{R1 + R2} + \frac{V_B \cdot R2}{R1 + R2}$$

$$V_B = I \cdot Rs \cdot Gain$$

$$V_o = -Gain \cdot Rs \cdot \left(\frac{R2}{R1}\right) \cdot I + \left(1 + \frac{R2}{R1}\right) \times V_A$$

Compare with

$$V_o = Slope \times I + Const$$

$$R2\big/R1 = \frac{Slope}{Gain \times Rs}$$

$$V_A = \frac{Const}{1 + R2\big/R1}$$

Once R2, R1, $V_A$ & $V_B$ known, the CC and CP dividers set for 2.5 V on LM431 reference pin.

**Figure 16.3**   A CV/CC/CP adapter control.

power region will give $V \propto 1/I$ which happens to be a rectangular hyperbola (not a straight line), but the approximation to a straight line is still very good over a limited region. This circuit was created by the author and it works under the following intuitive principle: if as the current increases we "fool" the feedback circuit into thinking that the voltage has gone up, it will reduce the output voltage. So we may end up getting the product $VI$ to be fairly constant. The interesting part about the circuit is that as we increase current, first the usual regulation reference (LM431) is active (CV section, grayed out), then at higher currents the CP section automatically takes over, and finally at higher currents, the CC section takes control. Once the reader has understood the principle he or she can actually get rid of the divider on the CC section by adjusting the gain of the op-amp.

## 16.5   Paralleling Bridge Rectifiers

Ever needed to increase the dissipation and current in an input bridge rectifier without looking for exotic packages? In Fig. 16.4 we show how two bridge rectifier packs can be successfully paralleled using lower-cost packages. Two diodes inside the same package can usually be assumed to be well matched and will therefore share current well. This is a standard technique used on high-power supplies.

## 16.6   Self-Contained Inrush Protection Circuit

The basic inrush protection circuit shown in Fig. 16.5 is fairly standard, though it assumes that PFC is present. There are two high-power wire-wound resistors which charge the bulk capacitor and are then bypassed by the silicon controlled rectifier (SCR). The problem with driving an SCR at this position is that if the voltage to do that is provided by the

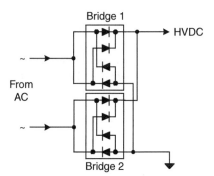

**Figure 16.4** Paralleling input bridge rectifiers.

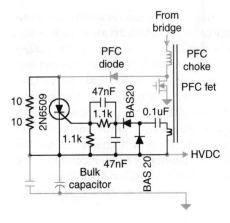

Figure 16.5   Inrush protection circuit.

auxiliary power supply, then we will need to do level shifting to reference the signal to the cathode of the SCR. But since the drive current for an SCR is not insignificant, we will actually lose a great deal of efficiency in the process through the dissipation in the level-shifting circuit. If we try to put a few turns around the PFC choke for the purpose we may have a problem because the PFC varies from low duty cycle to high. Therefore, a unidirectional winding may not work over the entire range. So here we have a charge pump and voltage doubler to do the job. It works well over 90 to 270Vac. It has been built in very large volumes.

## 16.7   Cheap Power Good Signal

A simple three terminal device at the output can provide power good indication. This is a fairly unique device from Mitsumi (at *www.mitsumi. co.jp*). Several variations of the PT series (e.g., PST591 to 595) are available for different applications. See Fig. 16.6.

## 16.8   An Overcurrent Protection Circuit

In Fig. 16.7 we have thrown a thinner gauge wire but with the same number of turns on the forward converter choke. Thus the two windings are magnetically equivalent and there should have been no voltage difference between them. But there is. The *dc resistance* (DCR) of the

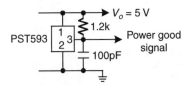

Figure 16.6   Easy power good indication.

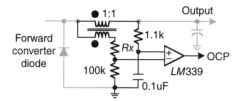

Figure 16.7  Overcurrent protection for a forward converter.

main winding creates a slight differential voltage which depends on the load current. This is detected by the op-amp and by adjusting $R_X$ we can set a current limit. Remember that this limit will be fairly crude because the resistance of copper increases by 4 percent every 10°C. However, this circuit has been used in large volumes.

## 16.9  Another Overcurrent Protection Circuit

In Fig. 16.8, $R$ is the DCR of a post $LC$ filter of a flyback, or could be an actual sense resistor, or even just a length of manganin (or constantin) wire. It has been built in extremely large volumes. It uses the drop across a diode like the 1N4148 ($\approx$ 0.6 V) as a reference to compare the drop across the sense resistor with. The resistor in series with the diode is calculated to operate the diode close to the "knee" in its $VI$ characteristics. The voltage divider $R1/R2$ is adjusted to set the current limit. Note that to avoid common-mode noise or interference problems, most experienced designers of flybacks try not to put any sense resistor in the return rail (ground).

## 16.10  Adding Overtemperature Protection to the 384x Series

The popular 3842/3844 series of controllers do not have a built-in *over temperature protection* (OTP) . In Fig. 16.9 we use the fact that in typical off-line applications, the error amplifier of the controller is not being

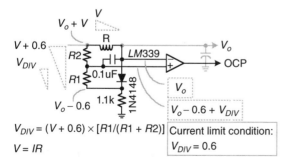

Figure 16.8  A cheap overcurrent protection circuit for flybacks.

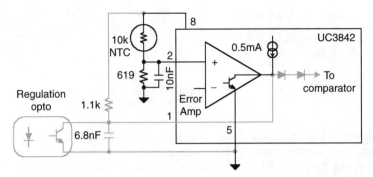

**Figure 16.9**  A cheap overtemperature protection add-on to the 3842/3844.

used because the error amplifier function is being carried out by the LM431 on the secondary side. So the feedback is usually directed to Pin 1 rather than Pin 2. Pin 2 is therefore vacant. Luckily, the output of the 3842/3844 error amplifier is an open collector type, and so it is possible to perform OTP too, besides regulation. The *negative temperature coefficient* (NTC) thermistor should be actually attached to the plastic body of the switching FET by a thermal glue for the most effective protection under overloads and short circuits on the outputs. This circuit has also been used in very large volumes.

## 16.11   Turn-On Snubber for PFC

In Fig. 16.10 we have a popular PFC turn-on snubber, using four additional diodes, two toroidal turn-on chokes, and a large electrolytic capacitor. This works by resisting the reverse recovery shoot-through that occurs in the PFC diode when the switch turns *on*. Then when the switch turns *off*, the energy gets recycled into the bulk capacitor after being temporarily stored in the large electrolytic capacitor. So it is not

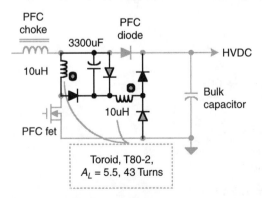

**Figure 16.10** Turn-on snubbers for PFC stages.

wasted. The author further modified this original (proprietary) circuit by reasoning that since the voltages across the two chokes are almost exactly equal, (that can be confirmed by analysis or measurement) why not wind them on the same core? So, in the schematic we can omit the two diodes shown with gray fill, but we should ensure that the dots on the windings are as shown. We can also halve the capacitance to about 1500 μF. The result is a nonproprietary circuit that works well, is also much cheaper, but perhaps needs to be more thoroughly tested.

## 16.12    A Unique Active Inrush Protection Circuit

This was developed by the author for a low-volume universal input 300 W telecom power supply. Since space was at a premium (it had to be in a 3U rack-mount profile), the two large 10 W resistors of more conventional inrush protection circuits were not possible. Here the in-rush control is by means of a FET (see Fig. 16.11). But the FET is *not* operated linearly and need not be mounted on a heatsink either. This circuit waits till there is a zero crossing in the input ac voltage waveform and only then allows the FET to turn *on*. So, the inrush current starts rising *with* the rising voltage, rather than turning *on* and finding the

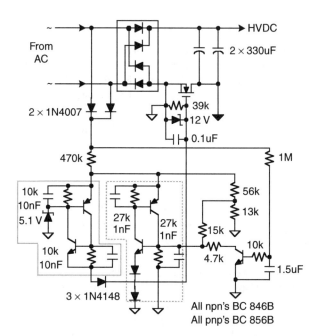

**Figure 16.11** An active inrush protection circuit that works on the zero-crossing principle.

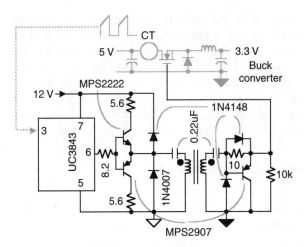

**Figure 16.12**   Floating drive from a 384x controller output.

instantaneous voltage at any arbitrary level. The inrush current therefore reaches a well-controlled peak (independent of parasitics like ESR) of about 42 A. The required specification on the inrush current was 45 A maximum.

### 16.13   Floating Drive from a 384x Controller

In Fig. 16.12, we have a method of floating the output of the 384x controller. This was implemented for an external dc-dc converter module providing 3.3 V from a power supply which had only 12 V and 5 V outputs. Note that slope compensation is also required for such a topology or it just won't seem to work. In a 384x this normally takes the form of a simple capacitor (around 33 pF to 100 pF) between the Clock pin (i.e., $R_T/C_T$) and the Isense pin.

### 16.14   Floating Buck Topology

This is shown for the simplest case of turns ratio 1:1 in Fig. 16.13. Here the IC floats on an auxiliary rail it creates, which also happens to settle down to exactly half the voltage of the main output rail. This decreases the voltage stress on the IC since even the SW pin of the IC sees a voltage less than $V_{IN}$ under steady operating conditions. The input pin of the IC also sees a voltage of only $V_{IN} - V_{AUX}$. The operating principle is as follows: during the switch conduction time the voltage across the main winding is $V_{IN} - V_O$. When the switch stops conduction, the voltage across the main winding is $V_O - V_{AUX}$. But the latter must be equal to the voltage across the auxiliary winding, which is clamped to $V_{AUX}$.

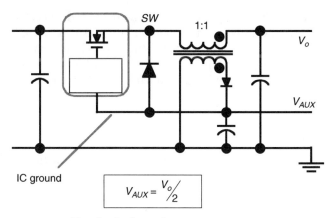

**Figure 16.13**  Floating buck topology.

Equating, we get $V_{AUX} = V_O/2$. The auxiliary rail can be used to deliver about 1/10th the load current without disturbing the energy balance. This was developed by the author and bears a U.S. patent number.

## 16.15   Symmetrical Boost Topology

In Fig. 16.14 we have a true ac-dc topology, as conceived by the author. No input rectification stage is required and the output is still a dc level. Note that since the diode drops appear in series with the (boosted) output, the impact on efficiency is less than if they were positioned at the input, as in an input rectifier stage. This also saves one diode drop actually. The circuit works by having only one FET switch at a given time, and so when the next ac half-cycle starts, the other FET

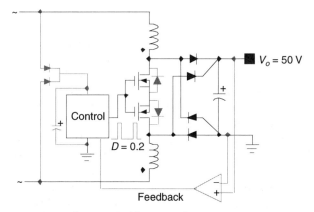

**Figure 16.14**  Symmetrical boost topology.

starts switching instead. Feedback is accomplished by a differential amplifier and the input ac must be rectified to provide a dc supply to the controller. This circuit can be very useful if we don't want to require any special precautions against damage by accidental reverse polarity at the input. This circuit will just continue working normally as if nothing has happened.

## 16.16   A Slave Converter

Consider the equation for the output of a buck-boost in discontinuous conduction mode

$$V_o = \frac{D^2 \bullet V_{in}^2 \bullet 10^6}{2 \bullet I_O \bullet L \bullet f} \text{ volts}$$

where $L$ is in µH and $f$ is in hertz.

This has the following proportionality

$$V_o \propto D^2 \bullet V_{in}^2$$

But the duty cycle of a buck converter in continuous conduction mode is

$$D \propto \frac{1}{V_{in}}$$

So, if we use the duty cycle of a buck in continuous mode to drive a buck-boost in discontinuous mode, we can get the dependency on $V_{in}$ to cancel out as follows

$$V_o \propto \frac{1}{V_{in}^2} \bullet V_{in}^2 = \text{constant}$$

This was achieved in Fig. 16.15.

We have also used the fact that the output voltage of a discontinuous mode converter at a fixed duty cycle depends on its inductance. So we have "tuned" the slave to have the required output level (at its expected maximum load current) by a careful choice of inductance. Within a valid range, this technique provides completely adjustable auxiliary output voltages, something we cannot normally expect from composites based only on continuous conduction mode.

Note that the zener on the output of this slave converter is almost completely nonconducting when the slave converter is working at its designed (maximum) load. The efficiency is therefore as high as we normally expect from any conventional switching power converter. However, if the load on the slave decreases, the zener comes into play and starts automatically shunting the balance of the current a way. It then behaves as a conventional shunt regulator. Therefore load regulation,

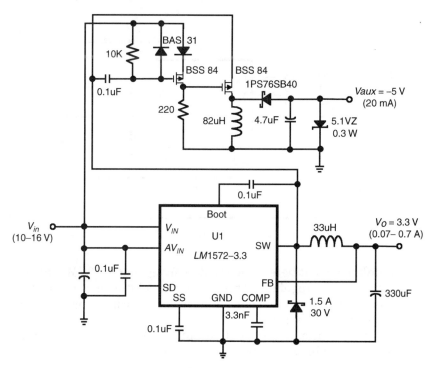

**Figure 16.15**  A slave buck-boost in DCM riding off a master buck in CCM.

which is taken for granted when dealing with single or multi-CCM stages, is not "automatic" here. It is being "enforced" by the zener, but luckily, if the inductance has been chosen correctly, this needs to happen only at less than maximum loads.

But we do have line regulation. As the input voltage increases, the feedback loop of the regulated buck converter commands its duty cycle to decrease to maintain output regulation. It so happens that this decrease in duty cycle is exactly what was required by the discontinuous-mode buck-boost to "regulate" its own output almost perfectly.

The schematic can probably be simplified a great deal. This was rather hastily developed by the author to prove a principle for a certain *request for quotation* (RFQ) but was later granted a U.S. patent.

## 16.17   A Boost Preregulator with a Regulated Auxiliary Output

This is shown based around a typical buck IC, the LM1572. The input range of the LM1572 is 8.5 V to 16 V and its output is set to 5 V. As shown in Fig. 16.16, once startup has been achieved, we can make it work down to a couple of volts input, while maintaining the output at 5 V. This turns it into a step-up or step-down converter. Boost preregulators

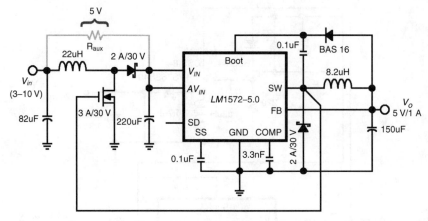

**Figure 16.16**  Boost preregulator with automatically-regulated intermediate output.

are not unknown, but here we see that no independent PWM control is required for the preregulator. This makes the solution more attractive. It is essentially a two-switch master-slave boost-buck cascade, with the buck stage being the master.

But here is the interesting input-output *transfer function coincidence* it is based on. We have for a boost

$$D = \frac{V_{O\_BOOST} - V_{IN\_BOOST}}{V_{O\_BOOST}}$$

where $V_{IN}$ is the input to this stage. The output of the boost forms the input to the buck, so,

$$V_{O\_BUCK} = D \times V_{IN\_BUCK}$$

that is,

$$V_{O\_BUCK} = D \times V_{O\_BOOST}$$

Eliminating $D$, we get

$$\frac{V_{O\_BUCK}}{V_{O\_BOOST}} = \frac{V_{O\_BOOST} - V_{IN\_BOOST}}{V_{O\_BOOST}}$$

that is,

$$V_{O\_BUCK} = V_{O\_BOOST} - V_{IN\_BOOST}$$

We see that the *strapped output* across $R_{AUX}$ is being automatically regulated. Though it is not ground-referenced it can provide power to a stand-alone circuit block like a light emitting diode (LED) display.

# Reliability, Testing, and Safety Issues

## 17.1 Introduction

Power supply engineers have several performance goals that they typically weigh against cost. One of these performance goals is reliability. As we will see, a few cents saved by a bad design choice is not worth it if we are going to see excessive numbers of field returns. That is why, among others, a *demonstrated reliability test* (DRT) is considered a must. Here we will briefly cover the terms in frequent use in the power supply industry as regards testing and qualification. We will also touch upon safety issues and show how they can be impacted by seemingly innocuous design choices.

## 17.2 Reliability Definitions

Reliability is the probability that a device will perform its specified function in a given environment for a specified period of time. In other words, reliability is quality over time and environmental conditions.

The exponential distribution is one of the most common distributions used to describe reliability. We assume that the reliability exponentially decays with time as

$$R(t) = R(0) \times e^{-\lambda t}$$

where $R(0)$ is the reliability at time $t = 0$ and is assumed maximum [that is, $R(0) = 1$]. Then the time constant is the point where the number of units still functioning is $1/e$ of the original number, i.e., 36.8 percent are left. This time constant is called the *mean time between*

TABLE 17.1 Failure Definitions and their Conversions.

| MTBF (hours) | Failure rate (per POH) (h$^{-1}$) | $\lambda$ (% per 1000 h) | ppm (hours) | FITs (failures per billion POH) |
|---|---|---|---|---|
| $10^9$ | $10^{-9}$ | $10^{-4}$ | $10^{-3}$ | 1 |
| $10^8$ | $10^{-8}$ | $10^{-3}$ | $10^{-2}$ | 10 |
| $10^7$ | $10^{-7}$ | $10^{-2}$ | $10^{-1}$ | $10^2$ |
| $10^6$ | $10^{-6}$ | $10^{-1}$ | 1 | $10^3$ |
| $10^5$ | $10^{-5}$ | 1 | 10 | $10^4$ |
| $10^4$ | $\Leftarrow 1/x$   $10^{-4}$   $\times 10^5 \Rightarrow$ | 10   $\times 10 \Rightarrow$ | $10^2$   $\times 10^3 \Rightarrow$ | $10^5$ |
| $10^3$ | $10^{-3}$ | $10^2$ | $10^3$ | $10^6$ |

*failure* (MTBF). 1/MTBF is the failure rate. Failure rate can be expressed as the percentage failures per 1000 device-hours of operation (usually called $\lambda$), or as the total number of failures in a million device-hours (expressed as parts per million, or ppm), or as the number of failures per billion device-hours (expressed as failures in time, or FITs). See Table 17.1 for the respective definitions and their relationships.

**Note:** We can have 1000 devices operating for 1 h, or one device operating for 1000 h. They are statistically equivalent and both correspond to 1000 *device-hours*. For power supplies we prefer to talk in terms of *power-on-hours* (POH). This is the cumulative hours of operation of several power supplies working together.

**Note:** The MTBF is almost the same as mean time to failure (MTTF), and the two are used almost equivalently. Technically, however, MTBF should be used only in reference to repairable items while MTTF should be used for nonrepairable items. However, MTBF is commonly used for both cases.

Now, since we have 8760 h in 1 year, a typical power supply MTBF of 250,000 h is equal to about 30 years. Yes, we could theoretically have one power supply working for 30 years, but we know it really doesn't (or can't). The reason is that we have wear-out failures starting to occur after about 3 to 5 years. We know that the chief culprits are the aluminum electrolytic capacitors (or the fan). So what an MTBF of 250 kh *does not mean* is that the power supply will work 30 years before we see a failure. We could however have 100 units in the field and they would accumulate 30 power-on-years in 0.3 calendar years only. That is when we will see the first failure (*on an average*). And that is what MTBF generally indicates. See also Figure 11.2 for an important clue.

The definition of MTBF applies only in the region where life or wear-out issues are not present. In fact the decay in the number of functioning units is exponential on the basis of the fact that the *failure rate is a constant*. So, if we start off with say 1000 units and 10 percent fail in the first year, we are left with 900 units. In the next year, another 10 percent fail so we are left with 810 units. The next year yet another 10 percent i.e., 81 fail so we are left with 729 units and so forth. This goes on till wearout starts occurring and the failure rate then climbs steeply. However, if we plot the values before wearout—1000, 900, 810, 729, and so on—we get an exponential curve. The time constant of this curve is the MTBF.

In the above example we took the failure rate as 10 percent in 8760 h i.e., the percentage in 1000 h is $10/8.76 = 1.142$. From Table 17.1 we see that this is $\lambda$ by definition. To get the failure rate per POH we need to divide this by $10^5$. Thus we get the latter to be $1.142 \times 10^{-5}$. From the same table we see that to get MTBF we have to take the reciprocal of this. Therefore, MTBF is $10^5/0.001142 \approx 9 \times 10^7$ h.

Similarly, a typical power supply MTBF of about 30 years (250 kh) means a failure rate of $1/250\,k = 4 \times 10^{-6}$. To get $\lambda$ we have to multiply this by $10^5$ (see Table 17.1) and so we get 0.4 percent per 1000 h.

## 17.3 Chi-Square Distribution

As mentioned, to verify MTBF we could power up one unit and wait for about 30 years for a failure to confirm that the MTBF is 250 kh. But the problem with that is that (1) within the test time we are not allowed to have any wear-out failures (2) we need to verify the MTBF *before* we release the product, not 30 years later, and finally (3) how do we know that that particular power supply was indeed a representative sample, or "typical?" We want to increase the number of units being tested so as to include several more production batches to see the effect of variations/spreads/tolerances, and also to increase the POH instead of the calendar years. Since the analysis clearly becomes statistical at this point, a typical power supply specification may ask for "an MTBF of 250 kh at 60 percent confidence level, at an ambient temperature of 55°C." This would mean that we should be able to assert with 60 percent confidence that indeed the MTBF is better than 250 kh. The means to do this is the $\chi^2$ (Chi-square) distribution and its table. This is described in the United States Military Handbook Mil-Hdbk 781A. The engineer doesn't need to know statistical theory in detail here but only to understand how to use it to compute the MTBF. Here are the steps

1. By definition, failure rate is

$$FR = \frac{\chi^2(\alpha, 2f+2)}{2 \times POH}$$

TABLE 17.2   Chi-square Lookup Table.

| No. of failures | $\chi^2$ at 60% CL | $\chi^2$ at 90% CL |
|---|---|---|
| 0 | 1.833 | 4.605 |
| 1 | 4.045 | 7.779 |
| 2 | 6.211 | 10.645 |
| 3 | 8.351 | 13.362 |
| 4 | 10.473 | 15.987 |
| 5 | 12.584 | 18.549 |

or

$$MTBF = \frac{2 \times POH}{\chi^2(\alpha, 2f + 2)} \ \text{hours}$$

where $\alpha$ is the *significance level* (or the acceptable risk of error) and is related to the confidence level by

$$CL = 100 \times (1 - \alpha)\%$$

$f$ is the number of failures.

While the engineer can look up a book on statistics to get the $\chi^2$ table, we have provided the results most frequently used for estimating reliability of power supplies directly in Table 17.2. The confidence levels usually used are the 60 percent and 90 percent levels. Further, the testing usually never goes for more than zero or one failure, as it would involve an even larger number of power supplies. Let us do some sample calculations.

**Example 17.1**   How many POH are required to demonstrate an MTBF of 250 kh at 90 percent confidence level (temperature specified: usually 55°C)?
    We should have zero failures when operated for

$$POH_0 = \frac{\chi^2 \times MTBF}{2} = \frac{4.605 \times 250000}{2} = 575,625 \ \text{h}$$

And we should have had only one failure when operated for

$$POH_1 = \frac{\chi^2 \times MTBF}{2} = \frac{7.779 \times 250000}{2} = 972,375 \ \text{h}$$

On completion of these many *power-on-hours* we can expect the second failure to occur.

**Example 17.2**   How many units are required for a 4-week test to demonstrate an MTBF of 250 kh at 60 percent confidence level, with one failure?
    For one failure we need

$$POH_1 = \frac{\chi^2 \times MTBF}{2} = \frac{4.045 \times 250000}{2} = 575,625 \ \text{h}$$

505625

In 4 weeks we have 672 h. Thus we need

$$\frac{575625}{672} = 752 \text{ units}$$

Note that these units will all need to be operated simultaneously at maximum load or 80 percent of maximum load (as specified) and at maximum ambient of 55°C (or as specified). Typically, some of these will be run at the customer's location and some at the power supply manufacturer's location.

An alternative form does not require the Chi-square table. This says that the demonstrated MTBF is

$$MTBF = \frac{1}{1 - (1 - CL)^{POH}}$$

where CL is the confidence level and POH the number of power on hours *before* the first failure occurs (i.e., for zero failures).

## 17.4  Chargeable Failures

When doing the demonstrated reliability test, some failures may be discounted. For example, if the failure is clearly a result of user mishandling, it may not be considered a potential cause of future field failures, and therefore is not relevant to a reliability estimate. Failures such as these are considered *nonchargeable*. On the other hand, *chargeable* failures are those which if not corrected will occur in the field. "Failure" may just mean that the power supply has gone out of its stated specifications. Mechanical dents and the like are not chargeable. Note that a chargeable failure may be the result of a workmanship problem during production, and if so, it has to be corrected by process improvement and/or training so that we can be sure that the failure will not occur in the field. When the supplier has provided adequate evidence that a particular failure mode has been understood and a fix implemented, only then will it become a *nonchargeable* failure (and will be eliminated from being counted in the ongoing demonstrated reliability test). Note that this may also involve a design change. Finally, it must be agreed to by all concerned as constituting a corrective action.

Here is a summary of these failure classifications

1. *Chargeable failures*
   - Any failure that degrades the performance and effectiveness beyond acceptable limits

- Any failure that could result in significant system damage, such as to preclude mission accomplishments
- Any failure that will need repair activities

2. *Nonchargeable failures*
   - Any failure that does not degrade the overall performance and effectiveness of the system beyond acceptable limits
   - Any failure that has been corrected and proven to be fixed
   - Any failure that occurs during run-in test or incoming inspection

## 17.5   Warranty Costs

Most engineers are surprised to find out how much it can cost to get back one unit, repair it, and then have it reinstalled. This was in fact posed as a prize-winning question for employees in the German factory the author worked in, a few years ago. The correct answer was around DM180 (approximately $120 at that time). But it was interesting to note that almost no guess from the hundreds of employees even made it into the ballpark.

 Engineers should also know the 10× rule which goes as follows: if a failure is detected at the board level and costs $1 to fix, then if discovered at a system level it will cost $10, and if it goes to the field and then failure occurs, it will cost $100 to repair and so forth.

We should know how the figures relating to warranty add up as a function of MTBF? Here is a sample calculation

**Example 17.3**   If there are 1000 units with an MTBF of 250 kh, how many are expected to fail over 5 years?

Let us assume that the failed unit is immediately replaced with a unit of similar reliability. Then the number of failures over 5 years (43.8 kh) is

$$No.\ of\ failed\ units = \frac{1000\ units \times 44000\ ^{h}/_{unit}}{250000\ ^{h}/_{failure}} = 175.2\ failure$$

If it takes $100 estimated to repair one unit in the field, this will cost $17,520. So cost per unit is $17.52.

This calculation is often performed in terms of *annualized* (or *annual*) *failure rate* (AFR) and gives almost the same result. Starting with $N$ power supplies, with a certain MTBF, we know that the sample size shrinks as

$$N(t) = N \times e^{-t/MTBF}$$

So at the end of 1 year (i.e., t = 8760 h) we are left with

$$N \times e^{-8760/MTBF}\ units$$

The number of units failing every year is therefore described by

$$Annual\ failure\ rate = 1 - e^{-8760/MTBF}$$

Over 5 years the cost is estimated to be

$$cost/unit = (Annual\ failure\ rate) \bullet (years\ of\ product\ life) \bullet$$
$$(per\ unit\ repair\ cost)$$

So, if MTBF is 250 kh, and the cost to bring back, repair, and return one failed unit is \$100, over 5 years we get the following cost/unit

$$Annual\ failure\ rate = 1 - e^{-8760/250000} = 0.034$$

So,

$$cost/unit = (0.034) \bullet (5) \bullet (\$100) = \$17.20$$

This is going to increase the selling price dramatically. Unless of course the warranty is drastically reduced (90 days?)!

## 17.6  Calculating Reliability

Until fairly recently the United States Military Handbook Mil-Hdbk 217F was commonly used to calculate reliability based on either a simple *part count* or a *part stress* analysis.

This particular standard may not be in common use today, but in the author's opinion the underlying philosophy should still be taken note of. This philosophy is in fact shared by many other reliability prediction methodologies, many of which are still in frequent use. In all of these the failure rate of the equipment is taken to be the sum of the failure rates of all the individual components. Each component has a specified *base failure rate*. This is then adjusted by multiplying it with a number of factors (called the *pi factors*) which are related to the environment $\pi_E$, application $\pi_A$, quality level $\pi_Q$, secondary stresses (e.g., voltage stress $\pi_V$) and the like. The $\pi$ *factors* in the Mil-Hdbk were based on statistical data gathered over several years and so they weren't always reflecting the recent and rapid improvements in component technology. Therefore it was common to get a calculated MTBF for a power supply of the order of only 100 kh by part stress analysis, and around 150 k to 200 kh by part count analysis (under *ground benign conditions*). But neither of these numbers came close to what demonstrated reliability (and field) tests showed (typically 400 kh for an off-line power supply for example). So, over the last decade or so most companies had established an unofficial and internal thumb-rule multiplicative factor (of around 4) with which to multiply the MTBF prediction from the part stress analysis of Mil-Hdbk 217F.

Though Mil-Hdbk 217F is now rarely used, *stress analysis must always be performed*. This involves testing and profiling *each* component inside the power supply. Voltage levels, current, and temperature should be recorded, and this exercise is also helpful in identifying the weak links in the design. Note that the part count analysis of Mil-Hdbk 217F always seemed somewhat of a "play with numbers." Some prominent industry personalities had even laid out their "pet peeve" as follows: if you remove several components corresponding to a certain protection block, the reliability would fall, though part count analysis would say it has improved because there are now fewer components which can fail. However, it was always understood that part count analysis was to be carried out in the initial bidding phases only since there was no prototype on hand yet to carry out a formal part stress analysis. But it has to be acknowledged that even the latter analysis would likely not account for most *abnormal* operating conditions, i.e., those in which the protection block would act to protect the power supply. Therefore engineering judgment should always be considered of prime importance when designing power supplies, especially when carrying out reliability enhancements.

As mentioned, there are several other reliability calculation methods, and some companies (e.g., Siemens) even have their own internal prediction methodologies. All are, however, very similar to Mil-Hdbk 217F in philosophy, though they do end up with very different final numbers, mainly because their $\pi$ factors are numerically different.

Demonstrated reliability is still clearly the best way to go, despite the fact that a very large number of units need to be tested. But we must also realize that even this test is usually done only under steady operating conditions. Thus, a good design engineer will also carry out several bench tests to confirm reliability under abnormal conditions.

## 17.7   Testing and Qualifying Power Supplies

A power supply may go through several tests before it is considered mature. Besides the obvious functional checks, thermal scans, and thermocouple measurements (and safety and EMI compliance tests) some of the other tests typically required fall into the following general categories.

### HAST/HALT, HASS, and ESS

HAST/HALT, HASS, and ESS stand for *highly accelerated stress/life test*, *highly accelerated stress screen*, and *environmental stress screening*, respectively. HALT is performed during the development phase of the product. It is intended to stress a few samples to find the product's limits before destruction occurs. Thermal soak, thermal cycling, vibration, and other stresses may be applied that will sometimes exceed the specified operating range of the product. Thus HALT will give us a quantitative idea of the design margins.

Typically, HALT will be performed several times on successive design iterations to measure improvement. On the other hand, HASS is conducted during the subsequent production cycles to measure the effect of normal process variations on product reliability. But we note that the stress limits set for HASS are sufficiently reduced below the maximum limits indicated by the HAST test so as to avoid causing premature ageing during the production cycle itself. In HASS we will apply all selected stresses simultaneously, with the product functioning and being constantly monitored during that time.

In ESS, various stresses are applied to each outgoing unit. The idea is to weed out early life failures (i.e., those with "infant mortality") in production rather than hear about it from the field. Various forms of stress can be applied. The most common are thermal soak (*burn-in*), thermal cycling, and vibration. Sometimes shock and input voltage margining are used, though humidity may also be increased (e.g., to 85 percent).

Some specific examples of power supply/module tests are:

1. *Burn-in.* A typical commercial power supply burn-in setup is actually very simple. Several power supplies are powered up for several hours in a small room with resistive loads on their outputs. The heat from this combined resistive load heats the room up. A thermostat operates an exhaust fan when the temperature rises above 55°C typically, and this constitutes a crude ambient temperature control. Each output has a light emitting diode (LED) across it so that the operator can periodically check to see if any unit has failed. Some limited form of thermal cycling, and simple tests like a timer operated power-on/power-off cycling may also be carried out.

2. *Overstress tests (e.g., on 4 units).* The purpose of all this is to determine if it is cost effective to take steps to improve reliability. For example, if the stress margins can be improved significantly by just raising the wattage of a certain resistor, it would be usually well worth it. This test is thus done during the design phase and is essentially exploratory in nature.

   - *Thermal.* One unit will run at full load in a thermal chamber at maximum rated temperature and nominal line. Thermal protection, if fitted, will be deactivated. The temperature is increased in 10°C steps with a 30-min operation at each temperature until failure occurs. Failure analysis and repair are performed. The test is repeated (total 2 runs) and during the second run, the temperature on the suspected weak link is monitored.

- *Line voltage.* One unit will run at full load in a thermal chamber at maximum-rated temperature and maximum line voltage. The line is increased in steps of 10 Vac, till failure occurs. Failure analysis and repair are performed. The test is repeated (total 2 runs) and during the second run the voltage and current of the suspected weak link are monitored.

- *Load stress.* One unit will run at full load, nominal line, in a thermal chamber at maximum-rated temperature and maximum load. The load is then increased in steps of 10 percent of maximum load (each output simultaneously), at 30-min intervals till failure occurs. Failure analysis and repair are performed. The test is repeated (total 2 runs) and during the second run the voltage and current of the suspected weak link are monitored. We again repeat the test, but now with load on only one output at a time ("corner conditions"). Note that any current limit protection on the outputs is to be deactivated for the purpose of this test.

3. *Thermal ESS (e.g., on 4 units).* Here normal operation is expected through 10 cycles of $-30°$C to $85°$C at a rate of $20°$C/min. The thermal profile is provided. Typically, there will be three dwell levels per cycle of 15 min each. One will be at the highest temperature, after which the unit is taken straight down to the lowest temperature where it remains for 15 min, and thereafter the unit is brought back to room temperature where it stays for 15 min before it starts the next thermal cycle. The chamber temperature rate may be varied from $20°$C/min to $60°$C/min. After this test, the random vibration test is to be done so as to provoke failures due to any damage during this test.

4. *Vibration test (e.g., on 4 units).* The test is conducted with the system energized, fixed onto the vibration table in its normal configuration. The frequency band is set from 5 to 500 Hz. Each system is run for 10 min at 3, 4, 5, and 6 g (rms), 5 min at 7 g (rms) in the $x$ axis then the system integrity is checked. The test is repeated for the $y$ and $z$ axes. Plots of the control accelerometer and the response accelerometers are recorded for the starting and maximum operating test levels.

## 17.8   Safety Issues

Here we will focus mainly on the issue of "clearances" and "creepages" as they relate to our design and topology. Some basic definitions are as follows:

*Clearance distance.*   This is the shortest distance between two conductive parts (or between a conductive part and the bounding surface of the equipment) measured through air. Clearance distance helps prevent dielectric breakdown between electrodes caused by the ionization of air. The dielectric breakdown level is further influenced by relative humidity, temperature, and *degree of pollution* (or *pollution degree*, defined below) in the environment.

*Creepage distance.* This is the shortest path between two conductive parts (or between a conductive part and the bounding surface of the equipment) measured along the surface of the insulation. An adequate creepage distance is meant to protect against *tracking*. This is a process that produces a partially conducting path of localized deterioration on the surface of an insulating material. The degree of tracking required depends on two major factors. One is the *comparative tracking index* (CTI) of the electrical insulating material (expressed in volts) and the other is the *pollution degree* of the environment.

To see how these need to be computed see Fig. 17.1. Note that EN 60950 is the commonly followed norm (originally IEC 950). This provides tables giving clearances and creepages depending on pollution degree and CTI (as applicable) and voltage. Here are some points we should remember and double check:

The peak voltage between two points affects the required clearance and therefore it must be measured. Creepage depends only on *working voltage* which is basically the rms of the voltage between the points.

For worldwide input power supplies for information technology (IT) equipment, a popular rule of thumb is to allow 8-mm creepage

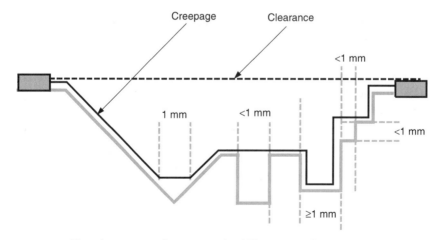

**Figure 17.1** How clearances and creepages should be measured.

between primary and secondary circuits and 4 mm between primary and chassis ground. Clearance usually gets automatically taken care of in the process. Thus 4-mm *margin tape* is used in safety transformers as this produces 8-mm creepage between primary and secondary windings. If the above spacings are used during the design stage, there is a high probability (over 90 percent) that the power supply will pass the relevant safety test. However, with power factor correction (where the high voltage dc (HVDC) rail is around 400 V), we have to be very careful as this rule may not hold. We should compute the required spacings (and margin tape width) in the manner indicated later in this chapter.

*Triple insulated* wire is an emerging choice and it may be used in a transformer if cost permits. Its advantage is that no margin tape is required. But the wire must comply with Annex U in EN 60950.

The optocouplers (especially surface mount packages) should be checked carefully to see if they comply with the mandatory separation requirements. As per IEC, for optocouplers, a working voltage measurement can be taken to determine creepage, but if the measured voltage is less than the mains voltage, the latter is to be used. If the working voltage of the optocoupler is more than 50 $V_{RMS}$, a minimum distance of 0.4 mm through the insulation is needed.

There is no requirement for the distance through insulation of sheet materials. We always use three layers of Mylar (i.e., polyester) tape at the isolation boundary of a transformer, and to reduce leakage this can even be 0.5 mil thick if cost permits. Though usually, three layers of 1-mil- or 2-mil-thick polyester are more common.

Manufacturers often introduce slots in the PCB to enhance the creepage between primary and secondary circuits. This depends on the CTI of the PCB material. But note that the slots must be wider than 1 mm to count (see Fig. 17.1)

The most commonly used PCB materials and polyester insulators have a CTI of 200 to 400 V. They thus fall into *Material Group III*. We should calculate creepages based on this assumption unless we know better.

As for pollution degree, the most common assumption in commercial power supplies is *Pollution degree 2*. This corresponds to a typical office environment where only temporary conductivity may occur by condensation. *Pollution degree 1* is for dry, nonconductive pollution, *Pollution degree 3* corresponds to a typical heavy industrial environment, and *Pollution degree 4* corresponds to persistent conductivity as by rain or snow.

In testing, every component can be subjected to a 10-N force, and the required spacings should still be maintained. This is one reason why we should prefer to lay components flat on the PCB, especially those near the edges of the PCB (near the chassis) and those near the isolation boundary.

Mandatory separations also need to be maintained if, for example, a single solder joint (anywhere) comes undone. This can cause a two-lead component to rotate, thereby possibly bridging the primary and secondary sides. Therefore, room temperature vulcanizing silicone (RTV) or hot melt glue is usually liberally applied on the PCB (this being also useful for clearing a typical vibration test).

## 17.9   Calculating Working Voltage

The impact of topology on correctly calculating working voltage is now taken up. This is required to calculate the width of the margin tape, for example. We will focus on the case of a single-ended forward converter with a PFC front-end and an HVDC of around 400 V. This case is special because the thumb rule of 4 mm/8 mm does not work and we actually need more than that. Rather than take it to the test lab and discover that the transformer needs to be completely redesigned, it is important that we know exactly what the rms voltage we are talking about here is.

What exactly are we measuring? The transformer is isolated so what do we mean by rms voltage *across* the transformer? To answer this question we have to trace the path backward as in Fig. 17.2. We can

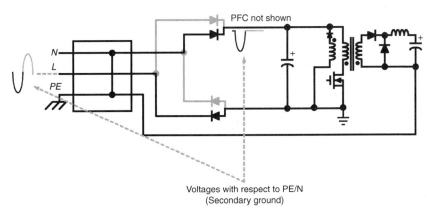

Voltages with respect to PE/N
(Secondary ground)

**Figure 17.2**  How the bus voltage looks with respect to the secondary ground.

see that since *protective earth* (PE) and *neutral* (N) are connected at the service entrance and the secondary ground is the system ground, the HVDC (i.e., bus voltage) has the shape shown. Note that the black part of the ac waveform occurs when L goes below N. During this time, the HVDC also falls with respect to system ground. On the gray part of the waveform, L is higher than N, so the HVDC is flat. Besides this low frequency undulation, we also have a high-frequency switching waveform on either side of the transformer.

We have two termination pins per winding. So, which primary-to-secondary combination gives us the worst case for the purpose of calculating or measuring the working voltage? The worst-case rms is actually recorded between the drain (dotted end) and the nondotted end of the secondary. That is because the nondotted end has the opposite phase. So when the drain goes high, this nondotted end of the secondary goes low and the *difference* between the two is the maximum. In Fig. 17.3 we have shown what the voltage waveforms look like.

What if we have multiple outputs? For calculating the maximum rms voltage across the transformer we need to consider the winding which corresponds to the *highest output voltage* (magnitude).

We have provided a Mathcad file to do the calculations in Box 17.1. For simplicity, we have avoided performing a summation over vectors (which is actually the best way to prevent Mathcad convergence problems) and preferred a simple integration instead. But because

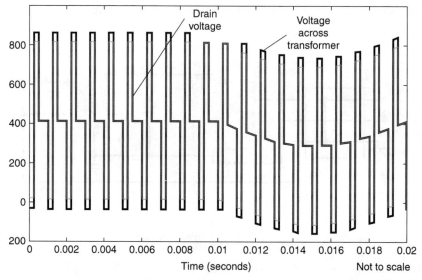

**Figure 17.3** Voltage on drain and across transformer for forward converter (Option 2).

Select := 1

**Enter 1** here if this is conventional Forward Converter (with Output Diodes in Common Cathode configuration), and output being $+Vo$ (return connect to PE)

**Enter 2** here if this is conventional Forward Converter (with Output Diodes in Common Cathode configuration), and output being $-Vo$ (return connect to PE)

**Enter 3** here if this is reversed Forward Converter (with Output Diodes in Common Anode configuration, and output being $-Vo$ (return connect to PE)

**Enter 4** if here this is reversed Forward Converter (with Output Diodes in Common Anode configuration, and output being $+Vo$ (return connect to PE)

$k := 10^3$   $m := 10^{-3}$   $u := 10^{-6}$   $t := 0, 1 \cdot u .. 20 \cdot m$

Enter inputs $Vac := 90$   $Vdc := 385$   $Vo := 12$

Turns Ratio $n := \dfrac{44}{5}$   $D := \dfrac{n \cdot Vo}{Vdc}$   $D = 0.274$

Switching Frequency

(enter $2k$ to avoid non-convergence) $f := 2.k$   $\omega := 2 \cdot \pi \cdot f$   $T := \dfrac{1}{f}$

Line Frequency $fl := 50$   $\omega l := 2 \cdot \pi \cdot fl$   $Tl := \dfrac{1}{fl}$

Construct Waveforms:

$Vrect(t) := Vac \cdot \sqrt{2} \cdot |\sin(\omega l \cdot t)|$

$Vpwm1(t) := \text{if } [\text{mod}(t, T) < D \cdot T, Vdc, 0]$

$Vpwm2(t) := \text{if } [\text{mod}(t, T) < (T - D \cdot T), Vdc, 0]$

$Vpwm(t) := Vpwm1(t) + Vpwm2(t)$

$Vsec1(t) := \text{if }\left[ \text{mod}(t, T) < D \cdot T, \dfrac{Vdc}{n}, 0 \right]$

$Vsec2(t) := \text{if }\left[ (1 - D) \cdot T < \text{mod}(t, T) < T, \dfrac{-Vdc}{n}, 0 \right]$

$Vsec(t) := Vsec1(t) + Vsec2(t)$

According to reference of measurement introduce required voltage difference

$vt := \begin{vmatrix} Vsec(t) & \text{if} & \text{Select} = 1 \\ Vsec(t) + Vo & \text{if} & \text{Select} = 2 \\ O & \text{if} & \text{Select} = 3 \\ (-Vo) & \text{if} & \text{Select} = 4 \\ \text{"Error"} & \text{otherwise} \end{vmatrix}$

$Vacross(t) := \begin{vmatrix} [Vpwm(t) - Vrect(t) + v(t)] & \text{if} & t > \dfrac{Tl}{2} < t < Tl \\ Vpwm(t) + v(t) & \text{otherwise} \end{vmatrix}$

Worst-case maximum rms voltage

$Vworking := \left( \dfrac{\displaystyle\int_0^{Tl} Vacross(t)^2 dt}{Tl} \right)^{\frac{1}{2}}$

$Vworking = 471.016$ Volts

**Box 17.1**  Working voltage of single-ended forward converter transformer.

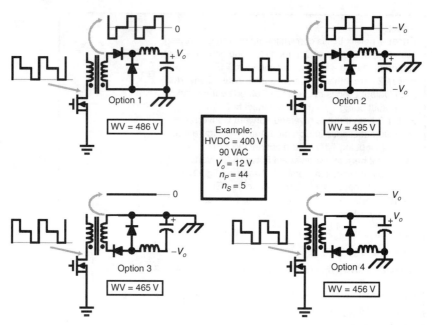

**Figure 17.4**  Different options referenced in the Mathcad file, for a single-ended forward converter.

integration is then required over the two very different frequencies involved, for certain output voltages the Mathcad file will not give desired results. In that case, the reader may then have to reduce the set PWM frequency down to 1 kHz just to make the program work. But once it does, we always get the right result, since the voltage does not really depend on the switching frequency. This program has actually been compared by the author to a far more detailed file (one that does involve a summation over vectors, and which also uses the exact switching frequency of the PWM), and the results from the simpler file were shown to be equally accurate. Note that we can enter four schematic options as displayed in Fig. 17.4. The voltage values shown boxed next to each schematic are the result of the attached Mathcad file. We see that if we want a positive output voltage rail, option 4 provides the lowest voltage across the transformer and thus possibly the narrowest margin tape. Mandatory creepage distances as a function of working voltage are displayed in Table 17.3. We have also provided an example of an interpolated creepage calculation within the same figure. So for an estimated working voltage of 450 $V_{RMS}$ we need margin tape of width 4.6 mm. *Note that the worst case occurs at 90 Vac not 270 Vac (when PFC is present)!*

**TABLE 17.3 Creepage Distances as per the xx950 Safety Standards.**

| Working voltage V (rms or dc) | Pollution Degree 2 | | |
|---|---|---|---|
| | Material group | | |
| | I CTI > 600 V | II 600 > CTI ≥ 400 | III 400 V > CTI ≥ 100 V |
| ≤50 | 0.6 | 0.9 | 1.2 |
| 100 | 0.7 | 1.0 | 1.4 |
| 125 | 0.8 | 1.1 | 1.5 |
| 150 | 0.8 | 1.1 | 1.6 |
| 200 | 1.0 | 1.4 | 2.0 |
| 250 | 1.3 | 1.8 | 2.5 |
| 300 | 1.6 | 2.2 | 3.2 |
| 400 | 2.0 | 2.8 | 4.0 |
| 600 | 3.2 | 4.5 | 6.3 |
| 800 | 4.0 | 5.6 | 8.0 |
| 1000 | 5.0 | 7.1 | 10.0 |

Linear interpolation permitted (round off to next higher 0.1 mm increment) e.g., 450 V requires

$$\frac{450 - 400}{600 - 400} \times (6.3 - 4) + 4 = 4.6 \text{ mm}$$

This is for basic insulation (i.e., primary to safety ground creepage, also width of margin tape). Double this distance for reinforced insulation (i.e., primary to secondary creepage).

## 17.10 Estimating Capacitor Life

With 500 V electrolytic capacitors becoming available, no other capacitor has the CV (capacitance times its voltage rating) capability of this popular capacitor. For cost reasons too, it is virtually indispensable in commercial off-line power supplies. People may not always want it, but usually can't do without it either. Even in modern dc-dc converters, where the "all ceramic" solution is much the buzzword nowadays, after finding unexpectedly severe input voltage overshoots (under hard application of input power) and the consequent instability and ringing, a common emerging recommendation is to put a *high-ESR* capacitor like an aluminum electrolytic in parallel to the ceramic input capacitor, in an effort to damp out the input resonances (lower the Q-factor).

In most commercial power supplies the aluminum electrolytic capacitors determine the eventual life of the product under operation. It is therefore important to get a much better understanding of this critical (and almost unavoidable) component. The most important aspect of designing with this capacitor is estimating its life.

The first parameter in a typical aluminum electrolytic capacitor's datasheet that we must understand is the *dissipation factor* (DF) or tan $\delta$. It is related to the *equivalent series resistance* (ESR) by

$$ESR = \frac{\tan \delta}{2\pi f \times C}$$

DF is thus the ratio of the resistance to the reactance $1/2\pi C$. Note that the ESR used here must be the ESR at the frequency $f$ (usually stated at 120 Hz). Clearly a low DF means a low ESR.

What is defined as "end of life?" Typically this is said to occur when the change in capacitance measured from the *initial* value is greater than ±20 percent and/or the tan $\delta$ has become more than 200 percent of its *initial* value. Considering the fact that the initial specified value of the capacitor lies within a certain standard tolerance band (usually ±20 percent), we need to include another 20 percent to account for that. For example, if our theoretically calculated holdup time requirement is 100 μF, we should start with a capacitor value greater than 156 μF *nominal* (check: 156/(0.8 × 0.8) = 100). This is really a more than worst-case estimate. Nevertheless, several manufacturers of reliable commercial power supplies do add a flat 40 percent to the theoretically calculated value to account for ageing and initial tolerance.

How bad can the ESR really be at the end of the useful life? Since DF is allowed to double, we see that the change in ESR can be

$$ESR \propto \frac{\tan \delta}{C} \Rightarrow \frac{200\%}{80\%} = 250\%$$

A higher ESR generally plays a helpful rule in the feedback loop, but we should still do a Bode plot to confirm the phase margin. We can hand-select (using actual measurements) lower capacitance, and higher ESR capacitors from the used parts bin in an effort to mimic end-of-life characteristics on our bench prototype.

We should also be aware that as the ESR increases, the heating in the capacitor will also increase toward end of life, raising its temperature further. However, manufacturers of quality capacitors like the Chemicon group (*www.chemi-con.com*) have clearly accounted for this in their life predictions, so we usually do not need to worry about this particular issue.

The next important datasheet parameter is the ripple current rating. It is typically stated in amperes rms at 120 Hz and 105°C. This essentially means that if the ambient temperature is at the maximum rated of 105°C, we can pass a low-frequency current waveform with the stated rms, and in doing so we will get the stated life. The life figure is typically 2000 h to 10000 h under these conditions. Yes, there are lower-grade 85°C capacitors available, but they are rarely used as they can hardly meet typical life requirements at high ambient temperatures.

The datasheet also provides certain *temperature multipliers*. For example, for the LXF series from Chemicon, the numbers are

1. At 65°C the temperature multiplier is 2.23.
2. At 85°C the temperature multiplier is 1.73.
3. At 105°C the temperature multiplier is 1.

This is easy to understand if we realize that such (long-life) capacitors are typically designed for a 5°C differential from ambient to *core* (deep inside the capacitor), and about 5°C differential from ambient to case. The temperature inside the capacitor is thus slightly higher than the case. But it is the temperature of the electrolyte that determines how quickly it evaporates and how long the capacitor will last. Misestimation of a few °C of this temperature could mean thousands of hours subtracted from the life of the product.

Now realizing that the amount of heating and the temperature rise are proportional to $I_{RMS}^2$, we can see that at 85°C

$$\frac{T_{CORE} - 105}{T_{CORE} - 85} = \frac{I_{105}^2}{I_{85}^2} = \frac{1}{1.73^2} = \frac{1}{3}$$

Solving for $T_{CORE}$, we get

$$T_{CORE} = 115°C.$$

Using this value we can confirm the 65°C multiplier

$$\frac{115 - 105}{115 - 65} = \frac{10}{50} = \frac{I_{105}^2}{I_{65}^2}$$

So, the multiplier must be $5^{0.5} = 2.236$, which agrees with the specified value. Therefore from the vendor's ripple current multipliers (for temperature) we can easily deduce the *design core temperature for this series of capacitors*. Then, with a little help from an empirical relationship toward the end of this section, we can also figure out the most important capacitor design parameter—the *difference from case to core* (or *can to core* as more frequently called).

Out of the 10°C rise in temperature, roughly 5°C is from ambient to case and another 5°C from case to core. Let us call these differentials symbolically as $\Delta T_{case\_amb}$ and $\Delta T_{case\_core}$. Note that these are the *design values* used by the manufacturer for the specific family of capacitors. They are not our actual measurements or estimates in an application. *If, however, we pass the rated ripple current (as specified at 105°C) through the capacitor, we do get exactly these design differentials in practice, and that is true whatever our actual ambient is.*

**Example 17.4**   A typical 85°C-rated capacitor has the following temperature multipliers: 1 at 85°C and 1.3 at 70°C. What is the designed core temperature?

$$\frac{T_{CORE} - 85}{T_{CORE} - 70} = \frac{I_{85}^2}{I_{70}^2} = \frac{1}{1.3^2}$$

Solving

$$T_{CORE} = 85 + 15 \times \left(\frac{1}{1.3^2 - 1}\right) = 107°C$$

So in this case, typically, $\Delta T_{case\_amb} \approx \Delta T_{case\_core} \approx 11°C$

We see that 85°C capacitors do have a much higher allowed temperature differential. This is an advantage, but not quite as much as it seems. The problem is that temperature multipliers provided in the datasheet are never used in practice since what they amount to is increasing the current so as to bring the core temperature back up to its maximum value. But we know that if the core temperature is at its maximum, the life we can expect is only the specified 2000 to 10000 h. But we want much more.

A typical power supply design requirement is 5 years or 44000 h at maximum load (or 80 percent of maximum load) operating at a room ambient of 40°C. Note that *the power supply is typically tested at a maximum of 55°C, but for life expectancy, a reduced temperature is usually specified.* But how can we achieve even that from a 2000-h capacitor? We use the doubling rule as derived from Arrhenius' theory

$$L = L_o \times 2^{\frac{\Delta T}{10}}$$

This effectively states that the life of a capacitor doubles every 10°C fall in temperature (of the core). $L_O$ is the guaranteed life (2000 to 10000 h) when passing the maximum specified ripple current at 105°C i.e., when its core is at 115°C, as calculated above.

**Hint:**   For semiconductors a similar rule of thumb prevails—the *failure rate* doubles every 10°C rise in temperature. Life and failure rate are actually separate issues. Life is a wearout effect at the end of the familiar "bathtub" curve, whereas failure rate is measured *between* the regions of infant mortality and wearout.

**Note:** The standard 44000 h life requirement is equivalent to 5 year operation at $24 \times 7 \times 365$. In reality this may never be the case. It is better to discuss this with the customer as it adds greatly to the cost of the power supply. A well-known PC-market competitor usually specified a life of only 15000 h for most products. This amounts to roughly 8 h a day for 5 years, and is probably a far more realistic goal, one which can also often be met with cheaper $85°C$ capacitors at certain key locations in the power supply. But we do have to use extremely good quality capacitor manufacturers. No cheap substitutes please!

**Example 17.5** If we pass the rated ripple current through a 2000-h capacitor (no temperature multipliers applied) at an ambient of $55°C$, what is the expected life (first pass estimate)?

At the rated current we can expect that the core is at $55°C + \Delta T_{core\_amb}$. So the temperature "advantage" we have gained (measured from the maximum-rated temperature) is $(105°C + \Delta T_{core\_amb})$ minus $(55°C + \Delta T_{core\_amb})$ i.e., $50°C$. Since this capacitor provides 2000 h at the maximum temperature, at the reduced ambient we may get a life of

$$2000 \times 2 \times 2 \times 2 \times 2 \times 2 = 64{,}000 \text{ h}$$

Note that in the above analysis $\Delta T_{core\_amb}$ eventually got canceled out. So, this amounts to writing the following simple equation for life

$$L = L_o \times 2^{\frac{T_{core\_rated} - T_{core\_application}}{10}} = L_o \times 2^{\frac{T_{rated} - T_{amb}}{10}}$$

In our example, $T_{core\_rated}$ is $115°C$, and $T_{rated}$ is the maximum-rated ambient of $105°C$. $T_{amb}$ is the actual ambient in our application. $T_{core\_application}$ is the temperature of the core in our application, which in our example is $65°C$. But note, however, that we would have got the same life prediction had the capacitor manufacturer used any other designed $\Delta T_{core\_amb}$. As we saw, it got canceled out, but that is only because in the example, we *followed the manufacturer's recommendations* and passed only the maximum-rated current through the capacitor.

In practice, we don't have a good way of knowing the local ambient (in the immediate vicinity) of the capacitor. Nearby components may also be heating the capacitor. Therefore a common and conservative industry practice is to cut the outer sleeving of the capacitor and to insert a thermocouple under the sleeve in contact with the metal case. This way, small air draughts don't affect the results. We then take this case temperature as the effective ambient, *unless we know better*. Suppose the case temperature is measured to be $70°C$. Then the estimate of the capacitor life is now

$$L = L_o \times 2^{\frac{T_{rated} - T_{application}}{10}} = 2000 \times 2^{\frac{105-70}{10}} = 22{,}600 \text{ h}$$

However, we have to be clear what the *source* of this heating is. If it is *not* heat from nearby components, the $\Delta T_{case\_core}$ may be actually much higher than we think. The life cannot be the same as compared to the case where the heat is purely from external sources, since that won't produce the harmful internal temperature differential (from case to core). Therefore a case temperature measurement is simply not enough. *We have to measure the ripple current* through the capacitor too, to at least confirm that we have not exceeded the maximum ripple current rating of the capacitor (which is equivalent to not exceeding the designed case to core delta). The relevant points are summarized below

> Capacitor manufacturers recommend that in general we don't pass any more current than the maximum-rated ripple current. This ripple current rating is the one specified at the worst-case ambient (e.g., 105°C). But even at lower temperatures we should not exceed this current rating. No temperature multipliers are to be used to buttress this rating. Only then is the case to core temperature differential within the design specifications of the part. And only then are we allowed to apply the simple doubling rule for life, since the core temperature rise is then considered factored into the life prediction figures provided by the manufacturer.

> If the measured ripple current is confirmed to be within the rating, only then can we take the case temperature measurement as the basis for applying the doubling rule, even if the heat is coming from adjacent sources. Again this is because the case to core temperature differential is within design expectations.

> However, in direct communication, Chemicon has in the past allowed a higher ripple current than rated, but the life calculation method given is then slightly different. This amounts to a special doubling rule which we will describe below using a practical example.

**Example 17.6**  We are using a 2200 µF/10 V capacitor from Chemicon. Its catalog specifications are 8000 h at maximum-rated 1.69 A, stated at 105°C and 100 kHz. The measured case temperature in our application is 84°C and the measured ripple current is 2.2 A. What is the expected life?

The life calculation provided by Chemicon was

$$L = L_O \times 2^{\frac{105-84}{10}} \times \overbrace{2^{\frac{5-\Delta T}{5}}} \Big| \text{hours}$$

where

$$\Delta T = 5 \times \left( \frac{2.2}{1.69} \right)^2 = 8.473°C$$

So,

$$L = L_O \times 2^{\frac{105-84}{10}} \times 2^{-0.695} = 21,000 \text{ h}$$

Let us understand the terms involved here. The $\Delta T$ calculation above essentially says what we already know

$$\frac{\Delta T}{\Delta T_{case\_core}} = \left( \frac{I_{application}}{I_{rated}} \right)^2$$

We know from the vendor's data that this family of capacitors was designed for a 5°C differential between case and core, and that differential is caused by passing the rated 1.69 A through it. So this $\Delta T$ calculation gives us the temperature differential when we pass 2.2 A through it. We then get a rise of 8.473°C rather than the designed 5°C.

The term $(17 - \Delta T)$ in the exponent of the life calculation gives us the temperature *in excess of the designed 5°C*. Let us call this $\Delta T_{excess}$. So the life equation is

$$L = L_O \times 2^{\frac{T_{rated}-T_{case}}{10}} \times 2^{\frac{-\Delta T_{excess}}{5}} \text{ hours}$$

The first term with the positive exponent causes the life to increase above $L_O$ and the second term exerts the opposite effect. We can also see that a temperature differential from case to core *in excess of the designed value* is considered more harmful than a normal temperature differential (i.e., one that is caused by staying within the current rating). Chemicon models this excessive temperature rise conservatively as *causing a halving of life every 5°C increase, rather than the usual 10°C.*

**Note:** This equation should not be used to predict life if the ripple current is less than rated. $\Delta T_{excess}$ is *not* allowed to be negative here.

**Note:** Capacitor manufacturers typically don't guarantee life under forced air cooling. The designer should either measure the capacitor without forced air cooling if possible, or add a judicious safety margin.

Rather than take the case temperature as the ambient temperature of the capacitor, which is more of a worst-case calculation, we could try to actually measure its local ambient. Assume that the general ambient is $T_{amb\_ext}$. The local ambient near the capacitor is $T_{amb}$. The procedure to deal with radiation from nearby components is as follows

1. Take the capacitor from the circuit board putting it on the underside, but still connected to the circuit. In this position we can measure the temperature on its case $T_{case\_1}$. This is

$$T_{case\_1} = T_{amb\_ext} + T_{self-heating}$$

2. At the same time we place an exactly similar capacitor at the position where the original capacitor was, but this has one lead "missing," so it is in effect not connected to the circuit. We measure its case temperature $T_{case\_2}$. This is

$$T_{case\_2} = T_{amb\_ext} + T_{ext-heating} \equiv T_{amb}$$

3. Therefore having measured the ambient in the surrounding air which is $T_{amb\_ext}$, we know all the required components of the temperature build-up

4. Also note that the following equation is recommended for a more careful analysis of the ratio that exists between $\Delta T_{core\_case}$ and $\Delta T_{case\_amb}$ (which was earlier stated to be $\approx 1$)

$$\boxed{\Delta T_{core\_case} \Big/ \Delta T_{case\_amb} = 0.0231 \times CaseDia_{mm} + 0.845}$$

This curve-fit equation was derived by the author from data provided by Chemicon. It is accurate to within 6 percent for capacitor outer diameters in the range of 10 mm to 76 mm. Above D = 40 mm, the error from the use of this formula is less than 1 percent.

**Caution:** A temperature measurement is not enough, nor is a ripple current measurement. Measuring ripple current with a view to at least confirm that it is below the rated ripple current of the component is certainly advisable, though the designer is cautioned against using the ripple current alone to estimate heating on the basis of some assumed coefficient of convection $h$ (see Chap. 13 for more on $h$), because that would ignore heating from nearby components, and we will thus overestimate the life expectancy.

**Hint:** When measuring ripple current through a capacitor, the normal procedure is to lift the lower terminal (the one going to ground) and to insert a loop of wire for inserting a current probe. But this reading is extremely hard to do without us affecting the current in the capacitor even as we are attempting to measure it. This may become an *invasive* measurement. An alternative is to insert a small noninductive and calibrated sense resistor (e.g., made of manganin or contantin wire) and measure the voltage across it. However, we should not place reliance on a direct rms reading of the sensed voltage since the noise will likely skew the results. We should record the waveform on an oscilloscope and then do a calculation based on the vertices as discussed in Chap. 1. Also, when measuring the rms current through *paralleled* capacitors (e.g., those on an output rail) it is not a good idea to lift the lead of only one of them to do a current measurement since the current will just happily redistribute into the remaining capacitors. We should cut a *common return trace* and then insert either the current probe or a sense resistor.

Lastly, the vendor may have directly provided a ripple current rating at 100 kHz in addition to the 120 Hz number. If not, the vendor would certainly have provided *frequency multipliers. A typical frequency multiplier is 1.43 at 100 kHz.* That means that if we are allowed 1 A ripple current at 120 Hz then at 100 kHz we are allowed 1.43 A. This by design will produce the same heating as 1 A causes at 120 Hz. Therefore this is also equivalent to saying that the ESR at 100 kHz is related to the ESR at 120 Hz by the following equation

$$\left( \frac{I_{100kHz}}{I_{120Hz}} \right)^2 = \frac{esr_{120Hz}}{esr_{100kHz}} = (1.43)^2 = 2.045$$

Thus the high-frequency ESR is about half the low-frequency ESR. Clearly, since the case to core temperature differential is unaffected in the process, *frequency multipliers can and should be used.* In Chap. 5, we can see how to apply these frequency multipliers to our advantage when the current waveform has both low-frequency and high-frequency components.

## 17.11   Safety Restrictions on the Total *Y*-Capacitance

In off-line power supplies Y-caps are usually connected from line to safety ground (that is, PE, or protective earth). The purpose is to bypass the high-frequency common-mode noise. But these don't just bypass noise, they also conduct some of the low-frequency line current. That is what the line-to-line capacitors (X-caps) also do, the difference is that the Y-caps carry this current into the protective earth (chassis). If the earthing is not good for any reason, the user could get electrocuted on touching the chassis (or housing). Therefore, international safety agencies limit the total rms current introduced into the earth by the equipment to a maximum of typically 0.25 mA, 0.5 mA, 0.75 mA, or 3.5 mA (depending on the type of equipment and its *installation category*—its enclosure, its earthing, and its internal isolation scheme). Somehow, 0.5 mA seems to have become the industry default design value, even in cases where 0.75 mA or 3.5 mA may have been allowed. *It is important to know how high one can actually go in terms of ground leakage current, as this dramatically impacts the size and cost of the line filter, in particular the choke.*

Keeping the discussion here at a theoretical level, we can easily calculate that we get *79 μA per nF at 250 Vac / 50 Hz*. This gives us a maximum allowed capacitance of 6.4 nF for 0.5 mA, or 44.6 nF for 3.5 mA. Typical configurations in off-line power supplies are four *Y*-caps, each being 1 or 1.2 or 1.5 nF, or only two *Y*-capacitors, each of value 2.2 nF. Note that there may be other parasitic capacitances or/and filter capacitances present, which should be accounted for in computing the total ground leakage current, and thereby correctly selecting the *Y*-caps of the line filter. However, we must keep in mind that if for improved EMI performance (CM noise rejection), a *Y*-cap is connected from the *rectified* input dc rails to earth (or from the output rails to earth) there is no ground leakage current through these capacitors. Therefore, in principle, there is no limit on their capacitance in this position either. However, we may need to put *two Y*-caps in series at these positions to comply with worldwide safety standards and regional deviations.

### 17.12   Safety and the 5-cent Zener

In the introduction to Chap. 1 we mentioned the 5-cent zener problem and promised to explain it here. In Fig. 17.5 we have actually presented several issues which should be kept in mind when driving FETs with controllers (in off-line applications). Here is why the engineer didn't get hired:

1. *R*1 is a gate pull-down resistor that is not just indispensable, it needs to be of low enough resistance too. The reason is that under a sudden (hard) application of input power at high-line conditions, as the voltage on the drain of the FET suddenly ramps up, it injects a pulse through the drain to gate (Miller) capacitance. This follows the simple equation $I = C \times dV/dt$. This injected current charges up the floating gate and thus has the potential to spuriously turn the FET *on*. This is further aggravated by the fact that the controller is not likely to have had time to be powered up fully at this moment. Most controllers (like the 3842/3844 series) have in effect a tristate output until the reference voltage on the pin of the IC becomes available. So, the IC cannot effectively provide a pull-down during this time. In actual tests, *R*1 had to be decreased to between 4.7 kΩ and 10 kΩ to ensure a safe input power application under all conditions.
   a) When abnormal tests are conducted by safety agencies, they can short or open any single component in the power supply. This is expected to usually lead to failure, but that is ok, provided the power supply "fails safe." This means that at no time should a hazardous voltage appear across the accessible output terminals.

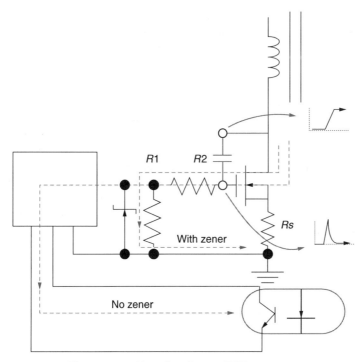

**Figure 17.5**  Things to consider when driving FETs.

The problem here is that when the FET fails, it almost invariably causes a large momentary surge of failure current flowing internally first from drain to source. But the metal oxide sense resistor $Rs$ invariably fails open shortly thereafter (before even the fuse can blow). However, the energy in the inductor is not yet spent, and it still demands a freewheeling path to flow through. So, now it "knocks" at the gate of the FET. A huge surge current finally goes through the gate, and follows a path through the IC (destroying it) and then onward into the optocoupler. In actual tests the opto package cracked open sometimes, thus potentially breaching the "sacred" primary to secondary boundary. That is certainly unacceptable to safety agencies. But, if an 18 V gate zener is placed as shown, the zener almost invariably fails in a shorted condition. It thus diverts the FET failure current away from the IC (until the fuse blows). This helps pass safety testing, but is also invaluable during prototyping because though FETs can be quickly replaced, desoldering tiny control ICs a few times destroys the fine copper traces on the PCB, rendering the board unusable in a short while. With the zener in place, the IC usually never gets damaged (nor any of the components connected to its other pins).

2. Some high-end designs used to put a transient voltage suppressor diode (TVS) in parallel to $Rs$ for the same reason. The TVS is basically just a higher peak energy zener that is guaranteed to fail in a shorted condition.

$R2$ is usually not required nor used. But a few years ago there was some suspicion (still unsubstantiated) that the small body capacitance of the zener was creating a resonant $C\text{-}L\text{-}C$ type of tank circuit with the inductance of the PCB trace going to the gate (including the internal bond wires) and the gate capacitance, thus leading to oscillations and "inexplicable" failures. To play safe, cagey designers add a 10 Ω resistor in between the zener and the gate with an intent to lower the $Q$ of the tank circuit, and to thus damp out any oscillations. We should certainly mount the zener as close to the FET as possible to avoid introducing more parasitic inductance.

NOTES:
- ELECTROLYTIC CAPS ARE MAIN CAUSE OF WEAR OUT, 3 TO 5 YRS.

MTBF p346

IF 3% FAILURE/YR + POWER ON TIME IS 8 HRS/DAY

= 2920 HRS/YR = 1%/1000 HRS ≡ λ

FROM TABLE 17.1 WE NEED MTBF = 10⁵ HRS

IF WE USE CHI SQUARED @ 60% CONF, 0 FAILURE

POH WOULD BE  $\frac{4.025}{2}$ × 10⁵ = 2.013 × 10⁵ IF WE

HAVE 6 MO TEST = 4380 HRS, 45 UNITS WOULD BE

NEEDED. EACH AT 3A PRIMARY = 135 A CIRCUIT!

20 UNITS WOULD TAKE 1.15 YRS

# Appendix:
# Components and
# FAQ

**1.** Why are film capacitors recommended for low-impedance applications?

ANSWER: Metallized film units most often fail because of electrode migration through holes left by film imperfections, but since the resulting metal-to-metal bond almost always exhibits high resistivity, heat resulting from the capacitor's fault current clears the electrode bridge, restoring normal capacitor operation. Such self-healing action prevents permanent film capacitor failure and is the reason why manufacturers of metallized film capacitors claim that their products are best suited to applications that can't tolerate catastrophic component failures.

But we should be aware that film capacitors' self-healing properties depend on the availability of sufficient clearing current; thus the probability of catastrophic failure is still present, and it increases with increasing circuit impedance. Moreover, the clearing current results in sudden voltage drops across the dielectric, and this can result in potential soft errors in critical applications.

**2.** Why are tantalum (Ta) capacitors not recommended for low-impedance applications?

ANSWER: Tantalums exhibit a self-healing effect similar to that of metallized film capacitors, but they require a *slow* buildup of heat in fault sites within the capacitor. Low-impedance circuits may allow too much current to flow through reduced resistivity zones, thereby accelerating internal heating and turning a small material imperfection into a catastrophic component failure. Tantalum capacitors therefore have inherent limitations on the surge current they can safely pass. This usually necessitates that we limit the applied voltage to 50 percent of the rated voltage, especially if the capacitor is to be used as a front-end (input) capacitor. We should also be aware that manufacturers usually have special families of more robust, surge current tested tantalum capacitors, and we should prefer these in front-end applications. Note that in a boost topology, the output capacitor also sees a high inrush surge current at power-on.

**3.** What are the pros and cons of tantalum capacitors versus aluminum electrolytic capacitors?

ANSWER:

1. The aluminum electrolytic has the advantage of recoverability of the oxide film, so catastrophic failure is avoided if there is an internal flaw. Thus the aluminum electrolytic fails open in most cases, whereas the Ta capacitor, because of its solid electrolyte, does not self-heal easily and can fail either with high leakage or even short.
2. The temperature and frequency characteristics are advantageous because the leakage of Ta is also usually much better than aluminum.
3. Solid capacitors like Ta have almost a "semipermanent" life (no wearout) and a decreasing failure rate with time. Lifetime issues have to be considered for aluminum because of the evaporation of electrolyte (see Chap. 17). Their failure rate can also climb steeply above 70°C.
4. The ripple current an aluminum electrolytic can withstand is usually far in excess of Ta. Aluminum electrolytics also have higher reverse resistance (ability to stand momentary reverse voltages). They can also typically withstand 1.2 times their rated voltage for a second or so (but verify with vendor).

**4.** Can SMD Al electrolytics and SMD film capacitors withstand soldering temperatures?

ANSWER: In the earlier days, vapor-phase reflow soldering was very popular for surface mount technology (SMT). This got largely replaced by the infrared reflow system. Subsequently, the most popular method is a forced convection furnace. For through hole devices, wave soldering (i.e., *flow soldering*) is used.

An obstacle to widespread leadless component acceptance stems from the high temperatures that surface mounted devices (SMD) must withstand during reflow and flow soldering. Although modern monolithic ceramic capacitors survive immersion in solder baths without degradation, more volatile electrolytics (especially aluminum) tend to outgas at high temperatures. Film devices can also suffer under extreme heat.

To address the problem of soldering at high temperatures, Nichicon offers leadless aluminum electrolytics which can withstand 260°C soldering temperatures for as long as 10 s under flow soldering and 230°C for 20 s under reflow (with 90 s at 150°C during preheat cycle). For hand soldering the bit temperature can be 350°C applied for up to 3 s.

An increasingly popular vendor of aluminum electrolytic capacitors and SMD film capacitors is Panasonic. Panasonic offers stacked film units which typically permit 260°C for 5 s maximum (as for flow soldering i.e., wave soldering) and allow more than 230°C (maximum 260°C) for 30 s as for reflow soldering.

**5.** What precautions are necessary when working with Al electrolytics at low temperatures?

ANSWER:  Though Al capacitors are available for operation down to $-55°$C, we should know that the capacitance will fall about 15 percent from its value at room temperature, and the ESR will increase by about 10 times. The former could affect output voltage ripple, holdup time, and the like, whereas the latter could also increase output voltage ripple besides causing a large change in the Bode plot. In voltage mode control in particular, the ESR of the output capacitor serves a useful purpose as it provides a zero in the feedback loop for ensuring stability. The zero should therefore be fairly fixed in its location (in the frequency domain), which in turn means that the ESR should also not change much, or the phase margin can get rather unpredictable and possibly unacceptable.

**6.** How do I achieve effective decoupling?

ANSWER:  In a system, software upsets can be caused by noise transients on any pin. However, the most vulnerable pin is the $V_{CC}$ pin since it has direct access to all internal parts of the chip. We also need to maintain the $V_{CC}$ specifications on the chip while still providing the input current spikes needed to keep it working. So, decoupling of $V_{CC}$ is necessary, especially in power supply controllers and switcher ICs.

There are two kinds of decoupling capacitors—board/bulk decouplers and chip decouplers. A board decoupler will normally be a 10 µF to 100 µF electrolytic placed near to where the input power enters the PCB. Its placement is not very critical, provided it is accompanied by a correctly placed chip decoupler(s). The purpose of the board decoupler is to refresh the chip decoupler(s), the latter being the small ceramic capacitors placed close to the IC. The chip decouplers provide the current spikes during crossover transitions. The "critical traces" (see Chap. 12) leading up to the IC from the decouplers must be of very low inductance, or voltage spikes will be induced in the traces which will enter the IC via the $V_{CC}$ pin. See also Chap. 2 dealing with dc-dc converters and their configurations on how topology can affect decoupling requirements.

A general thumb rule is that the bulk (reservoir) capacitance should be at least 10 times that of the high-frequency decoupling capacitance.

It is often said that the chip decoupler should not have too *large* a capacitance. There are two reasons for this statement. One is that some capacitors, because of the nature of their dielectrics, tend to become inductive or lossy at higher frequencies. However, mica, ceramic, glass, or polystyrene work well to several hundred megahertz (though some of these materials are obsolete now). The other reason cited for not using too large a capacitance has to do with lead inductance. Sometimes it is wrongly assumed by engineers that a larger capacitance will be physically bigger and so the decoupling area on the PCB will automatically be larger. Figures quoted on the series resonant frequency of a 0.01 µF capacitor run from 10 MHz to 15 MHz, depending on the lead length. Above the resonant frequency the capacitor would be essentially inductive and hence useless as a decoupling element. The resonant frequency is given by $f_o = 1/2\pi(LC)^{1/2}$, which indicates that for the same effective lead length (including effects of PCB layout too), increasing $C$ by 100 lowers the resonant frequency by a factor of 10. So, a 1 µF capacitor at the same PCB location would have a resonant frequency of 1 to 1.5 MHz only, and it may do more harm than good as a decoupler.

The effectiveness of a decoupling capacitor also has a lot to do with the way the power and ground traces connect this capacitor to the chip. In fact the area formed by this loop can be even more important than the value of the capacitance itself. We need to minimize the area if we want to achieve effective decoupling. Given that the area is minimal, and ignoring lead inductances, it can generally be said that the larger the value of the decoupling capacitance, the more effective it is as a decoupler since the voltage across it is then relatively smoother. But we know that the effectiveness has also a lot to do with the major ripple component which we are trying to smooth out. High-frequency ripple components do not respond well to physically large capacitors, particularly because of the higher lead inductances. This is why some designers try to improve even the chip decoupling by placing two ceramic capacitors side by side, one larger and one small.

Finally, it should be also kept in mind that good filtering and good decoupling are not necessarily the same thing. For example, we can introduce a resistance $R$ between the bulk and the chip decoupling capacitors. That clearly improves filtering characteristics, but it will affect the refresh rate of the chip decoupling capacitor and that can severely affect the overall performance by causing the $V_{CC}$ pin to droop out of specification (from the viewpoint of the IC).

**7.** Why is the temperature coefficient (*tempco*, TC, or TCC) of the material important for bypass/decoupling?

ANSWER: Even though bypass applications don't require a tight tolerance on the capacitance, we must pay close attention to the minimum component values. At 25°C a 0.15 µF of *Z5U* capacitor may be adequate, but at 10°C the capacitance falls to only 0.07 µF. To have proper bypassing even over the limited range of 10°C to 85°C requires choosing a *higher initial value*. In this case we should have used 0.33 µF of *Z5U*. We get what we pay for. We must only be sure we are not paying for performance we don't need. Therefore, based on cost considerations and the application we may also consider a smaller *X7R* or film-type capacitor. See question 9 hereafter on estimating the actual in-circuit capacitance.

**8.** What do *X7R*, *Z5U*, *Y5V*, and the like mean?

ANSWER: The categories of ceramic capacitors are:

*Class I.* A temperature compensating capacitor. Very stable, but will usually have a low dielectric constant ($K$) and therefore larger size. Most popular formulation is the COG, also called "*NP0*" (i.e., *negative positive zero*, referring to the almost flat temperature coefficient of capacitance). The tan $\delta$ (and ESR) of COG is also relatively stable, changing by only about 25 percent over its operating temperature range. The ESR does increase somewhat with frequency, though the capacitance does not change significantly with frequency. COG has no ageing characteristics.

*Class II.* Medium $K$ types, for example, $X7R$ with $\tan \delta = 0.03$, or $Y5V$ with $\tan \delta = 0.025$ at room temperature. In both cases, $\tan \delta$ decreases significantly with temperature. For example, from $25°C$ to $-40°C$, the $\tan \delta$ of $X7R$ will increase by about 300 percent.

*Class III.* Even smaller than Class II (higher $K$) but will usually have a lower $Q$ (higher $\tan \delta$ typically about 0.05 to 0.08 at room temperature). So, the $\tan \delta$ of $Z5U$ is worse than for $X7R$ or $Y5V$).

Class II and Class III are further sub-classified according to the table below.

| Low temperature limit of range (°C) | Upper temperature limit of range (°C) | Maximum allowable change in capacitance from 25°C (at 0 V dc, over entire operating temperature range) |
|---|---|---|
| $X = -55$ | $4 = 65$ | $F = \pm 7.5\%$ |
| $Y = -30$ | $5 = 85$ | $P = \pm 10\%$ |
| $Z = -10$ | $6 = 105$ | $R = \pm 15\%$ |
| | $7 = 125$ | $S = \pm 22\%$ |
| | $8 = 150$ | $T = +22, -33\%$ |
| | | $U = +22, -56\%$ |
| | | $V = +22, -82\%$ |

A variation of $\pm 15$ percent ($R$ above) corresponds to $\pm 150000$ ppm. If the temperature range is, say $X7$ (i.e., $125 + 55 = 180°C$), a $\pm 15$ percent variation is equivalent to a TCC (*thermal coefficient of capacitance*) of $\pm 150,000/180 = \pm 833$ ppm/°C. For COG ceramic capacitors the TCC is expressed as $0 \pm 30$ ppm/°C applicable over the $X7$ range. Note that temperature range is important in expressing TCC because the latter is an average value.

**9.** I have bought a 1 μF capacitor. What could be the capacitance it actually presents in my circuit?

ANSWER: Capacitance as specified in a datasheet is usually measured at an applied voltage of 1 V rms, at 1 kHz and 25°C. In an actual circuit we could see the following worst-case spread (after 100 kHz, with the possibility of both ac and dc voltages being considered):

**1.** *For COG.* Consider initial tolerance ($\pm 5\%$), TCC ($\pm 0.15\%$), voltage stability (0%), frequency stability (0%), ageing (0%). Combining, we get $C = 1$ μF **+5.16, −5.14%**

**2.** *For X7R.* Consider initial tolerance ($\pm$10%), TCC ($+2$, $-10\%$), voltage stability ($+15$, $-10\%$), frequency stability ($+5$, $-15\%$), ageing ($-3\%$). Combining, we get $C = 1\,\mu\text{F}$ **+35%, $-$40%**

**3.** *For Z5U.* Consider initial tolerance ($\pm$20%), TCC ($+2$, $-54\%$), voltage stability ($+22$, $-56\%$), frequency stability ($+5$, $-15\%$), ageing ($-25\%$). Combining, we get $C = 1\,\mu\text{F}$ **+57%, $-$90%**

Note that, in general, the tolerance reading for COG capacitors may already include its temperature drift.

**10.** Why are chip capacitors prone to cracking?

ANSWER: SMT is now the preferred assembly method for many types of electronic equipment. Surface mount capacitors are also seen as the most effective means of eliminating bypass interconnection inductances. But we have to be careful, especially in power supplies where the components can be exposed to extreme temperatures and repeated thermal cycling.

First, the differential *coefficient of thermal expansion* (called CTE or TCE) between ceramic capacitors (the most frequently used leadless units) and standard epoxy boards (PCBs) can lead to unreliable solder connections. Because of mechanical stresses arising from dimensional variations, ceramic chip capacitors with their inherently brittle crystalline microstructure are best suited for relatively small PCBs that exhibit minimal absolute thermal expansion/contraction *and minimal sag*. Many power supply manufacturers still avoid large ceramic capacitors completely above a certain size, preferring to parallel several smaller capacitors instead. We also should know that the laminate material with the closest matching CTE to that of surface mount components is FR4 (see Chap. 12). We should also be cautious when using chip capacitors on the underside of a *mixed* board (i.e., along with through hole components). In this case the surface mount devices will usually need to be glued, turned over, and completely immersed in molten solder during the wave soldering process. This not only adds to the cost, but may also affect long-term reliability. Chip capacitors are also often damaged simply due to mechanical stresses incurred during production and the faults can show up even weeks or months later. These cracks have often been traced to the claws of the pick and place machines, or when individual boards are being cut out from a large tile after soldering.

**11.** What are the typical TCCs of film capacitors?

ANSWER:

**1.** Polyester (Mylar) +600 to +900 ppm/$^\circ$C (capacitance increases with temperature).

**2.** Polypropylene $-$200 ppm/$^\circ$C (capacitance decreases with temperature).

**3.** Polystyrene $-$125 ppm/$^\circ$C (capacitance decreases with temperature).

**4.** Polycarbonate $-$100 ppm/$^\circ$C (capacitance increases with temperature, now obsolete).

(These values are approximate/averaged numbers over the rated operating temperature range. They should not be considered as linear)

12. What are the essential differences between the most well known film capacitor types and what are their alternative names?

ANSWER: Typical values in order of polypropylene (*KP*), polystyrene (*KS*), and polyester (*KT*) are:

1. Maximum working temp: 100, 85, 85 °C
2. Tan $\delta$: 0.4, 0.2, 3 % at 100 kHz
3. Tempco: $-300$, $-120$, $+1200$ ppm/°C
4. Metallization possible?: yes (*MKP*), no (*KS*), yes (*MKT*)
   - Polypropylene may sometimes be called "FKP" for film and foil construction. Double-metallized polypropylene maybe called "MMKP."
   - Polypropylene has much higher $dV/dt$ rating (about 10,000 V/µs) than polyester (100 V/µs) and is more suited to pulsed applications or at the front-end of a converter.
   - Note that polycarbonate (*KC*) is virtually extinct now (as is polystyrene). Polycarbonate film capacitor users are being asked to switch to polyphenylene sulphide (PPS) film capacitors that are available in both SMD and through hole versions. These can operate up to 150°C (with some specified voltage derating) and their tan $\delta$ is around 0.2%. Their capacitance is very stable with temperature and frequency.

Alternate names of some film capacitor materials are:

1. Polyester: Mylar, polyethylene terephthalate (or PET)
2. Polystyrene: Styroflex

13. I need 470 pF. Should I use a capacitor with the marking "470" on it?

ANSWER: According to the standard electronics industries alliance (EIA) codes, for ceramic, film, and Ta capacitors, the significant digits are only the first two digits, whereas the third digit is the number of zeros (capacitance expressed in pF). Therefore 470 stands for 47 pF not 470 pF. A capacitor marked 471 is 470 pF. Note that by the same system, 4*R*7 is 4.7 pF. For aluminum electrolytic, the same rule applies but the capacitance is usually expressed directly in µF.

14. What are the common standard capacitor tolerances?

ANSWER: Here we are talking only about the tolerance on the initial specified value. We can have $J = \pm5\%$, for example, for COG, $K = \pm10\%$, for example, $X7R$, $M = \pm20\%$, for example, Z5U or X7R, S = $\pm22\%$, $N = \pm30\%$, $Z = +80\%, -20\%$, $W = +100\%, -10\%$, $T = +50\%, -10\%$. For film capacitors we may specify tighter tolerances like $F = \pm1\%$ and $G = \pm2\%$.

**15.** How could the voltage rating of a capacitor be indicated on the case?

ANSWER: There is a single-digit code system in which we have $A$(10 V), $B$(16 V), $C$(25 V), $D$(50 V), $E$(100 V), $G$(200 V). There is also a two-digit code in which we have 0$G$(4 V), 0$J$(6.3 V), 1$A$(10 V), 1$C$(16 V), 1$E$(25 V), 1$V$(35 V), 1$H$(50 V), 1$J$(63 V), 2$A$(100 V), 2$D$(200 V), 2$E$(250 V), 2$V$(350 V), 2$G$(400 V), 2$W$(450 V). But, high-voltage disc capacitors may simply be marked 1kV(i.e., 1000 V), for example.

**16.** How is the polarity indicated on an SMT polarized capacitor?

ANSWER: On a plastic packaged Ta (or polymer) capacitor, the solid band is the positive terminal (anode). On conformal coated Ta capacitors, the wire nib (protrusion) is the anode. On the aluminum electrolytic, the solid band is the negative terminal (cathode). Remember that on a diode, the solid band is again the cathode!

**17.** What is the transient voltage capability of ceramic capacitors?

ANSWER: A flash test is usually conducted for testing the dielectric withstand voltage (DWV). Almost all ceramic capacitors are tested by applying a DWV in excess of 2.5 to 3 times their working dc voltage (WVDC) for 1 to 5 s, the current being limited externally to less than 50 mA.

**18.** Why are there no exact ferrite material equivalents from vendor to vendor?

ANSWER: It has been aptly stated that ferrite manufacture is "a blend of art and science." That is why no standard materials are available throughout the industry.

**19.** What are the typical tolerances of ferrite cores?

ANSWER: The process of making ceramic ferrites involves *sintering*, which produces volume shrinkages of 30 to 39 percent. Thus, parts are made with initial volumes 40 to 63 percent larger than the expected final volume.

Still, freshly sintered parts result in mechanical tolerances of $\pm 1$ to $\pm 3$ percent. But basic tolerances for permeability, saturation, residual, coercive force, and other such characteristics can be as high as $\pm 25$ percent. The temperature coefficient is even more liberal at about $\pm 50$ percent.

**20.** How does the air gap affect the $A_L$ tolerance?

ANSWER: The initial permeability $\mu_i$ as defined by vendors, is only for magnetically *enclosed* objects, for example, toroids. For many reasons, such as the accuracy of weighed quantities or the resolution of the X-ray fluorescence analysis used to check composition, the initial permeability referred to a toroidal core can only be narrowed down to a tolerance range of $\pm 20$ percent.

Since $\mu_i$ is given only for a closed toroid, the *effective permeability* $\mu_e$ for nontoroidal (cut) cores is a function of various factors such as quality of polishing, mechanical tolerances, number of surfaces in contact, positioning of core halves, dirt on the surfaces to be mated, and the like. So, cores supposedly "with no air gap" also have a permeability $\mu_e$ which will be less than $\mu_i$. In theory, for materials with $\mu_i \cong 2000$, a $\mu_e/\mu_i$ of about 96 percent is possible. In practice this number is only 75 percent. For highly permeable materials ($\mu_i \cong 10{,}000$), the highest ratio of 65% by calculation is obtained for mirror surfaces with (residual) air gap of 1 µm to 2 µm. Note that "no air gap" for cut cores should always be taken to mean at least *1 µm of air gap* (default value). If we introduce a deliberate air gap, then as the air gap becomes large, the initial permeability plays a smaller and smaller effect on the $\mu_e$. For example, if the air gap is 20 µm, the difference in $\mu_e$ for the above two classes of materials may be only about 20 percent. If the gap is greater than 100 µm, the $\mu_e$ is virtually the same. Clearly, in deliberately gapped cores, the surface texture is of little consequence. But for large cores with very small air gaps, the tolerances can be even worse than calculated, because the relationship between the grinder disk width and core width (edge effects) also comes into play.

To sum up, the tolerance of the $A_L$ value can be in general expressed as follows:

$$\frac{\Delta A_L}{A_L} = \frac{\Delta \mu_i}{\mu_i} + \frac{\Delta l_g}{l_g} + \frac{\Delta\{X\}}{\{X\}}$$

where, $l_g$ is the air gap, and $\{X\}$ stands for all the geometric influences previously indicated. However, these influences are difficult to grasp and quantify exactly.

Typically, the air gap is toleranced by vendors (e.g., Epcos), as follows:

$l_g \le 0.1$ mm, $\Delta l_g = 0.01$ mm, that is, $\Delta l_g/l_g$ starting at 10%

$0.1$ mm $< l_g \le 0.5$ mm, $\Delta l_g = 0.02$ mm, that is, $\Delta l_g/l_g$ is as high as 20%

$l_g > 0.5$ mm, $\Delta l_g = 0.05$ mm, that is, $\Delta l_g/l_g$ is less than 10%

We, therefore, should be very cautious when dealing with small air gaps, as we can have large tolerances for the inductance.

**21.** What are the pros and cons of specifying $A_L$ versus air gap?

ANSWER: The designer has the following choices:

**1.** Specify air gap tolerances
**2.** Specify $A_L$ tolerances
**3.** Choose between symmetrical/unsymmetrical grinding for the mating halves

The basic question is whether to specify the ferrite in terms of the $A_L$ value, or the dimensions of the air gap. The two are related. If the air gap increases,

$A_L$ decreases. In power applications such as storage chokes and flyback converters, the object is to store energy in the air gap, so the air gap (which is basically where we store the energy as seen from Chap. 7) must be defined fairly precisely. The $A_L$ value by itself is not critical here. Nevertheless, we probably do have an allowed (or recommended) range for inductance for various other reasons. For example, if the inductance is too low and we are operating close to maximum power, we may hit the peak current limit and thus be unable to deliver the required power. If the inductance is too low, we will also enter discontinuous conduction mode earlier as we decrease load and may end up with undesirable early pulse skipping eventually (see Chap. 9). Too low an inductance in current mode converters with slope compensation may also pose problems (see Chap. 14). So, a *minimum $A_L$* may need to be specified. Too large an inductance ($A_L$) may lead to a non-optimum design, and cause some strange problems as mentioned in Chap. 1. Therefore a *maximum $A_L$* may also need to be specified. However, we should always strive to make our design itself relatively intolerant to variations in $A_L$. But we also note that the magnetics design for a switching application is further complicated by the fact that the actual operating values may not be close to those at which the $A_L$ value is specified in the datasheet (usually 10 kHz, 0.25 mT, and $23 \pm 3°$C). In contrast, typical power transformer circuits work at much higher frequencies, at flux densities around 300 mT and temperatures up to 120°C. So, the correlation between the material's datasheet and our actual application may not be so trivial.

As indicated, $A_L$ and air gap are related. It can be shown by calculation that the air gap for a given $A_L$ value can only fluctuate within certain limits, and that a worst-case estimate is possible for the span of this fluctuation. If the $A_L$ values are very low, and the acceptable tolerance range for both air gap and $A_L$ sufficiently wide, it is conceivable that *both $A_L$* value and air gap could be simultaneously guaranteed. But from the manufacturers' point of view, they feel that it is safer and more precise to specify *either $A_L$ or* air gap tolerances.

Actually, prescribing an air gap is simple, as only geometrical tolerances are involved (though there is a problem with larger core forms). Specification of the $A_L$ value is more complex since there are specified test conditions like a specified coil, specified temperature, specified mating pressure (however only pressures distinctly below 10 N/m$^2$ or unrealistically high pressures can falsify measurements), and the like, all of which may also be very different from the conditions that the designer is to use in his or her application. Hence, ultimately the user's $A_L$ value needs to be correlated and effectively converted to an equivalent manufacturer's $A_L$ value. The magnetic material vendor should be approached for help in this matter.

Historically, the early $P$ and $RM$ cores were specified with only an air gap, and were also supplied in pairs for maintaining tight tolerance limits. In the late 1970s a change was made with $E$-cores, and they were sold only in unit delivery (and specified air gap). The idea was to give customers the choice of mating different air gaps for different applications. Nowadays, manufacturers have started specifying only $A_L$ tolerances, while maintaining unit delivery. Now, for most applications, it is sufficient to select *symmetrical* cores (i.e., identical halves) from the *same pallet*, so that the end $A_L$ value has the *same* tolerance. If we randomly select unsymmetrical core halves, then since the $\mu_i$ value of the two halves may be different, a statistical dual distribution can occur (as in overlapping unbalanced/unsymmetrical distributions). If unsymmetrical halves are required for some reason, the manufacturer can be approached to provide an asymmetrical form of packing, with appropriate packing codes to ensure that the asymmetrical halves are from the *same sinter batch*. Then the final $A_L$ tolerances will be as expected and quality problems will not occur.

Ultimately, it is advisable to create a user's test setup to correlate with standard test parameters with the help of the manufacturer, and in critical applications, to resort to 100% testing for relevant parameters.

**22.** Should we measure the $L$ on an LCR meter set to *series mode* or *parallel mode*?

ANSWER:  Theoretically, we can model an inductor either as an ideal inductor in series or in parallel with a resistor. Here, as with capacitance, the equivalency between series- and parallel-mode values holds true *only at the conversion frequency*, because the *quality factor* used in the conversion is frequency dependent. However, as with capacitance, the $Q$-factor is the same regardless of the measurement technique, that is,

$$Qp = Qs.$$

If the inductance is large, parallel mode is more suitable, since the reactance at a given frequency is large and the indicated inductance is closer to the effective inductance. Also, the parallel resistance becomes more significant than the series component. Conversely, for a low inductance, the series mode is preferred. For midvalues, a more exact comparison of reactance to resistance is called for to determine which equivalent circuit mode to use. The rules given above are obviously dependent on test frequency too. So, if the test frequency is 1 kHz, the thumb rule is that below 1 mH, we should use series mode, whereas above 1 H we should use parallel mode. In between these values, we can use the manufacturer's recommendations and/or our judgment in relation to expected resistances. Note that $Q$ is inherently lower at lower frequencies.

Distributed capacitance, if any, decreases the effective inductance, and more importantly, increases the frequency dependence of that effective inductance. A large number of turns will increase the effective series resistance and lower the $Q$, and a distinct difference in series-mode measurement and parallel-mode measurement will be obvious (especially at lower frequencies where $Q$ is even lower).

**23.** What is the $100 \times 100 \times 100 \times 100$ rule of thumb for ferrite loss?

ANSWER:  Most ferrites have a core loss of about 100 mW/cm$^3$, (that is, 100 kW/m$^3$) at 100°C, 100 kHz, and 100 mT. Better materials improve this number by about 20 percent, whereas at temperatures around 40°C the losses increase by about 20 percent. Note that as per convention, the 100 mT just quoted is $B_{AC}$ or half the peak-to-peak swing.

**24.** Is a sleeve-bearing fan a better choice than a ball-bearing type?

ANSWER:  The ball bearing fan is certainly more expensive. It has a longer life when operating at elevated temperatures. But it gets noisier with running time. Hence, if the useful life of a fan were defined as ending when the fan became noisy, the ball-bearing fan would have a smaller life than the sleeve-bearing fan. The sleeve-bearing fan is quieter in the long run and accepts any mounting attitude. Its life is comparable to a ball-bearing fan if temperatures are not too high. It is more rugged and can sustain multiple shocks without any effect.

**25.** My flyback converter is behaving rather strangely. What is the most likely cause?

ANSWER:  One of the most likely causes of severe and inexplicable malfunction, especially with a new build, is the transformer. We have to confirm the polarities of the windings are as we expected them. Two quick methods are provided:

We can use the probe test signal prong available on the front of the oscilloscope and connect it by a jumper to one end of a selected winding. We then connect CH1 probe tip to the same end of the winding, while connecting its ground clip to the other end. We then connect CH2 probe tip to one end of the other winding, and the ground clip to its opposite end. We compare the waveforms seen of both channels. Though they may be looking "squashed"/distorted, we can still check if the edges on both channels rise and fall *in-phase* or not. If they are in-phase, the ends we have connected to the two probe tips are both "dotted" (or equivalent) ends. If not, one end is dotted, the other is "nondotted." Note that the volt/divisions of both channels will need to be adjusted for visual clarity.

Another way to do this is to make a starting assumption about polarities. Then we should connect/solder the presumed "nondotted" end of one winding to the presumed "dotted" end of the other, putting the windings effectively in series. If we then measure the inductance across the two free ends that remain, we should get a reading higher than the inductance of either winding (measured individually). If not, our initial assumption was false. Either way, we will now know which ends are "dotted" (equivalent).

**26.** My flyback converter's zener clamp is getting very hot and blows up. Why?

ANSWER:  We have to estimate the zener dissipation by using the correct formula given in Chap. 6. We also need to make an *in-circuit* check of the actual net leakage inductance as seen by the switch, in the manner described in the same chapter.

27. My dc–dc switcher IC keeps behaving strangely. What is the most likely cause?

ANSWER:  Each topology has certain "critical trace sections" (and thereby critical components too). See the table provided in Chap. 12 that deals with PCB layout. Spikes from these traces can enter the control sections of the IC causing it to misbehave. Also ensure effective bypassing at the input by means of a ceramic capacitor (typically 0.1 µF to 1 µF) very close to the IC.

28. I am using an integrated buck switcher IC with a through hole diode mounted on a heatsink. I have adequate decoupling and small trace lengths but I still see some strange behavior. Why?

ANSWER:  With through hole devices, though we can reduce the trace lengths, we cannot usually do anything about the lead lengths. Even this small length can pose problems if this belongs to an identified *critical trace section*, and if the crossover time of the switcher is very small (<50 ns). In this case the switching node trace has become inadvertently long by the use of the through hole package. We need to put in a snubber (typically 470 pF to 4.7 nF in series with 10 Ω to 100 Ω). This should be mounted very close to the switching node (SW) pin of the IC and the ground pin of the IC.

29. I am using an integrated buck switcher IC which is in a through-hole package for mounting on a heatsink. I have adequate decoupling and small trace lengths, but I still see strange behavior. Why?

ANSWER:  Now the lead lengths of the IC may be posing a problem. Clearly, we cannot mount the snubber or IC decoupling capacitor any closer to the IC chip. The only option now is to slow down the device *crossover transitions*. A common technique when using *N*-channel FET switches is to insert about 10 Ω to 100 Ω in series with the bootstrap capacitor. Since the bootstrap capacitor is basically the input capacitor of the supply rails powering the floating driver stage, putting in a resistor limits the current into the gate of the FET and slows the transition (though, in principle, only the *turn-on* transition). This should help reduce the amplitude of the inductive spikes in the layout, which are the basic cause of the anomalous behavior (see Chap. 12).

30. On my flyback power supply, the series pass regulator of my auxiliary output is running very hot. What should I do?

ANSWER:  Firstly we should think of removing the series-pass stage if possible. One way to do this is *double point sensing*. Here we essentially compromise some regulation of the main output somewhat, and use that freedom to better regulate the auxiliary output. We, thus, introduce a large resistance value from this auxiliary output to the reference pin of the LM431 (or TL431) while simultaneously slightly increasing the resistance from the main regulated output to the reference pin.

We can also improve the cross-regulation by connecting the lower end of the auxiliary winding to the cathode of the output diode of the main winding, rather than to ground. Of course, we now need to adjust the number of turns of the auxiliary winding accordingly.

If the series pass stage is considered unavoidable (as, for example, if the output regulation needs to be very tight), we should note that a popular and very cost-effective series FET (for about 2 to 3 A auxiliary load current) is the MTP3055, with an $Rds$ (drain to source resistance) of 150 m$\Omega$. If the FET is running very hot, there is little to be gained by decreasing the $Rds$ much lower, as that is not the root cause of the heat. What we probably are seeing is that the voltage at the input of the series pass stage is a little too high. We can reduce this by two techniques: (a) We can wind a *half-turn* for this winding as discussed in Chap. 6. (b) We can trim the voltage down further by inserting a manganese–zinc (Mn–Zn) ferrite bead (or more expensive amorphous material) in series with the winding. Here we must ensure that this bead is not too big as it is only meant to "chisel away" part of the leading edge current trapezoid. We may need to *increase* the number of turns on the bead if necessary so as to increase the ampere-turns sufficiently to saturate the bead (just a little past the crossover transition). We should recheck that we are still meeting the regulation specification on this output, especially when the main output is at its minimum load condition.

**31.** The customer is talking about *EVT*, *DVT*, and *MVT* samples. What are these?

ANSWER: These acronyms are by now fairly well accepted buzzwords inside the power supply industry. The first prototypes coming out from design are called *engineering verification test* (EVT) samples, or *first articles*. They will be put through functional tests and a typical expectation is that they will meet 80 to 90 percent of the specified functional requirements. After that they will go back to engineering for the necessary remaining work, and this will lead to the *design verification* (or *validation test* samples (DVT). These will ultimately go into pilot-production and then full-production. But before that happens, we may be asked for the final *manufacturing verification test* (MVT) samples, which will include compliance to safety, EMI, and testability.

# References

1. Collett, P. C. E., "Investigations into Aspects Affecting the Design of Mains Filters for Frequencies in the Range 10kHz-30MHz," ERA Report No.82-145R, ERA Technology Ltd., England, 1983.
2. "Capacitors for RFI Suppression of the AC Line: Basic Facts"; 4th ed., Evox-Rifa Application Notes, Evox-Rifa, Lincolnshire, IL 60069, USA.
3. Snelling, E.C., *Soft Ferrites, Properties and Applications*, 2d ed., Butterworths & Co, 1989.
4. "Power Factor Corrector, Application Manual," 1st ed., SGS-Thomson Microelectronics, October 1995.
5. "Data Handbook, Aluminum Electrolytic Capacitors," PA01-A, N.A. ed., Philips Components, 1993.
6. "Understanding Aluminum Electrolytic Capacitors," 2d ed., United Chemi-Con, 1995.
7. "Micro Linear Databook," Micro Linear Corp., 1995.
8. "Fair-Rite Soft Ferrites," Databook, 13th ed., Fair-Rite Products Corp., NY 12589.
9. "Magnetics Designer," Supplementary Information, Intusoft, 1997.
10. "UC3842/3/4/5 Provides Low-cost Current-mode Control," Application Note, U-100A, Unitrode Integrated Circuits.
11. Billings, K, H., *Switchmode Power Supply Handbook*, McGraw-Hill, 1989.
12. Pressman, A. I., *Switching Power Supply Design*, McGraw Hill, 1991.
13. McLyman, W.T., *Transformer and Inductor Design Handbook*, 2nd ed., Marcel Dekker, 1988.
14. Unitrode Power Supply Design Seminar, SEM-500, Unitrode Integrated Circuits.
15. *3C85 Handbook*, Ordering Code 9398 345 90011, Philips Electronic Components and Materials, 1987.
16. Sum, K. K., "Intuitive Magnetic Design," Electronic Design Workshops, Penton Media, Nov 15–16, 2000.
17. Bloom, G. E., "DC-DC Switchmode Power Converters, Circuits and Converters," April 25, National Semiconductor Corporation Seminar Presentation; Bloom Associates, CA., 2002.
18. Mulder, S. A., "Application Note on the design of Low Profile High Frequency Transformers, a New Tool in SMPS Design," Ordering Code 9398 074 80011, Philips Components Corporate Innovation Materials, 1990.
19. Ahmadi, H., "Calculating Creepage and Clearance Early Avoids Design Problems Later," *Compliance Engineering Magazine*, March/April 2001.
20. Redl, R., "Low-Cost Line-Harmonics Reduction," Seminar in Bremen, Germany, Power Quality Conference, 1995.
21. Carsten, B., "Calculating Skin and Proximity Effect, Conductor Losses in Switchmode Magnetics," PCIM Conference, 1995.
22. "Magnetics® Ferrites," Databook, Magnetics, Division of Spang and Company, 1999.
23. Lee, S., "Thermal Management of Electronic Equipment," PCIM Conference, 1996.
24. Middlebrook, R. D.; Cuk, S., "Advances in Switched-Mode Power Conversion: Volumes I, II, and III;" TESLAco, Irvine, CA.
25. "Data Book and Design Guide;" Power Integrations.

# Index